COLLIER'S
WONDER BOOK

Edited By

Waldemar Kaempffert

Editor of Popular Science Monthly

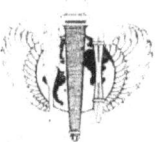

P. F. Collier & Son Company
New York

The original cover illustration for *Collier's Wonder Book,* 1920 Edition.

The illustration used for the cover of this Raspberry Ridge Reprint is part of the original *Collier's Wonder Book.* The title graphics were taken from the original cover and then combined with the illustration to form a new and brighter cover from the old elements. All interior pages are exactly reproduced from the original book.

Prepared for reprinting by Janice Harbaugh

Raspberry Ridge Publishing
United States of America

Contents

ASTRONOMY

What Holds the Stars Together?

Gravitation, the all-pervading force

By Alfred J. Lotka

A HEALTHY man weighing three grains? Ridiculous! Nevertheless, it is true.

You see, it all depends. Weight is a relative thing. The earth pulls you down with a force of, say, 150 pounds weight in New York. But, at the same time, the moon in the sky is pulling you up, and so far as the moon is concerned you weigh, at New York, about three grains—about as much as a fly.

The Moon Constantly Pulls Us

The reason why you weigh so little relatively to the moon is that the moon is only about one eightieth of the earth in mass, and that it is so far away—about sixty times as far from New York as that city is from the center of the earth. Still, even so, and at that distance, the moon is constantly pulling you toward her with this weight of three grains. You do not notice it because the earth is holding you down with so much greater force. But if the earth and the sun were suddenly removed and all motion were arrested, leaving you hanging in space, you would immediately begin to fall to-ward the moon. It would not be a very hasty fall, it is true—at least, not at first. You would have plenty of time to think about your misfortune, for it would take you over two minutes to fall one foot! You would, in fact, be quite unconscious at first that you were falling. Still, when, after a long time, you reached the surface of the moon, you would have gained speed, and you would land there at the rate of 7,780 feet per second. (So, after all, it is well the earth does not suddenly abandon you to the attraction, gentle but firm, of the moon.)

As you approached the moon not only your speed but also your weight would increase. You would not regain your full earthly dignity of 150 pounds, but at least you would be saved the humiliation of those miserable three grains. By the time you reached the surface of the moon, supposing you were somehow able to break your fall, you would step about lightly in your new surroundings with a weight of 25 pounds. You would astonish the natives (if there were any) by jumping over their houses and performing other unheard-of acrobatic feats.

However, the moon is not a healthy place to stay on. Temperature conditions are very uncomfortable, and there is no air. So we are not much interested in any athletic prowess which we might develop there.

The case is a little different with the planet Mars. According to Percival Lowell, Mars is probably inhabited. Now, on Mars a 150-pound earth-man would weigh about 53 pounds, say in round numbers about one third of his earth weight. Here is a problem: How big could nature afford to build a Martian man without putting a greater load on every square inch of the soles of his feet than a 150-pound earth-man rests on his? The answer is that the Martian could be made three times as tall, say 17 feet 6 inches. For if he stood on the earth his weight would then be 3x3x3x150 = 4,050 pounds; but on Mars it will be only 1,350 pounds.

If an ordinary man touches the ground with his feet over an area of about 50 square inches, our Martian would stand on 3x3x50 = 450 square inches. If, then, we figure out the number of pounds borne by each square inch, we find for the ordinary man on earth 150÷50 = 3 pounds. For the Martian 17 feet 6 inches high we find 1,350÷450 = 3 pounds, just the same. We see, therefore, that the Martian could be built on three times the scale in height, breadth, and depth as an ordinary man; and though he would, on earth, weigh more than two tons, this would not put any greater tax on his feet on Mars than that which is normal for us on earth. Yet this Martian giant, though in no way encumbered by his own weight, would be twenty-seven times as powerful as an earth-man. In certain special operations where the work consists in overcoming gravity, such as digging canals, he could accomplish 3x27 = 81 times as much as an earth-man on earth, since Martian gravity is only one third of that of the earth.*

On the moon and on Mars gravity is less than on the earth. On the planet Jupiter it is more than 2½ times as great as here. A 150-pound earth-man on Jupiter would be weighed down with a load of 390 pounds. Walking would be a very tiring process. On the sun gravity is more than 27 times that on earth. If the sun were solid and the temperature such as to permit the presence of life, an earth-man on its surface would not only be utterly unable to rise into a standing position, but even when lying flat on his back he would be crushed by his own weight of more than two tons. If there were sun-men they would not exceed 2½ inches in height.

Influences Every Phase of Life

So then the size of man, and of every living creature, is determined, among other things, by gravitation. But that is only a minor detail.

Gravitation has a fundamental influence on every phase of life. What makes a plant shoot its stem upward, its branches sideways, its roots down into the soil? Gravitation—at least, in part. What keeps the earth shrouded in the atmosphere from which you draw your life-breath? Gravitation. What keeps the earth from flying off at her speed of 18½ miles per second into the dark recesses of space, where, far from the sun's warm rays, not only would all life perish from cold, but even the air would freeze solid? And what holds

*This estimate is somewhat excessive because we have used the round figure 3 instead of the accurate 2.65; but it will serve to illustrate the principle.

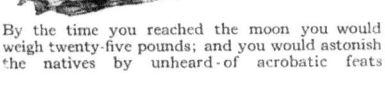

By the time you reached the moon you would weigh twenty-five pounds; and you would astonish the natives by unheard-of acrobatic feats

the moon in its course around the earth? Gravitation. And so on, indefinitely. How does it all come about? A partial answer at least can be given to some of these questions.

The response of plants to gravitation is strikingly illustrated in the life-plant, so called perhaps because of its extraordinary vitality. A single one of its leaves may be cut into thirty pieces, and from each piece, when sown on the moist ground or suspended in moist air, a new plant will grow. If a piece of the stalk of this plant is hung up horizontally in a moist atmosphere it starts to grow. But it grows all on one side, the lower face only, with the result that it acquires a curved outline. If we paint black marks or rings around the stem, it can be seen how the space between these remains unchanged at the upper face, but how the rings spread apart on the lower face in consequence of the growth. A seedling planted horizontally in a vertical surface of soil will very soon curve into an S shape, the stem growing upward, the root downward. The explanation of all these effects probably is that certain substances collect by their weight toward the bottom of certain parts of a plant, causing increased or diminished growth.

Not only plants but animals also display geotropism, as this property of being directed by the earth's gravitation is called. A certain aquatic animal known as the sea-cucumber, if placed on a flat plate, will keep on climbing vertically upward no matter how the plate is turned. The only thing that guides it is gravitation. There is no other inducement for it to seek the highest point in the plate.

Is man geotropic? He most certainly is, though in his case gravitation does not direct the course of his steps but merely regulates his erect position.

If you suppose that this does not require any special faculty, just call to mind the actions of a drunken man. How is the regulation accomplished? By means of a system of "spirit levels" carried in the head, the so-called semicircular canals of the inner ear. Injury to these canals causes more or less acute symptoms of dizziness or inability to maintain equilibrium.

Spirit Levels in Our Ears

The greatest freak in the matter of geotropism is a little crustacean known as Palæmon (the prawn). This creature has a most extraordinary method of solving the problem of keeping on an even keel. It possesses an apparatus similar to the spirit level of our inner ear, but within its liquid are little loose particles of grit, socalled otoliths or "ear-stones." It appears that these, by their pressure against the bottom or the sides of the space containing them, indicate to

A sun-man would be about 2½ inches tall. On the sun you would become an unwieldy giant, unable to rise to your feet or move in any way without the assistance of puppet-strings manipulated by high-power cranes

their owner whether he is steering an even course, or whether he is tilted this way or that. Now, the remarkable thing is this: At the time of moulting, Palæmon loses his otoliths or ear-stones. So what does he do? He coolly picks up sand from the bottom and puts it in his ears; and he goes on his way rejoicing, with the confidence of the man who has not touched a drop for a year. Even that does not exhaust Palæmon's repertoire of tricks. One ingenious biologist conceived the idea of giving Palæmon iron filings to put in his ears instead of sand. The obliging creature promptly inserted them, and lo and behold, Palæmon had developed a new sense: he responded to a magnet somewhat after the fashion of an animated compass needle! He had become magnetotropic.

To understand how the atmosphere is tied to the earth by gravitation, we must form a mental picture of a gas, such as the air. Can you call to mind the appearance of a swarm of gnats dancing in a sunbeam? How they flit to and fro and up and down with irregular motion? If your powers of sight could be increased about ten million times the air might present to you an appearance not unlike that swarm of gnats. For the air consists of innumerable particles (molecules) of diminutive size, flitting about and jostling each other, now colliding like billiard balls, now flying apart, now hitting against the solid objects of which the visible world is composed.

Just as would be the case with a jumble of billiard balls rolled at random on the table, the individual molecules have all kinds of different speeds. But the average speed of a large number of them is definite, and depends on the temperature. At 60° F. the average speed of the molecules of the air is about 1,500 feet, or something over ¼ mile, per second.

Now, if you read the article on page

7 on "Hurling a Man to the Moon," you will see the statement that to shoot a body off the earth so that it will never return requires a velocity of about seven miles per second. You will therefore see that a molecule of the air, at the average speed at 60 degrees, can never leave the earth.

This is not saying that some of the more rapidly moving molecules might not do so. However, it can be shown by a complicated calculation that, if the earth loses any of its atmosphere at all, the loss is so slow that even after millions of years it would not be noticeable.

So then the inhabitants of our globe are guaranteed against an air famine for many generations to come. How about the other planets? Computation shows that they too are provided for, though Mars is losing or has lost its hydrogen and helium, both of which are much lighter than air. The moon, on the other hand, is quite unable to hold an atmosphere, and is well known to be devoid of any.

Weight Is a Relative Thing

The earth's gravitational pull falls off as the distance increases. A stone weighing one pound at the earth's surface, four thousand miles away from the center, will weigh only one quarter of a pound if taken to a point out in space four thousand miles from the surface or eight thousand miles from the center. In other words, if the distance from the center of the earth is doubled, gravitative attraction is reduced in the proportion of 1 to the square of $\frac{1}{2}$. This is the law discovered by Newton.

But if, instead of going out into space, we make a deep bore-hole in the earth, how will gravity vary as we go down?

If the earth were a uniform perfect sphere, gravity would decrease as you go down into it. But, as a matter of fact, the central portion of the earth is more than twice as dense as the crust, and for this reason the weight of a body increases as you go down; though, if you could go far enough, a point would undoubtedly be reached where gravity would begin to decrease, becoming zero at the earth's center.

Weight is a relative thing. It changes not only as you burrow into the earth or fly out into space, but it even changes from point to point as you travel on the earth's surface. If your weight were 150 pounds at the North Pole, it would decrease by three quarters of a pound as you traveled to the Equator. There are two reasons for this change in weight. One is that, owing to the flattening of the earth at the poles, you are nearer its center there than at the Equator. The other is that at the Equator, owing to the earth's rotation, you are carried around in a circle with a velocity of about seventeen miles a minute, and a part of the earth's pull is spent in keeping you from flying off at a tangent. This part of the earth's attraction, about one three-hundredth of your weight, is not indicated by any kind of balance.

Supposing the earth were all alone in space, you could not tell by looking at the stars whether it was rotating or not. You would not know whether your weight was 150 pounds on a stationary earth, or 300 pounds on an earth revolving with such speed that one half your weight was taken off your feet by centrifugal action. In other words, you would have no way of distinguishing between gravitational force and the effect of inertia.

This inability to distinguish between gravitational and inertia effects is the root of the modern theory of gravitation. It is impossible to enter here into the details of a rather abstruse mathematical argument. But perhaps some indication of the nature of the theory can be given by stating that according to this theory the presence of a portion of matter, such as the earth, produces a kind of strain in space.

The theory has, so far, yielded one important concrete result. It accounts for certain observed irregularities in the motion of the planet Mercury which could not be explained in any other way.

The Moon's Grip on the Earth

Although, as we saw, the moon's gravitation is insufficient to hold the elusive molecule, it gives her a grip upon the earth, the strength of which may be illustrated as follows: If, instead of gravitation, we had to rely upon a steel bar to tie the moon to the earth, this bar, 240,000 miles long, would need to have a diameter of 225 miles to sustain the pull required to keep the

The same force that keeps the earth from flying off into space makes a plant shoot its roots down, its stem up, and its branches sideways. This response of a plant to gravitation is particularly marked in the life plant, so called because from a single leaf thirty new plants may grow

A Martian giant could be over seventeen feet high without putting any undue strain on his feet. In the way of digging he could do the work of fifty men on the earth. On the sun you would have to be shrunk to two and a half inches if you wished to retain an upright posture without strain

Earth man   Sun man   Mars man

moon circling around the earth! If the moon is being constantly pulled with this great force, why does it not fall to the earth?

The answer may be a little surprising. The moon does fall; it falls about one nineteenth of an inch every second. This does not mean that it is approaching the earth. But, instead of traveling on in a straight line, as it would without the deviating effect of gravitation, it is as it were, drawn aside by one nineteenth of an inch every second and thereby kept circling in its path around the earth.

By the way, why should the moon complete its cycle between two full moons in 29½ days? Has the lunar month always been of this length? And where did the moon come from?

These are not idle questions. The astronomer has an answer ready for them. The clue to the mystery is found in the tides. The moon, by her gravitational pull, raises on the earth two tides, one on the near side, the other on the far side. Between these two the earth rotates as between two brake-shoes. The obvious result is a very gradual slowing down of the earth's rotation. A little less obvious is the reaction on the moon. As the moon in its revolution lags behind the faster-moving earth, the first effect would be for the earth to hasten on the moon in its travel. But, instead of giving way directly to this impulse, the moon recedes farther away from the earth, and actually slows down. The net result is a gradual lengthening of the day, and at the same time a more rapid lengthening of the month, while the moon is slowly, very slowly, getting farther and farther away from the earth.

But if the moon is moving away from the earth, then formerly it must have been nearer to it; and if we go back far enough in time, what then?

Now, this is not a matter of guess-work. Astronomers can compute these things. When the day was about three hours long, the month was also three hours, and the moon was grazing the earth. In other words, the moon was originally part of the earth, and was thrown off by some catastrophe. What was the cause of this catastrophe?

The Birth of the Moon

Here also the astronomer is ready with an answer. The earth is more or less plastic. In those days it may have been more so than it is now. It may have been largely liquid. Now, a plastic or liquid body of the shape of the earth is capable of vibrating, somewhat like a church-bell. Such a bell has a definite period or frequency, so many vibrations per second, according to its size and pitch. In the same way, the earth has a natural period of vibration, which can be approximately computed. It is found to be about 1½ hours for one vibration.

Now the sun raises two tides in a day. In the three-hour day it raised a tide every 1½ hours. But this is the natural period of vibration of the liquid earth. The globe, under these conditions, was thrown into gigantic pulsations of ever-increasing violence. Year by year the vast wave stormed to greater and yet greater heights. This world of ours, so solid now under our feet, was throbbing as if its heart-beats would rend it.

And then, one fateful day, the gathering flood rose as a great mountain to the sky, and out of the surging tide the far-flung moon was born.

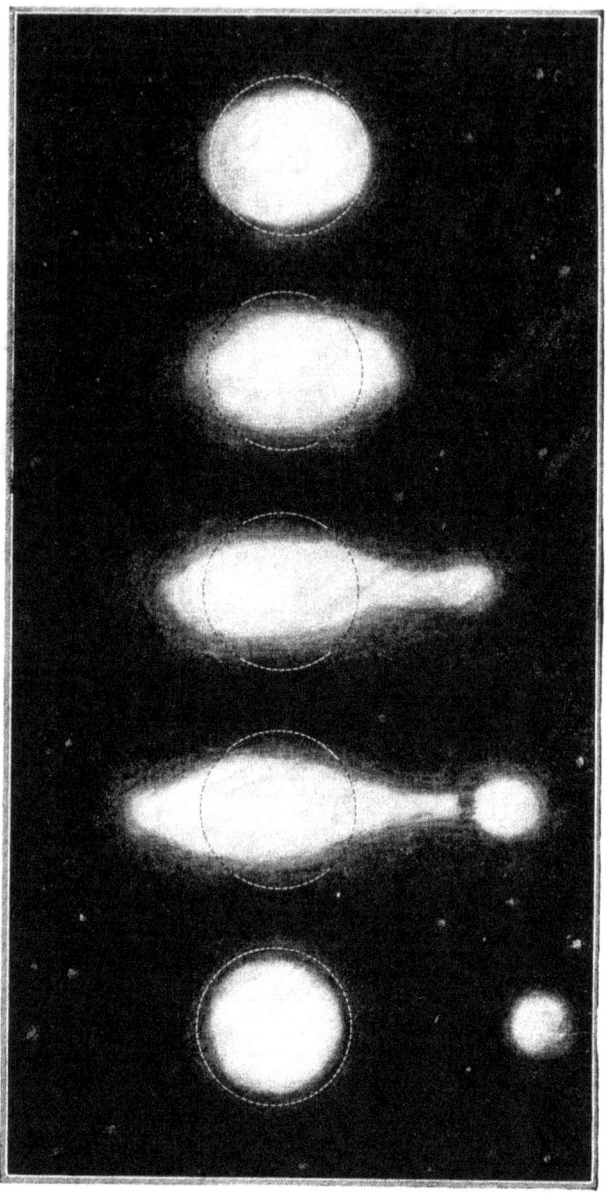

The Birth of the Moon

This illustration shows in general principle (though not in scale) the process by which, many millions of years ago, the moon was thrown off from the earth by tidal action. The several pictures represent successive stages in the vibration that culminated in the throwing off of the moon.

"The day was at that time about three hours long. Now the sun raises two tides a day. In the three-hour day it raised a tide every ninety minutes. But this is the natural period of vibration of the liquid earth.

"The globe, under these conditions, was thrown into gigantic pulsations of ever-increasing violence. Year by year the vast flood stormed to greater and yet greater heights. This world of ours, so solid now under our feet, was throbbing as if its heart-beats would rend it. And then, one fateful day, the gathering flood rose as a great mountain to the sky, and out of the surging tide the far-flung moon was born"

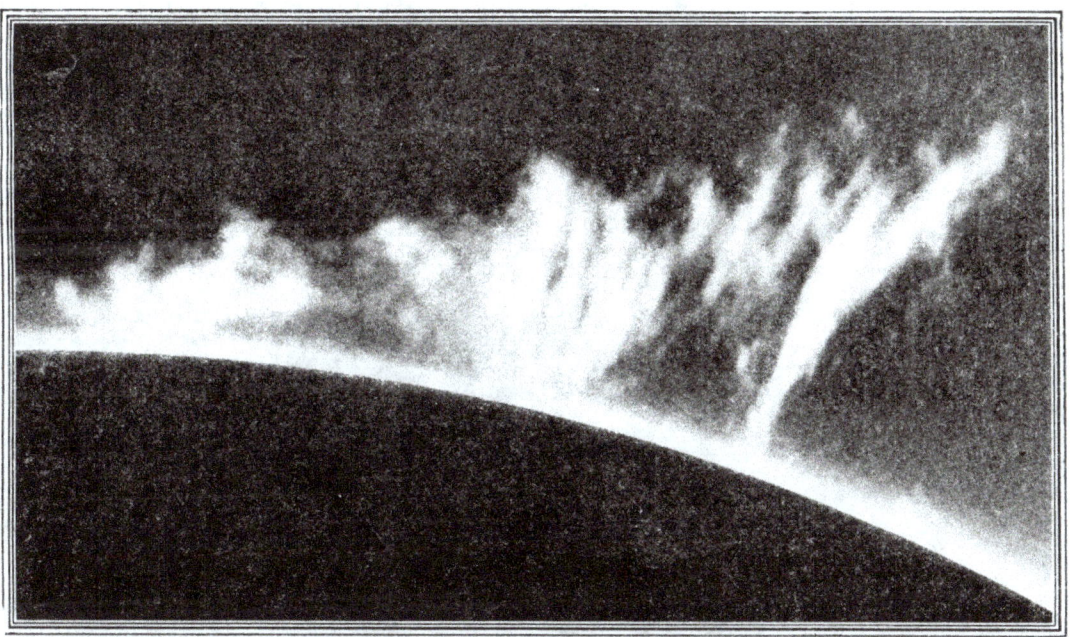

When sunspots are numerous, we also observe an excessive number of "prominences." These are red flames of hydrogen, which may shoot up at a speed of two hundred miles a second to a height of hundreds of thousands of miles

Blame It on the Sunspot

Solar upheavals that make mischief on earth

THAT favorite scapegoat, the War, was not responsible for all the abnormal events of the memorable year 1917, and the German spy, as a versatile mischief-maker, has a rival who is none the less formidable because his base of operations is 93,000,000 miles away from the nearest munition plant.

Let us go back in imagination to the 9th of August, 1917. Last night you may have taken pleasure in watching the ghostly streamers of a fine auroral display shift and shoot in the northern sky. Your morning paper tells you that telegraph lines all over the country were paralyzed by "earth currents" of extraordinary intensity. There were, in fact, hours together when not a single message could be sent by wire anywhere in the United States—an event without precedent in the history of American telegraphy.

Cause Still a Mystery

Torrents of natural electricity, sweeping through the earth, swamped the relatively feeble currents by means of which the circuit of a telegraph line is completed underground. At the same time, compass needles everywhere behaved as if bewitched, oscillating back and forth, to the bewilderment of mariners whose faith is pinned on them.

The more delicate magnetic needles installed at observatories underwent even more pronounced disturbances. In short, a violent magnetic storm was in progress.

What is the meaning of these uncanny events? After breakfast you seek out your friend the Professor, to inquire whether the end of the world is at hand. The Professor happens to be an astronomer. By way of answering your question he hands you a pair of colored spectacles and bids you take a look at the sun. Right in the middle of the solar disk you notice a rather conspicuous fly-speck.

Now you enter the observatory. The astronomer opens a long slit in the dome, revolves the latter until the sunshine streams through, takes the cap from the tube of his telescope, sets the driving-clock in motion, and swings the instrument around until a magnified image of the sun falls upon a white screen attached below the eyepiece. Thus enlarged, the fly-speck turns out to be a cluster of ragged black spots with gray borders.

"We are now," says the Professor, "in the midst of a period of remarkable solar activity, of which sunspots are one of the symptoms, though by no means the only one. The Sunday Supplement will undoubtedly tell you that the aurora of last night, the earth

currents, and the magnetic storm were 'caused' by the large group of sunspots which you see near the center of the sun; but it would be nearer the truth to say that, while all these things are intimately connected with one another, the ultimate cause is still a mystery.

"Astronomers have learned that the sun passes through periods of perturbation, or storminess, which recur on an average every eleven years, and that certain terrestrial phenomena tend to run through a similar cycle.

The Sun's Periodic Variations

"Our sunspot records show the eleven-year period unmistakably, although it is by no means a regular one, and a great many minor fluctuations are superimposed upon the general curve of 'spottedness.' A few years ago there were days and weeks together when not a single sunspot could be seen. Now it is not uncommon to see a hundred at a time.

"That sunspots denote a state of disturbance in the solar atmosphere is shown, first of all, by the fact that when spots are numerous we also observe an excessive number of the discharges known as *prominences*. These are the red flames which may be seen along the edge of the solar disk during a total eclipse. While some prominences are seen to hang quietly over the same place for several days, others shoot upward at a speed of from one to two hundred miles a second to a height of thousands and even hundreds of thousands of miles.

"Measurements made with the pyrheliometer of the amount of solar heat received by the earth show that the sun is hottest when spots are most abundant, although very delicate observations have shown that a spot is itself a relatively cool part of the sun's surface. Lastly, there is a marked difference in the amount of electricity which the sun gives forth at different times. Like all hot bodies, the sun discharges a host of electric corpuscles, or electrons. It is probably the bombardment of our atmosphere by these corpuscles that causes the aurora. Striking the rarefied gases of the upper air, they illuminate them, as an electric discharge lights up a vacuum tube.

Photographing the Sun

"Now let us look at a photograph of a sunspot. Here is one made at the Mount Wilson Solar Observatory. The spot has the appearance of a great hole opening into the sun's interior. Around the black center, which we call the *umbra*, is a broad border, or *penumbra*, which might be compared to a grating of bright bars with dark spaces between them. The bright shell in which these spots appear, called the *photosphere*, is a layer of incandescent clouds, consisting of metals and other substances, floating in an atmosphere of the same substances in a state of gas, and having a temperature of between 6,000 and 7,000 degrees centigrade (about twice that of the electric furnace).

"Even though we think of sunspots as holes torn in the solar clouds, their appearance through the telescope or in an ordinary photograph hardly suggests the idea of violent motion. This is much better brought out in photographs taken with an instrument called the *spectroheliograph*, an invention of Dr. George Ellery Hale.

How the Sun Affects Our Weather

"The final proof that a sunspot is a whirlwind or tornado in the solar atmosphere was furnished just ten years ago. Assuming that the electrons discharged by the sun are caught in the whirl around a spot, their revolution should produce the same magnetic effect which we get from the electric current flowing through the coil of an electromagnet. A sunspot should, therefore, be intensely magnetic. Laboratory experiments have proved that when light passes near a magnet it shows, when examined with the spectroscope, a splitting up of the spectral lines, known as the 'Zeeman effect.' In 1908, Hale, at Mount Wilson, tested the light from a sunspot for this, and found it at once.

"Will the study of the sun help us predict the weather? Probably it will; but at present nothing more definite is in prospect than predictions of the general character of entire years and seasons over broad areas of the earth. For more detailed predictions we can get no help from the sun until we know just why there are twice as many spots on the sun today as there were a month ago, and just how many there will be by the middle of next week. When we can predict the sun's weather, we shall have taken a long step toward predicting weather.

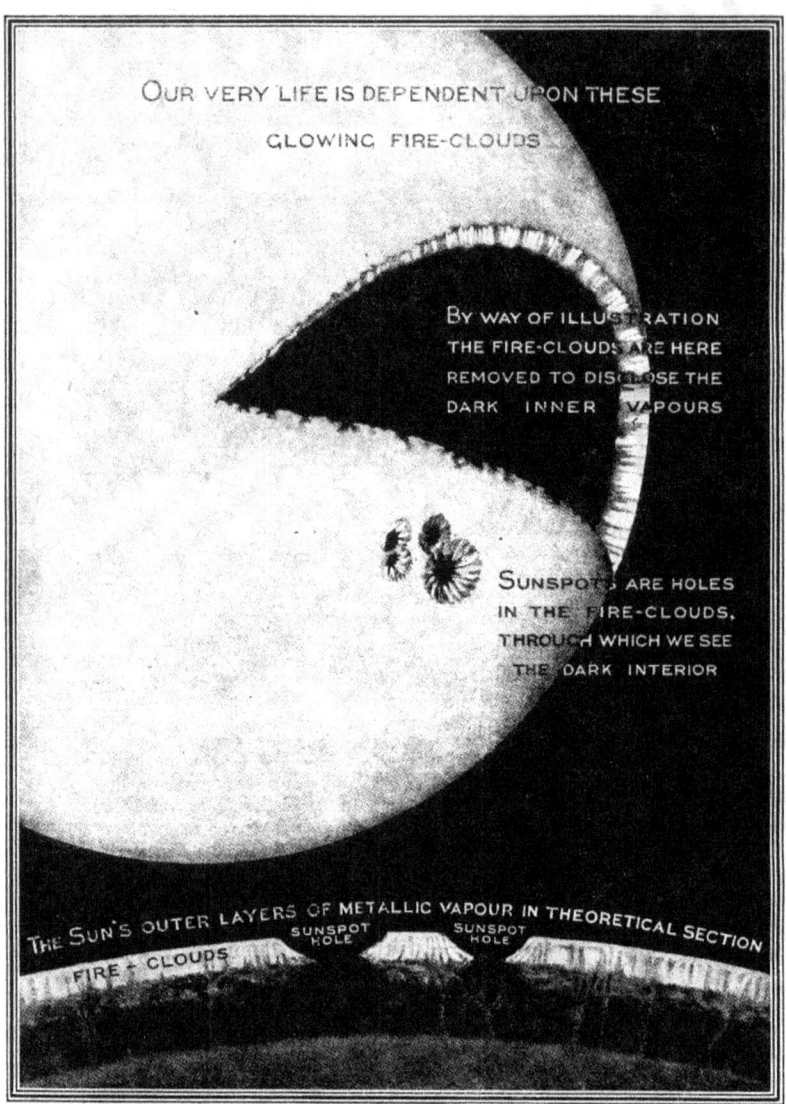

From a drawing by Scriven Bolton in the *Illustrated London News*

Sunspots are holes in the solar clouds. A sunspot is a whirlpool which sucks in the light, overlying hydrogen. Because it is a tornado in violent revolution, a sunspot produces the same magnetic effect that we get from a current flowing through the coil of an electromagnet. The electrical and magnetic effects of the sun on the earth are of far-reaching importance. Curiously enough, we are likely to have our coolest weather on the earth when the sun is most active, and therefore at its hottest

Hurling a Man to the Moon

How could a lunar Columbus break the grip of gravitation and reach the nearest heavenly body? What kind of motor would he use? How much power would it take?

By Waldemar Kaempffert and A. J. Lorraine

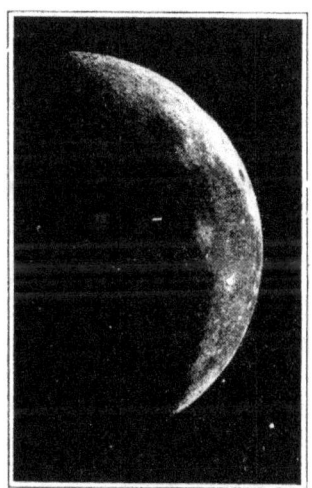

There in the heavens shines the moon—two hundred and forty thousand miles away. How can I bridge that awful chasm?

THERE in the heavens shines the moon—240,000 miles away. How can I bridge that awful chasm? Others have asked that question of themselves before me. Most of them have been novelists or scientific romancers. I am not a novelist, but an engineer, and it is as an engineer that I ask myself the question.

Two hundred and forty thousand miles! The distance alone seems hopeless. And yet, the distance that I must traverse gives me little concern. If I could take an express train I could easily reach the moon in six months. If I could use a fast airplane, a 120-mile-an-hour fighting-machine, it would be a matter of only three months before I beheld at close range the scarred and pitted surface of the moon.

The Distance Is Not the Important Thing

Two hundred and forty thousand miles are nothing. The captain of the fastest transatlantic steamer covers as many miles once every three years; the conductor of a speedy railway train in even less time.

No, it is not distance that perplexes me, but a barrier more formidable than

all the ice and snow, all the gnawing cold that guards the approach to the earth's poles. What is it—this barrier? Nothing—absolutely nothing. I am not trying to dismiss it by calling it nothing; for when I refer to "nothing" I mean the utter void of interstellar space. To be sure, the physicists say that the space between the stars is occupied by the "ether." That does not make the void any less a void for my practical engineering purpose.

You will understand my predicament better if you consider the means that we employ to move ourselves about mechanically on the earth. Every transporting vehicle that we use reacts in some way on its environment. The locomotive grips rails by friction. The balloon is buoyed up by the surrounding atmosphere. I find that the greatest height ever attained by a man in a balloon is 35,420 feet. Beyond a certain height a gas-bag will not rise; there is not enough air.

It can no more leave the atmosphere and float about in empty space than a fish can leave the water and sail in the air.

I am trying merely to make the point that, just as the locomotive needs rails or an automobile a road, so a balloon

Suppose it were possible for you to stand between two stars. The side that you presented to the sun would be exposed to his blaze, while the dark side would be frozen

needs air. Between the stars there is no air—only that baffling void. The airplane serves me no better than a balloon. It, too, reacts upon its environment. What keeps it up? Pressure beneath its wings—air-pressure. Hence, like the balloon, it can go so high, and no higher. I find that the altitude record for an airplane is 30,500 feet, made recently by an American aviator, Captain Lang. Without air a plane cannot fly. I repeat: there is no air between the stars.

It is clear that all the propulsive engines that man uses to transport himself from place to place on the earth are useless for my purpose. Like man himself, they are earth-bound. I must cast about for radically different means of locomotion—some mechan-

ism that is independent of any reaction on its environment.

Jules Verne once wrote a novel, "From the Earth to the Moon," in which he tried to reason with something like scientific sobriety that a gun might propel a man to the moon. He literally shot his hero through 240,000 miles of space. Will a huge gun answer my purpose?

Is There Any Limitation to Gravity?

In the first place, I must consider gravity. It ties us to the earth as if by an elastic band. Throw a stone up into the air, or fire a shell from an anti-aircraft gun. The stone or the shell rises to a certain point, and then it falls back as if it were pulled down by a stretched spring.

Now, an elastic band or a stretched spring has its limitations. If we strain it beyond a certain point it snaps. Is there anything corresponding in the case of gravity? I know that the analogy must not be pressed too closely. But this is true: If I could shoot a projectile from a gun with a greater and greater velocity, a point would ultimately be reached when, instead of returning to the earth, the projectile would continue to travel on an independent course through space.

A Shot Fired from the North Pole

I am dealing now with measurable forces. Huge guns have been built, and their muzzle velocities are known. Suppose I were to mount a gun at the North Pole and fire my shot at a sufficient altitude to clear all obstacles. What muzzle velocity would I have to attain so that if I were shot out, safely encased in a projectile, I would never fall back to the earth?

The calculation is simple enough. What is the result? Twenty-six thousand feet per second would have to be the speed of the projectile in which I am housed when it leaves the gun. I would keep on traveling in a circle around the globe. Indeed, I would become a satellite of the earth—a living moon.

The Projectile Would Become a Satellite

If I could illuminate the projectile in which I sit, you would see me rising every one and a half hours at the horizon; you would see me traveling across the sky in about forty-five minutes, and setting at the opposite point of the horizon. I would be a busy little moon; for I would rise and set as many as eight times in a twelve-hour night.

But I do not want to travel perpetually around the earth. I want to reach the moon. Hence my muzzle velocity must be greater than 26,000 feet per second. If it is somewhat greater, my path, as I sit in my projectile, would at first be drawn out into an oblong, or ellipse. Still, at regular intervals, I would always return to my starting-point.

I find that I must increase my velocity to about 37,000 feet (seven miles) a second if I am to leave the earth in my projectile for good and all and to travel forever farther and farther away from it. It is as if the elastic band were torn: gravity no longer confines me and my projectile to an orbit around the earth.

Thirty-seven thousand feet a second! It is stupendous. Can I attain it? Let us see what has actually been done by some of our big modern guns. What are some of the records? Our fourteen-inch naval guns, which in the great world war were mounted upon railway trucks and leveled at the German army, fired shells with a speed of 2,800 feet a second. A six-inch wire-wound experimental gun tested at Sandy Hook attained a muzzle velocity of 3,850 feet a second. It had but a short life. No doubt the highest muzzle velocity ever attained was that of the 75-mile gun with which the Germans bombarded Paris. Artillery officers have estimated that the initial velocity of its shell must have been from 5,000 to 5,500 feet a second—only one seventh of that required to hurl me to the moon. That German gun represents the limit of our present technical resources, so far as gun-making is concerned.

But not yet will I abandon the idea of reaching the moon by shooting myself from the earth. If one

gun carries seventy-five miles, how far will two guns carry? I am not trying to joke; this is no facetious application of the rule of three. I find in the patent records of the United States and other countries more than one such method of thus multiplying the power of guns in order to increase the range.

Over and over again inventors have suggested that it may be possible to fire a gun within a gun. Such is the timing arrangement that when the projectile—that is to say, the inner gun—has reached a certain point, it in turn is fired, and the shell projected from it proceeds with a new velocity to its destination. If two guns, one within the other, should prove practicable, why not three, five, eight, or a dozen?

This is precisely what a French army officer, writing in the *Revue du Ciel*, proposes (not very seriously, it is true) as the expedient to be adopted in an imaginary war between the earth and the moon. He assumes an initial velocity of 12,300 feet a second, which, he thinks, is within the limits of possibility in the future. His projectile consists of twelve shells, one within the other—and the final bullet.

Will guns within guns transport me to the moon? How big must the projectiles—the guns—be? How much powder must I use? I make the calculation. What is this? In order finally to land a paltry 16-pound shot on the moon, no less than eight tons of projectile must be fired! There is no need to calculate the explosive necessary.

A Ray of Hope in the Sky-Rocket

I see that guns within guns will avail me nothing. Moreover, only the last shot, that hopelessly inadequate 16-pound piece of steel, would reach the moon. The rest of the eight tons would come tumbling back to earth with a terrific velocity. Look out for your head!

It is very evident that I cannot reach the moon by the gun route. And, even if it were possible, I would be crushed to a pulp by the terrific impact at the moment of departure.

A machine-gun mounted on a small railway flat-car would be able to kick itself backward, away from its target, by firing a steady stream of bullets. So a sky-rocket might kick itself to the moon

But I am not yet discouraged. A distinguished French aeronautical engineer, Robert Esnault-Pelterie, suggests that I use a sky-rocket. What is it that makes a sky-rocket leave the earth? It is not its reaction upon the air, as many suppose, but rather the recoil from its discharge. A machine-gun mounted on a small railway flat-car would be able to kick itself backward, away from its target, by firing a steady stream of bullets. A sky-rocket might kick itself to the moon.

Some years ago, Rodman Law, the venturesome brother of the famous woman flyer Ruth Law, actually tried to fly with a sky-rocket of huge proportions. His was a successful failure. He did not get very far, but he lived to tell the tale.

Round Trip to the Moon in One Hundred Hours

In a gun the explosive force is applied very suddenly; the blast lasts scarcely 1-100 part of a second, and it deals a crushing blow to the projectile. Esnault-Pelterie calculates that in a sky-rocket big enough to reach the moon the motive force is applied for about twenty-five minutes. In that time an elevation of about 3,600 miles is reached. After that the car travels on by its own momentum without further application of power. This second stage of the journey occupies something over forty-eight hours. If the sky-rocket principle is practical at all, I ought to be able to reach the moon and to come back again in one hundred hours.

How much power does it take? This time I do not have to make the calculation myself. Esnault-Pelterie has already done that. And what does he tell me? No less than 414,000 horsepower are required! This is allowing a weight of about one ton for the car and its contents. The work of carrying me away from the earth's attraction by means of a sky-rocket figures out as nearly 7,000 calories for each pound lifted.

That is quite beyond the possibilities of any known explosive—dynamite, nitroglycerine, or T.N.T. The most terrific compounds used in the great war are not terrific enough to propel even their own weight over the 240,000 miles that must be covered to reach the moon, not to mention my weight, the weight of my food, and the weight of all the machinery in my car.

But there is one substance that does contain sufficient energy for my purpose, and that is radium. Locked up within a few pounds of it is all the energy that I need.

What Happens When Gravitation Ceases to Act?

"What would be the effect upon me if gravity were removed, and removed it would be at a point some thousands of miles away from the moon? . . .

"This is certain: as I approach the moon I will have the sensation of falling headlong through space—a terrifying sensation, which it might be impossible to counteract by the force of the will. . . . In the absence of hand-rails or straps, I should very likely sooner or later find myself sprawling in mid-air, unable to move my body from place to place in the ca. "

But how can I release it at the rate that I demand? It stubbornly refuses to give out even one half of its energy in less than 2,200 years! If I could hasten the process! Man has harnessed steam, electricity, waterfalls. Can he *unharness* the energy of radium? Some day he will, perhaps, and then a bold man may go sky-rocketing to the moon.

I can picture to myself some of the excitement of that journey, and some of the strange technical problems that must be solved. Being an earthly creature, I must breathe air. Since there is no air between the earth and the moon, with the exception of what little we find in our own atmosphere, I must carry my air with me in the form of compressed oxygen. That will be the least of my difficulties: oxygen-tanks are now an indispensable part of every submarine's equipment.

It will be far more difficult to cope with the cold. Brave explorers have frozen to death in the Arctic and Antarctic wilderness. But the cold that they suffered is as nothing compared with the cold that I must endure. Where there is nothing there can be no temperature. But for practical purposes it is safe to say that the temperature of interstellar space must be several hundred degrees below zero.

But that is not all. Heat, too, must be overcome—the heat of the sun. It seems paradoxical thus to mention intense cold and intense heat in the same breath. But heat, in this case, manifests itself when the radiant energy of the sun strikes its target. Suppose it were possible for you to stand between two stars. The side that you present to the sun would be exposed to his blaze, the dark side frozen. To maintain a suitable temperature, you will have to provide covers on the shady side to keep in the heat, and absorbing surfaces on the sunny side to 'absorb all you can. Esnault-Pelterie believes that this object can be attained by providing the car with a proper arrangement of light-absorbing (black) surfaces turned toward the sun and light-reflecting surfaces (mirrors) turned away from the sun.

I seat myself in the sky-rocket car.

How Does It Feel When Gravitation is Removed?

Like every other living organism, man has adapted himself to his environment. Remove me from my environment, and I perish. Without air I cannot breathe. Can I do without gravity? What would be the effect upon me if gravity were removed—and removed it would be at a point a few thousand miles away from the moon?

I cannot walk, or stoop, or write, or lift even so much as an eyelid without being influenced

by gravitation. I weigh 150 pounds, which means that gravitation is pulling me down by that amount. Can I safely endure the removal of a force which every moment of my life ceaselessly acts

Rodman Law actually tried to fly with a sky-rocket of huge proportions

upon me, and to which I have become so accustomed that I no longer heed it?

For aught I know, the organs of my body—my lungs, my heart—are mysteriously dependent on gravitation. It is certain that they would have developed differently if I were not the plaything of that ever-present force. If gravitation were to cease to act, would this not play havoc with my system? This is certain: As I approach the moon I will have the sensation of falling headlong through space—a terrifying sensation, which it might be impossible to counteract by the force of the will.

Rodman Law's attempt to go up in a sky-rocket was a successful failure. He did not go very far, but he lived to tell the tale

In any case, the ordinary methods of moving about would forsake me. Walking would be impossible. Were the car provided with suitable straps and hand-rails, I could use these to pull myself from place to place. In the absence of such aids, I should be strangely helpless. I should very likely sooner or later find myself sprawling in mid-air, unable to move my body from place to place in the car, having nothing to hold on to, and no gravitation to draw me back to the floor. It is true that if there is air in the car I might propel myself backward by blowing with my mouth, or forward by actually swimming in the air by moving my hands and feet.

Tables, chairs, instruments, tools which were not fastened down would be jumbled in chaotic fashion. The wrench with which I tighten a nut, if I release my grip, would not fall, but hang uncannily in space. It might perhaps be necessary to arrange things so that I should at least have the sensation that gravitation was still acting. This could be done by regulating the speed of the projectile, but only at the expense of much power.

After All, I Belong on the Earth

Lastly, there is the difficulty of stopping this sky-rocket, which has been hurled from the earth with a velocity of over seven miles. Brakes of some kind must be applied. Esnault-Pelterie calculates that in the last four minutes of the journey to the moon the motor must be reversed.

And, when I reach the moon, dare I leave the car and walk about on that airless, dead body—that planetary cinder? Even with a tank of oxygen strapped to my back and with a breathing-mask on my face, I might court death. The moon is cold—bitter cold—cold with the coldness of interplanetary space. May I leave my heated car and brave a temperature low enough to freeze the lightest gas? And the blinding, fiercely glowing sun—I must brave that too; for a man on the moon is exactly in the predicament of a man poised in interplanetary space. There the sun hangs in the sky, a sky not like ours, white and blue, but a terrible sky, jet-black even at midday, a sky pricked even at high noon with the stars that we on the earth see only at night.

I am walking in a land of dazzling high-lights and black shadows. It seems like some impossible nightmare.

One minute on the moon would be so harrowing an experience that I would surely rush back to my car, pull the starting lever of my radium engine, and hurl myself back to earth, where I belong.

Talking Across 34,000,000 Miles to Mars

How can the President congratulate some Martian Republic on the celebration of its Fourth of July?

By A. J. Lorraine

MARCONI, pioneer of radio communication, believes that it is possible to communicate with inhabited planets —if the newspapers report him correctly. Mysterious disturbances are recorded by sensitive radio receivers. Are messages coming from Martians frantically trying to communicate with us? Perhaps. At any rate, Marconi predicts the possibility of sending out electric waves to the planets. And if a great radio engineer believes in communication between worlds the matter is not to be lightly dismissed.

It is older than radio-telegraphy, this idea of talking to the stars. Before Marconi sent his first wireless message, it was suggested that we might flash signals into space or draw geometric designs on large surfaces in electric lights. Flammarion, Schiaparelli, Lowell, all of whom passed a life-time in studying Mars, have sponsored the scheme.

How Much Power Would It Take?

Now, the electromagnetic waves used in radio-telegraphy have this advantage over light waves: Atmospheric dust and clouds, both of which obscure light, offer no obstacles to the long waves that Marconi uses. What about the electric eyes, the receivers, that see these radio waves? The human eye is probably more sensitive —about ten times as sensitive, according to recent measurements. But the receiving area exposed to the incoming waves can be adjusted within wide limits, unlike the pupil of the eye, the size of which is fixed by nature.

Suppose, then, we try to send a radio signal to Mars. How much power would it take?

We ask the engineers who designed the radio stations used on the earth. About 400 kilowatts (540 horsepower) are required, they tell us, to communicate under normal conditions over 4,000 miles. Greater ranges are obtained when the conditions are favorable.

It looks black for wireless telegraphy. Still, Mr. Marconi may have reason for optimism. He may see possibilities in the use of directed waves, possibilities that are not obvious to others not so well informed.

Perhaps we shall have to fall back on light, after all. We shall have to establish a code based on the use of electric lamps or searchlights. But how?

One suggestion is that we should make a huge diagram in some cloudless part of the earth of a well known theorem in geometry. "The square on the hypothenuse of a right-angled triangle is equal to the sum of the squares on the other two sides."

That might be a good one to begin with. Every school-boy knows it; an educated Martian ought to recognize the diagram at once. To show that they understood, the Martians might reply by drawing in light some equally well known Euclidean proposition. Picture would follow picture, until at last perhaps some kind of a code would be established. Or, more simply perhaps, something like the Morse system could be applied to flash signals of long and short duration.

What can be seen on Mars? More than you probably suspect. A circular or round spot fifty miles in diameter can be distinguished with a good telescope under favorable conditions— a spot, for example, as big as London.

A dark line three quarters of a mile in width would also be visible. Bright points of light at night, or a line of lights, might be seen, even if narrower than three quarters of a mile. Searchlights have been made for which a brilliancy three times greater than that of an ordinary carbon electric arc lamp is claimed. They are about two thirds as bright as the sun. One of these installations is described as having a total of 1,200,000,000 candlepower. How blindingly dazzling it must be! How far into space it must penetrate! Yes; but Mars at its nearest is 34,000,000 miles away.

Talking by Light Flashes

Talking to Mars by flashes or by lines of light would at best be expensive. Generally speaking, only about two or three per cent of the energy that drives the dynamo by which the light is produced appears in the form of visible radiations. Seven per cent, it is true, is claimed for the mercury vapor lamp and the flaming arc. But how pitifully small is even that seven per cent! Again, the eye, marvelous though it may be, has its limitations.

It is true that even without a telescope you could, in absolute darkness, see a light of one candle-power at a distance of sixteen miles, according to some recent and very accurate measurements that have been made—but only just see it and no more. If energy thus received could be utilized to heat a pound of water, it would require about 267,000,000,000 years to raise its temperature one degree Fahrenheit. But this extreme sensitiveness is needed

Would, perhaps, some enterprising Martian with an eye to business establish an interstallar correspondence school, and patronizingly announce "Elementary Courses" in mysterious subjects for Earthmen and other undeveloped intellects?

If the Martians are superior beings, as Professor Lowell has argued, what shall we say to them? Something simple, something fundamental, to begin with. Thus we might present to them in electric lights strung in the desert of Sahara, as we have shown here, the proposition that "the square on the hypothenuse of a right-angled triangle equals the sum of the squares on the other two sides." The Martians ought to recognize that Euclidian theorem, and reply with another. Picture would follow picture, until at last some kind of interstellar code would be established

because the eye has very serious limitations in another way: The pupil, even when widely dilated, measures only about one tenth of a square inch in area. It therefore catches only a very small part of the light sent out by a distant object. A telescope helps, for it is a light-gatherer.

But suppose that we could send signals of some kind to Mars. What about the Martians? How could we make them understand? Lowell tells us that Mars is very much older than the earth. Therefore, one might reason, the Martians have had time to evolve into creatures intellectually far superior to ourselves.

The Superior Martians

How can we talk to this superior being? Difficult as it would be to establish a basis of common understanding, the obstacles are not insuperable. Helen Keller, blind, deaf, and dumb, has learned how to communicate with her fellow beings. How do we know whether a Martian has eyes? We do not know. But, at least, we can make some deductions based on what we know about nature.

The human organism is not a mere accident. It is the result of a definite set of circumstances. The sun sends out light: so we have developed eyes to see. Sounds are transmitted by the atmosphere: therefore we have ears to hear. Heat is sent to us by the sun: we are equipped to feel it. All this applies to innumerable species of animals. So, too, are ears common to many creatures.

The conditions on Mars are different from the conditions on the earth. But it receives light and heat from the sun, just as we do. Therefore the Martians, we may well suppose, are able to see light and to feel heat. Mars has an atmosphere that carries sound; therefore the Martians have ears. In other words, if under the conditions described many kinds of creatures on the earth have developed eyes, ears, and other senses, it is reasonable to conclude that the Martians have developed senses like ours if their planetary conditions are similar enough to ours.

Now, if the Martians are superior beings, as Lowell has argued, what shall we say to them? In the beginning, something simple, something fundamental. By way of casual conversation, let us flash to Mars the news that twice two is four. Irrelevant, you say? Not nearly so irrelevant as saying "How do you do?" to a friend on the street who is obviously doing very well. The mere fact that we of the earth have actually bridged the immense chasm that separates us from Mars ought to produce an effect so dramatic that the Martians would condescend to speak to us. Perhaps they would regard us with mingled curiosity and patronizing kindness, as if we were rather clever children.

What Would They Tell Us?

What would the Martians have to tell us? Their world is dying—dying for lack of water. Lowell tells us. Would they paint terrible pictures of the tragic end that is in store for them?

Or would they furnish us with thrilling descriptions of their marvelous achievements in science?

Would, perhaps, some enterprising Martian establish an Inter-Stellar Correspondence School for "Earthmen and other undeveloped intellects"?

Who knows? At any rate, if it amuses you, you may speculate about it.

Flying Above the Clouds to Catch the Eclipse

Astronomer would mount his camera on a seaplane and soar to meet the sun

A TOTAL eclipse of the sun can never last more than eight minutes. Usually it lasts much less. An astronomer will travel thousands and thousands of miles to an out-of-the-way place, in order to make the most of a few precious minutes. The actors in a play are no more carefully rehearsed than are astronomers stationed at the various instruments. No one member of an eclipse expedition sees the eclipse as a whole; each one performs the special duties assigned to him.

What if cloud or fog should steal between the earth and the sun? What if it should rain? All these elaborate preparations, all this tedious traveling, go for nothing.

But fogs are always low-lying—never more than a thousand feet thick. Therefore, if cloud or fog creep in between the earth and the sun, the solution is to climb above them and see the eclipse in all its uncanniness.

No wonder, then, that astronomers are interested in the experiment undertaken by Professor David Todd, of the Amherst College Astronomical Observatory, of using a seaplane in which to rise high above the clouds to view the eclipse.

Professor Todd's Experiment

With the assistance of United States Naval officers and a seaplane, Professor Todd set out to take photographs of the sun's eclipse which occurred on May 29. It was planned that the steamship on which the expedition sailed would stop at a point near the equator off the South American coast, launch the seaplane, and then stand by while the astronomer tried out his plan.

It might have been expected that Professor Todd would be the first to carry astronomy into the air. He is the most enthusiastic, indefatigable, and ingenious of eclipse observers. He even went so far, some years ago, as to devise a method of operating a whole battery of astronomical instruments from a central point, but was unable to employ his invention for the observation of the particular eclipse because the sky was at the time obscured.

Although at the time of going to press the results of Professor Todd's experiment have not been reported, it may be doubted if the plan of using a seaplane is practicable. Such is the vibration caused by a seaplane's engine that the steady platform that must be provided for all telescopes becomes a shaking base hardly suitable for Professor Todd's purpose. To be sure, it was his intention to offset the vibration by an elastic mounting of the telescope; but anyone who knows anything at all about the inertia of movable parts will admit that absolute steadiness can hardly be thus obtained.

A More Practical Scheme?

Professor George E. Hale, of Mount Wilson Observatory, has a far more practical scheme, to our mind. His plan is to send an unmanned balloon above the clouds, and to steady the cameras, which the balloon will carry, by means of a gyroscope. Professor Hale plans to study the corona—that ghostly appendage which surrounds the sun, and which is visible from the earth only during an eclipse—at any time.

As we ascend in the atmosphere of the earth we finally reach a point,

perhaps at an altitude of thirty miles or more, where the sky is not blue, but jet-black.

The sky is blue because the air is filled with countless billions of dust particles that diffuse the light of the sun. In the inky canopy of the sky above the region of dust particles, where the air is extremely thin, the stars appear in their proper places even in broad daylight. And the sun is a great blazing ball hung in the blackness. Its wonderful corona, the chief object of study during a total eclipse, gleams in all its pearly beauty.

Should Professor Hale Succeed

If Professor Hale succeeds in realizing his plan, we need not wait for a total eclipse in order to study the corona but we can photograph it whenever we please and study it day by day.

"Hello, Mars—This Is the Earth!"

THE more imaginative modern astronomers are inclined to believe, with the late Professor Percival Lowell, that Mars is inhabited. Assume that Mars is inhabited. How can we talk to the Martians? What a world-wide sensation there would be if we were to receive from Mars a flash in response to a signal of ours!

But how will the scientists signal to Mars? At its nearest, the planet will be about thirty-five million miles away in 1924. Various proposals have been made by Professor Pickering, Professor Wood, and the imaginative Professor Flammarion. In order to visualize and explain how these distinguished astronomers will communicate with Mars, Mr. Max Fleischer has directed the preparation of a motion-picture film for the Bray Studios. Through the courtesy of Mr. Fleischer and the Bray Studios we are enabled to present on these two pages excerpts from the film

The well known French astronomer, Professor Camille Flammarion, to popularize the notion of Mars' habitability, suggested that an electric lights. It would be a costly experiment. A huge tract of Sahara, for instance—would have to be "planted" with the lamps would have to be generated in a power house big once said that he hated to die rich. Here is a chance to

The distinguishing features of Mars are its "canals," first discovered by Schiaparelli and later minutely studied by Lowell. It was Lowell who advanced the theory that the canals are great irrigation ditches. Mars has almost dried up. Water is to be found only at the poles. If vegetation is to flourish, the water of the melting polar snows must be conducted to the arid regions. Hence the canals. Several hundred were charted by Lowell. The canals are the only evidences of intelligent life that we have

To the right we have the earth flashing a message to Mars. Who knows but some day we may tell the Martians all about our great war, all about the struggle for democratic ideals, all about the terrible upheaval through which we have just passed! Perhaps we will learn from an older and wiser planet how we ought to run the Earth

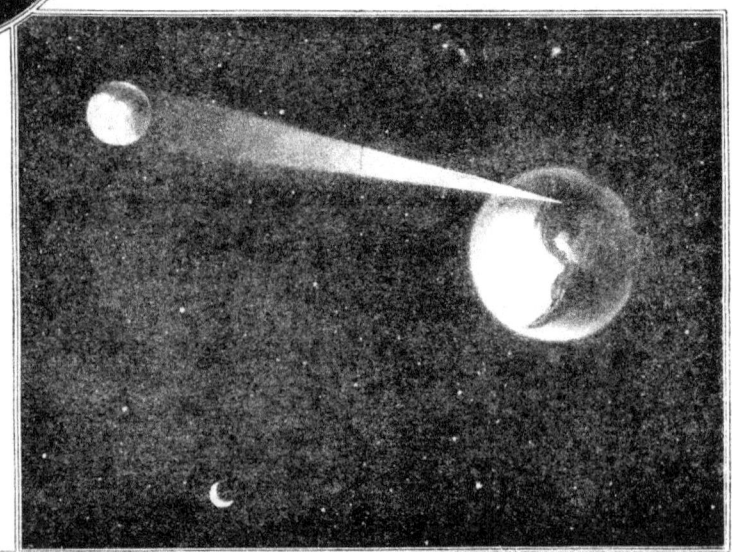

Will the Martians Answer Us?

who has done more than any other man in Europe enormous area on the Earth should be covered with of land—a considerable portion of the Desert millions of lamps. The current to illuminate enough to run a railway. Andrew Carnegie get rid of several million dollars at one swoop

Professor Pickering, of the Harvard College Observatory, suggested that a huge mirror be mounted, to swing on an axis so that it can reflect the sun's rays to Mars from any angle. It would take an enormous power plant to flash dots and dashes to the wondering Martians. And what would they see? As the picture on the left shows, a luminous spot which alternately glows and extinguishes itself

The picture at the left looks like a neatly cut-up farm. It represents Professor R. W. Wood's proposed method of communicating with Mars. The Professor would cover some huge white space on the earth, a portion of the Desert of Sahara, for instance, with strips of black cloth. These strips he would wind and unwind by means of electric motors. The result would be a series of winks. When the black strips are wound up, the white sand below reflects the sun's rays; when the strips are unrolled, the white area is covered. This is probably the cheapest method of optical signaling yet proposed

Drop a cork in fresh water, and very little of it will sink below the surface—to be exact, just twenty-four per cent of it will go under. Cork is very buoyant—much more so than Saturn, for instance

But Saturn has the honor of being more buoyant than any of its fellow planets. It is the only one that would float if they were all thrown into a great body of water. Twenty-eight per cent of it would stay above water. Saturn's lightness as compared with its great size has led astronomers to believe that it is not solid but merely a vast envelope of clouds surrounding a heated interior. Saturn's rings, too, are cloud-like, and so thin that they disappear when seen edgewise

You may not think so, but a human being, breathing properly, can walk into fresh water up to his nose and then find himself floating. He's not so buoyant as Saturn, but at least eleven per cent of him will bob above water. If you have not yet learned how to float on your back, bear this in mind: only eleven per cent of your body will remain above water. If you try to keep your whole head above water, you will find it won't work. Just your eyes, nose, mouth, and chin should emerge

Saturn—It Will Not Sink

SHOULD our eight planets suddenly be dropped into a great body of water, all except Saturn would sink like stones. Saturn, on the contrary, would float gracefully on the surface—rings and all—like a cake of much advertised soap.

That is because Saturn possesses only about three quarters of the density of water. The other seven—Mercury, Venus, Earth, Mars, Jupiter, Uranus, and Neptune—are all denser than water. Our earth leads them all, being about five times as dense. Mercury, Venus, and Mars follow close behind. But the larger, distant planets are far less dense than the nearer, smaller ones. Jupiter is only about as dense as the sun, and Uranus and Neptune have but one fifth of our earth's solidity.

Were our earth suddenly to possess the consistency of Saturn, and were we to retain our present densities, we would promptly sink in and disappear beneath the surface—that is, if the earth were in a fluid condition. If it still remained solid we would not sink beneath the surface, but would feel as if we were walking on something having the density of cherry wood.

Saturn is very beautiful. When seen through the telescope it has a yellowish color and its three strange rings are quite visible. It is seven hundred and sixty times as large as the earth in volume, and it travels very slowly around the sun. The earth's speed is eighteen and one half miles a second, whereas Saturn makes only six miles a second. That is one reason why it takes Saturn so long to revolve around the sun. Instead of doing it in a year, as we do, Saturn makes a revolution in thirty years.

In the course of each revolution Saturn's rings appear and disappear from time to time. They are very thin, and when they are in the same plane with our line of vision we cannot see them at all.

Hitting the Moon with a Rocket

Professor Goddard of Clark College plans to fire off a cracker right under the nose of the man in the moon

By E. F. Richards

A FEW months ago the plausibility of reaching the moon was considered in these pages on the basis of some calculations made by the French aeronautical engineer Robert Esnault-Pelterie, who showed that the most compact explosive known did not carry within itself sufficient energy to convey it to the moon by its own power.

Now Professor Goddard of Clark College, Worcester, Mass., comes forward with computations and experiments that cast an entirely new light upon the situation. In the first place, it is to be observed that, in order to reach the moon, it is not necessary that the explosive employed should possess sufficient energy to carry the whole of its weight to our satellite. For, as a rocket proceeds on its course, it continually discharges a part of its mass, so that only a fraction is carried the whole distance.

Secondly,—and this is where the significance of the recent computations and experiments appears,—it has been found by calculation that the velocity of the gases issuing from the rocket, by which velocity the kick is produced, has an extraordinarily great influence upon the amount of explosive required. Professor Goddard has succeeded, in his experiments, in raising the velocity of the gases discharged by the rocket from 1,000 feet a second, the best performance of ships' rockets now on the market, to 7,000 feet a second. The best rifle hurls out its bullet out of the barrel with an initial velocity of less than 3,000 feet a second.

The full significance of this can be appreciated only when we consider some actual figures. So, for example, Professor Goddard's computations show that in order to kick one pound from the earth to the moon requires, under the most favorable circumstances, an explosive charge of 602 pounds. This is assuming a velocity of 7,000 feet a second for the gases discharged by the rocket. But if we were restricted to a velocity of 1,000 feet a second, as in the ships' rocket, the charge required, per pound carried to the moon, would be the seventh power of 602—that is to say, 14,290 million million tons!

It will thus be seen that Professor Goddard's improvement in the design of the sky-rocket has, at a single step,

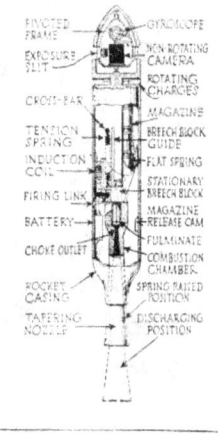

Here are two forms of Professor Goddard's rocket. Note in each case the tapering nozzle, the fundamental feature of the invention, by which the efficiency of the rocket is raised from the best previous record, namely two per cent, to sixty-four per cent. The drawing on the right shows a double rocket, the fuse of the small upper rocket projecting into the charge of the large lower rocket, so that the small rocket is fired off at the moment the charge in the large rocket is exhausted. The drawing on the left shows a type of rocket in which the charge is divided into a number of separate cartridges, which are fired off one after the other. The ignition is effected by a battery and induction coil. In the head of the rocket is placed a gyroscope which keeps the rocket pointed in one direction

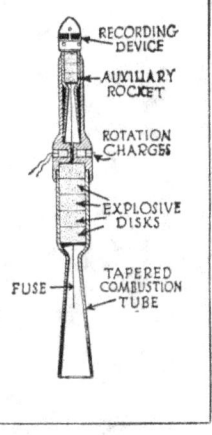

transferred the enterprise of hurling a missile to the moon from the class of utterly impractical **dreams** to the domain of entirely feasible and even comparatively light tasks.

The principal feature of Professor Goddard's improved type of sky-rocket is the tapering nozzle, designed on the principle of the turbine nozzle, so as to utilize the work of expansion of the hot gases. This design has increased the efficiency of the rocket from two per cent, the best performance attained hitherto by any rocket, to sixty-four per cent, exceeding by far the record of even the very best of internal-combustion engines.

For the present we may leave out of consideration the plan of anything like a personal visit to the moon—the chief difficulties here are physiological, not physical. But we can, if we want to, reach out a long arm and tickle the moon, as it were. Here a new problem arises. Suppose we send a rocket to the moon. How shall we know whether our aim has been true and the shaft has gone home? Pro-

fessor Goddard has not only worked out the problem on paper: he has conducted experiments to furnish the experimental data required.

It would, of course, be impossible to follow the course of such a small body through the 240,000 miles that separate us from the moon. But the rocket could be made to carry a charge of flash-powder arranged to go off when it hit the moon's dark surface, the event being brought off about the time of new moon. And the amount of flash-powder required can be easily determined by a simple experiment. Professor Goddard found that one fifth of a grain of powder made a flash plainly visible at a distance of two and one fourth miles. To produce an equally visible effect at the distance of the moon would require, accordingly, a charge of about fourteen pounds. Assuming that the total weight of flash-powder, plus accessory apparatus, were four times this amount, the total charge of explosive required would be about seventeen tons. Professor Goddard's invention is intended to carry aloft registering instruments and even cameras.

No telescope can follow the course of the rocket; but a charge of flash-powder could be carried which would explode when it hit the moon's dark surface

To Hit the Moon with a Rocket

Professor Goddard, of Clark College, has invented a rocket that operates on entirely new principles, and that would make it possible to hit the moon. It has been estimated that with ordinary rockets it would take 14,290 million million tons of explosives to reach the moon. Professor Goddard's rocket requires only 602 pounds, because the gases are discharged, not at the usual rate of 1,000 feet a second, but at 7,000 feet. The fastest projectile hurled from a rifle has a velocity of less than 3,000 feet a second, from which it is seen what a marked improvement Professor Goddard has made. The total charge of explosive required to reach the moon would be seventeen tons, equal to the total weight of ammunition discharged by a battleship when it shoots off all its guns at once. In other words, Professor Goddard's improvement at a single step transfers the enterprise of hurling the missile to the moon from the class of impractical dreams to the domain of comparably simple tasks

What We Don't Know About Volcanoes

by Calvin Frazer

LET us be honest and confess that what we do not know about volcanoes will, a hundred years hence, fill a good-sized library.

Volcanology is in the making. For that very reason it is all the more fascinating. A young man who wishes to make a name for himself (and little or no money) could hardly choose a more promising career than that of volcanologist, provided he has an aptitude for scientific research, the physique to endure rough mountaineering, and the courage to face the risk of being engulfed in fiery lava or buried under a shower of red-hot rocks. There are plenty of brilliant discoveries to be made, and the field is not a bit crowded.

Our grandfathers knew a great deal more about volcanoes than we do, but what they knew was nearly all wrong. They declared that the interior of the globe was a "molten mass." It is, almost certainly, a dense solid, more rigid than steel. How do we know? By several facts—

Long before war made them famous, gas-masks were faithful friends to students of volcanoes. These masked experts, working on the temporary crust of a hot lava lake, are collecting samples of gases

Photograph by Dr. A. L. Day

The weird beauty of the lava as seen at night when waves of living fire break against the mountain crags

its weight (as determined by astronomers), the way it transmits earthquake waves, the resistance it offers to the distorting effect of the sun's and the moon's attraction, etc. But the interior of the earth, as far as we can determine, is so intensely hot that if at any place far below the surface the stupendous crushing force of the rocks above is, by any means, somewhat reduced, the solid must change to a liquid or even a gas.

Earth Tides Cause Eruptions

Various circumstances do, in fact, bring about an easing up of this pressure, here and there and from time to time. The whole globe is shrinking as it cools, and its surface is warped into folds and creases. Moreover, there are regular tides in the earth's crust—a rise and fall of about a foot, on an average, twice a day. These things cause variations in pressure underneath, and the melting of deep-seated rocks probably takes place on a vast scale. It is only occasionally, however, that a little of the liquid rock, known as *magma*, rises to the surface and gives us a volcanic eruption.

How does the magma get to the surface? That is one of the things we do not know about volcanoes, but we have a strong suspicion that what happens is this: Imagine a large mass of melted rock miles underground. It is subject to tidal and other stresses, pressing it this way and that. In the brittle crust of the earth there are many fissures, through which the magma will be squeezed out as far as possible. Suppose one of these fissures, directed upward, to be nearly filled with the magma. Through the magma stream gases, generated from the more volatile elements of the rocks, and these gases accumulate at the top of the column, where they are under such great pressure that they become exces-

Just before the overflow of the lava lake of Kilauea emerging from the lake, which is flowing rapidly from right to

sively hot, according to a well known law of physics. It is probably this hot gas which, by melting the adjacent rock, eats its way through the strata, forming a vent for the magma. The liquid rock may come forth in a quiet flow at the earth's surface, as in the case of the volcanoes of Kilauea and Mauna Loa, in Hawaii, or it may burst forth in a great explosion. The explosive effect is probably much intensified by the presence of steam. Specimens of volcanic gas collected in the crater of Kilauea and pumped from the lava into collecting-tubes, without coming in contact with the air, were found to be heavily charged with steam.

Lakes of Lava with Tides

It is easy to see that if a narrow vertical channel is connected with a large subterranean reservoir completely filled with liquid rock, a small amount of

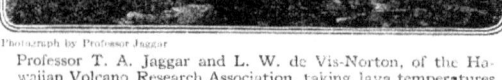

Professor T. A. Jaggar and L. W. de Vis-Norton, of the Hawaiian Volcano Research Association, taking lava temperatures

Photograph by Professor Jaggar

Looking out over the lava lake. The foreground is composed of lava less than an hour old, already breaking up

volcano. The crags are the head of the main lava column left in the picture, while a small fountain is playing at the right

pressure on the latter will cause a rapid rise of the liquid in the channel.

Apparently we have a case of this kind in the Hawaiian volcano of Kilauea, with its remarkable lava pit of Halemaumau. This pit is a small crater within the big one, and it contains a lake of lava, which rises and falls from day to day. Sometimes its surface is hundreds of feet below the summit of the pit, while at other times it rises quite to the top and overflows.

For some years past daily measurements of the level of this lava have been made with surveying instruments by the staff of the Hawaiian Volcano Observatory, under Dr. T. A. Jaggar; and one of the most interesting of volcanological discoveries is that the lava lake exhibits very

pronounced tides, depending upon the relative positions of the earth, the sun, and the moon. The lava is highest in summer and winter and lowest in spring and autumn; and there are minor oscillations twice a day and twice a month. Dr. Jaggar believes

Photograph by Professor Jaggar

Surveying the lava lake, 200 feet below the observer. The usual method of reaching it is by rope ladders

that the intermittent lava flows, on much larger scale, from the neighboring crater of Mauna Loa, are also due to tidal forces, and he justified this belief by predicting, some years in advance, the eruption of that volcano in 1916.

Can Eruptions Be Predicted?

Can the eruptions of other volcanoes be predicted? If so, then volcanology is one of the most important of the sciences in terms of human welfare. This practical problem is not yet fully solved, but it is in process of solution. A sudden succession of earthquakes at a volcano portends an eruption, and the few volcano observatories of the world are all equipped with seismographs — instruments that record the slightest earth tremor. When an eruption is in progress the chemical composition of the volcanic ash gives some indication as to how long the eruption will last.

That our lamentable ignorance of the subject of volcanoes is gradually giving place to valuable knowledge is due to a small number of investigators, some of the most brilliant of whom are Americans. The foremost volcanologists of the world today are the Americans Jaggar and Perret, and their field work has been admirably supplemented by the researches of the Carnegie Geophysical Laboratory, in Washington, under Dr. A. L. Day.

Shortly before the European war began, an international institution of volcanology was established at Naples by Immanuel Friedlaender. It is now temporarily located at Schaffhausen, Switzerland. The leading volcanologic observatories are the one on Kilauea, maintained by the Hawaiian Volcano Research Association, and one on Vesuvius

Tracing Earthquake Records by Rays of Light

A new recorder of earth tremors, which eliminates friction, has been invented at Kilauea Observatory

By L. W. de Vis-Norton, of the Kilauea Volcano Observatory

BY the construction of an instrument that has succeeded in eliminating the element of friction, a very distinct advance in the recording of earth tremors must be credited to Dr. T. A. Jaggar, volcanologist in charge of the Kilauea Volcano Observatory in Hawaii, and to his assistant, Dr. Arnold Romberg.

By doing away with the usual method of employing a stylus to trace a line upon a revolving drum covered with smoked paper, and by employing a system of photographic registration, some very remarkable results have already been attained, though the instrument is still susceptible of considerable improvement.

Record Made by Light Ray

The principle of the new idea in recording is that of an accurate record, made with a ray of light from a tiny straight-filament electric bulb. It was necessary to work with material at hand in the observatory, and one of the four regular seismographs in use was therefore dismantled and reconstructed on the spot.

The lamp was set up at a distance of 150 centimeters from the end of, and in line with, though slightly above, the arm of an Omori 100 - kilogram horizontal pendulum, the tip of the arm being fitted with a magnetized horizontal needle. An ordinary light

mirror, with a diameter of twelve millimeters, was then firmly fastened to a vertical taut silk fiber, held on a post standing on a concrete table; while a second magnet was attached to the back of the mirror in such a manner that it lay at right angles, with its north pole adjacent to the south pole of the arm magnet.

Projected in Ordinary Camera Lens

In this position it will be seen that horizontal earth displacement would move the pendulum and the mirror supports, while the frictionless magnets rotated the mirror around a vertical axis in response to inertia of the mass.

The light ray was projected with an ordinary camera lens, of about thirty meters infinity focus, upon a long strip of sensitized film clamped to the drum, which was speeded up to revolve at the rate of thirty-three millimeters per minute.

Attached to the long arm of the pendulum

was an openwork metal box suspended in a vessel filled with oil, whose effect is to strongly damp out mechanical vibrations, but which leaves the mirror extremely sensitive to earth movement, while a clockwork timing arrangement shuts off the light once each minute, with a longer period at the end of each hour, thus making a perfect timing record upon the photographic paper.

While, owing to imperfections in the available mirrors, the optical work is as yet in its earlier and crude stages, the magnification and openness of the records obtained are truly remarkable, and a vast improvement over the old stylus method of registration. While the records are, in some respects, similar to those of the Galitzin instruments, the finest microtremors are shown in a continuous wavy line, while the slower microseisms group themselves rhythmically, the whole series showing, for the first time, longer wave-movements, which will be made the subject of later investigation, these movements, under the old method, being indeterminate, irregular, and practically illegible.

Many beautiful photographs of even the most rapid movements of small local earthquakes have been obtained, together with a very beautiful seismogram of a strong, distant shock. The difficulties of damping, friction, magnification, opening the record, and time-marking have already been overcome, and the resulting seismograms are clear, transparent negatives.

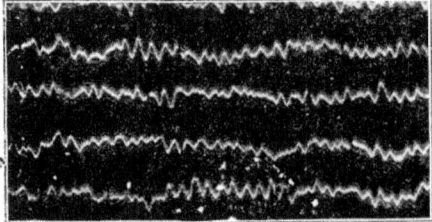

A record by the old method, made by a writing-pen on a drum covered with smoked paper. The perpendicular marks show intervals of one minute. The magnification is small, the phases of tremors indistinct

In this instrument devised by Dr. A. T. Jaggar and Dr. Arnold Romberg, of the Kilauea Observatory in Hawaii, earth tremors are recorded by a system of photographic registration; remarkably accurate records have been made, although the machine is still susceptible of improvement

In the new method by photography the time marks are indicated by small breaks in a continuous line. A study of this record of a normal period of usual earth tremor is of great interest

Jostling the Clouds to Change the Weather

Rain-makers, hail-preventers, and their strange devices

MARK TWAIN'S immortal remark that, though much has been said on the subject of the weather, very little has ever been done about it is one of those epigrammatic assertions that will hardly bear comparison with the facts. It is true that most people accept the weather philosophically as something that cannot be cured and most therefore be endured; but the aggregate of those who, from the earliest historic times down to the present, have declined to assume this acquiescent attitude would certainly populate a New York or a London.

Attempts to reverse the law according to which man is subject to the caprices of the weather have taken an immense variety of forms, and have not always been unsuccessful. When a farmer plants a grove of trees around his house to serve as a windbreak he is in a fair way to modify the weather over a small area of the earth's surface. When a horticulturist lights smudge fires in his orchard on a frosty night he certainly produces a kind of weather in the neighborhood of his trees that would not have prevailed there if Nature had been left to her own devices. Every human dwelling provided with a fireplace, stove or furnace alters the weather within a few hundred cubic feet of the atmosphere, producing an artificial summer in the midst of wintry blasts.

We did not, however, set out to describe these small-scale experiments in weather-making. It is literally true that hundreds of thousands if not millions of human beings have undertaken to control such major phenomena of the elements as rainstorms and thunderstorms, hail, tempest and sunshine.

Weather-Conjuring Is Not New

In that wonderful record of human delusions, "The Golden Bough," Dr. Frazer devotes about ninety pages to a rapid enumeration of the superstitious methods of controlling the weather that have found credence among the various races of mankind. The art of rain-making was practiced by the Chaldean astrologers, who passed on some of their ideas and practices to the Greeks and Romans. To this day there is hardly a savage tribe to be

"Electric Niagaras" are popular in France. A "Niagara" is a large copper lightning-rod, installed in a church steeple or any tall building. It draws down "torrents" of electricity

found anywhere in which rain-making, as well as other forms of weather-conjuring, is not one of the regular duties of the shamans or medicine men

Even the War Has Not Checked the "Hail-Rod" Man in Europe

That faith in weather-making lingers even in highly civilized communities is easy to understand, if one considers the enormous value of good weather, in terms of dollars and cents, and the tendency of the wish to become father to the thought when evidence is being weighed. Not a year passes without the announcement of some experiment in the production of rain in the semi-arid regions of our country and others. In Europe even the great war has not checked the erection of "hail-rods" and the discharge of precious ammunition against that perennial enemy of the farmer and vine-grower, the hail-storm.

The campaign against hail constitutes the most striking chapter in the history of weather-making. The practice of "hail-shooting" ante-

dates the invention of gunpowder. Even in remote antiquity it was the custom to shoot arrows or hurl javelins at the gathering hail-clouds, and the firing of cannon to avert hailstorms was in vogue centuries ago in Styria and northern Italy, and was well established in France before the Revolution.

Early in the nineteenth century a more popular expedient was the erection of hail-rods, or *paragrèles*, consisting of tall metal-tipped poles connected with the ground by a cord or wire. These devices were imitated from lightning-rods, and were intended to draw off the electric charge from the overhanging clouds, the idea being that without electricity no hail would form. At least a million of these rods were set up in various parts of Europe.

Gassing the Hail Storm

In recent times history has repeated itself, and both hail-cannon and hail-rods have been revived. The new vogue of the hail-cannon began in 1896, when a special form of mortar was introduced for the protection of

The gun is mounted on a tripod. Its mouth is fitted with a long sheet-iron funnel. No projectile is fired, but a whirling ring of gas and smoke is sent aloft by the discharge. This is supposed to produce the desired hail-preventing effect

Gunning for the Weather

"Hail-shooting" is older than gunpowder; for even in remote antiquity it was the custom to shoot arrows or hurl javelins at the gathering hail clouds. Nowadays the clouds are bombarded with special artillery. The old idea that cannon-fire precipitates rain dies hard. In Europe the new fashion of megaphone-like cannon began in 1896. By 1900 at least 10,000 cannon had been installed in Italy and 2,000 in Austria-Hungary, not to mention the hundreds erected in France. This picture shows an exhibition of hail cannon in Italy

The most pretentious of all "electric Niagaras" was that installed in the famous Eiffel Tower some years ago

the vine-growing district of Windisch-Feistritz, in Styria. The gun was mounted on a tripod, and its mouth was fitted with a long sheet-iron funnel. No projectile was fired, but a whirling ring of gas and smoke was sent aloft by the discharge and this was supposed to produce such disturbance in the atmosphere as to interfere with the normal mechanism of the storm.

The use of these guns spread rapidly, until by the year 1900 at least 10,000 of them had been installed in Italy and 2,000 in Austria-Hungary, while they had also been extensively adopted in France. More recent explosive devices consist of bombs and rockets, which are claimed to be more efficacious than cannon because they concentrate their effects at the level of the clouds.

Scientific commissions, appointed by the Austrian and Italian governments to investigate the various methods of hail-shooting, pronounced them worthless. Apart from the results of actual experiments, there were two good reasons for arriving at this conclusion: one was that in no case did the explosions produce any appreciable agitation of the air at the great altitude where (as proved by its icy character) hail is formed; and the other was that even if the explosions should occur at such altitudes there was no reason to suppose the formation of hail could be prevented thereby.

They Have "Electric Niagaras" in France

About the year 1899 a new form of hail-rod was introduced in France, and this has since become the favorite means of protection against hail in that country. It is essentially a large copper lightning-rod, installed in some cases on church steeples or other buildings, but more often on tall steel towers erected for the purpose. This device is fantastically called an "electric Niagara," because, according to its promoters, it draws off "torrents" of electricity from the clouds.

About one hundred Niagaras have been installed in France, some of them being set up in rows across the habitual paths of hailstorms, when they constitute so-called "barrages." Their construction and the observations of their effects have been carried on under the direction of a special committee appointed by the French Government.

Similar devices have been erected in Argentina, and plans for introducing them in South Africa were near consummation at the time the war broke out. The consensus of scientific opinion, however, strongly discredits their utility, and a propaganda against their use has also been waged, for obvious reasons, by the numerous hail insurance companies.

Climate of the U. S. Worries Weather Wizards

Other weather-making schemes have been numerous and varied. In the year 1845 a pioneer American meteorologist, James P. Espy, proposed the use of great fires in the western part of the United States in order to regulate the rainfall and winds to the eastward. The fires were to extend in a line of six or seven hundred miles from north to south and were to be set off once a week throughout the summer. Another genius proposed to destroy blizzards by a line of coal-stoves along the northern boundary of the country.

Several plans have been put forth for producing wholesale alterations of climate. An early scheme of this character contemplated the damming of the Straight of Belle Isle, in order to improve the weather of New England and the Canadian provinces; while a few years since a proposal to build an enormous jetty eastward from Newfoundland for the purpose of "protecting the warm north-flowing Gulf Stream from the onslaughts of the ice-cold south-flowing Labrador Current" actually received serious attention from Congress.

Jostling the Clouds to Make Rain

The rain-maker we have with us always. The ancient and apparently immortal delusion that rain can be produced by jostling the clouds with the aid of explosives received much encouragement in 1890-92, when experiments of this character were carried out near Washington and in Texas by R. G. Dyrenforth, with the aid of a Congressional appropriation. The experiments themselves were inconclusive, but they helped keep alive an unfortunate fallacy. In 1911 and 1912 the late C. W. Post, of breakfast food fame, revived the gentle art of bombarding the clouds to produce rain and became involved in a lively controversy with the Weather Bureau on the subject.

In 1891 L. Gatham, of Chicago, patented a method of producing rain, which consisted of liberating in the upper air, by means of bombs controlled by a time fuse, quantities of liquefied carbonic acid gas. His idea was that the instantaneous evaporation of this substance would chill the air over an extensive area to such a degree as to condense its moisture and precipitate it upon the earth.

This is but one of several proposed chemical methods of rain-making. From 1891 onward much was heard of one Frank Melbourne, known as the "rain wizard," whose procedure consisted of locking himself in a barn, freight car, or the like, and burning or evaporating certain chemicals. Sometimes rain followed these experiments and sometimes it did not; but in any case the mysterious doings of Melbourne and his imitators furnished a lucrative attraction for county fairs and an advertisement to the railway companies, who have frequently subsidized the exploits of the rainmakers. The very latest rain-making scheme is reported from Australia, where the government of New South Wales has financed the experiments of a man named Balsillie. His

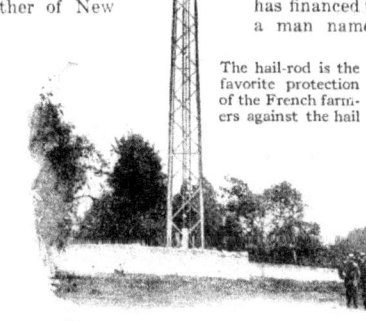

The hail-rod is the favorite protection of the French farmers against the hail

plan is to send up captive balloons to an altitude of a mile or more equipped to produce electrical discharges which are supposed to ionize the air sufficiently to promote condensation.

MT. EVEREST
29 000 FT.
760 YEARS

MT. WASHINGTON
6 300 FT.
166 YEARS

FIVE
WOOLWORTH BUILDINGS
3750 FT. 100 YEARS

WOOLWORTH BUILDING
750 FT. 20 YEARS

If it rained everywhere as it does at Cherra Punji, the water would be over the Woolworth Building in twenty years, and Mount Everest would be submerged in 760 years

The Wettest Place on Earth

OF course the very bottom of the Pacific Ocean (approximately six miles below its surface) is an exceedingly wet spot; but the "wettest place" upon earth, according to the usual meaning of this term, is Cherra Punji, in the Khasia Hills of Assam, India. Here the annual rainfall averages 458 inches, or about 38 feet. This annual average is from January to January; but during the summer months Cherra Punji is

The star shows the wettest place on earth

deluged with about 300 inches of rain. This is a summer average of over 3 inches per day, but more than 30 inches per day have been recorded for five successive days, approximately 150 inches falling in 120 hours. Thirty inches in one day would certainly be more than enough rain for any place on earth, except the Sahara Desert, where the rainfall is zero; but almost 41

inches descended upon Cherra Punji during June 14, 1876. And in the year of 1861 more than 900 inches, or about 75 feet, of rain fell there.

Now, let us see what the average annual rainfall upon Cherra Punji really means. The nearest approach to its 458 inches is at Maranham (277 inches), while at Vera Cruz 180 inches have been recorded. As for New York City, that has about 45 inches yearly, or about one tenth of the rainfall at Cherra Punji.

If the average annual rainfall all over the world for the past two thousand years has approximated 50 inches—this yearly average has been variously estimated — then since the beginning of the Christian era there has fallen from the clouds an amount of water not far from 100,000 inches in depth, or what would be equal to about 8,000 feet— that is, about one and one half miles. And supposing that, instead of an average yearly rainfall of 50 inches, there should have fallen from the clouds 458 inches, then the land-surface of our world—had all this water remained upon it— would have been covered by an ocean some 70,000 feet in depth. In other words, this land-ocean would have extended approximately 8 miles above the 29,000-foot summit of Mt. Everest in Asia.

Certainly Cherra Punji deserves the title of the "wettest place."

Mars Has a Moon that Sets in the East

OUR own moon rises, of course, in the east and sets in the west. So do all the other moons belonging to the other planets, except one of the two moons of Mars. This peculiar Martian satellite, named Phobos, rises in the west and sets in the east.

This seems very mysterious until we are told that Phobos travels around Mars faster than Mars rotates. That is, Phobos is revolving toward Mars' eastern horizon faster than Mars is rotating eastwardly, and therefore Phobos disappears, or sets, presently in the east, and reappears, or rises, in the west.

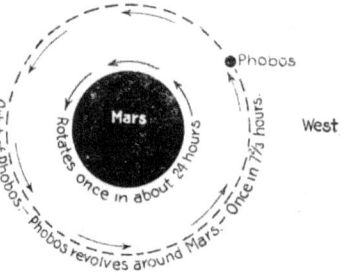

How Phobos, speed demon among the moons, outpaces Mars and rises in the west

He Makes the Weather Pay

The industrial meteorologist advises farmers and aviators

By Lee McCrae

Dr. Carpenter's balloon starting on a voyage of scientific discovery. Air lanes and landing places for aviators are determined on these air trips

MANY a man lives "with his head in the clouds," but none more literally than does Dr. Ford A. Carpenter, of Los Angeles. Yet he is a practical scientist and has been appointed industrial meteorologist of the Los Angeles Chamber of Commerce.

"'Free air' signs here in California are supplemented with free air information," Dr. Carpenter laughed, referring to his work. "I am the advisor of aviators, orchardists, engineers, farmers, surveyors, doctors, and manufacturers, and I am about the most industrious industrial manager you ever saw. I have recently charted the wind currents of southern California, established fourteen safe landing-places for aircraft, given a series of twenty-four lectures on climatology and kindred themes, located soap factories, a rubber plant, town sites, sanatoriums, and——"

"Buzz!" interrupted the telephone at his elbow. A moment later, in answer to a question, he was saying: "No, don't go up today. Your pictures would not be good taken under present conditions. Wait until we have had a rain. Call me after that and I will fix your elevation. Glad to tell you! So long!"

Then he went on: "That is a small sample of the work. I saved that young man a lot of expense trying to photograph Pasadena from the sky. It would have been time, trouble and money thrown away to go up today."

"Then you are a Weather Bureau *plus*."

"Exactly. Plus years of first-hand experimentation. I spent thirty-one years in Government Bureaus from Portland to San Diego, and on that experience I have built. Aviation now demands wind charts and definite air lanes. I am the first observer who has made systematic meteorological studies while in actual flight, and so can instruct flyers to go along at a certain altitude to—say San Francisco, for instance—and return by another altitude in order to have the winds in their favor, and avoid mishaps and possibly death. Recently we spent the night in a balloon noting the movements of night air-currents."

Few people know of the powerful influences of air-currents upon agriculture. Dr. Carpenter displayed photographs of Los Angeles and its suburbs, and told of the climatic survey made of the big Vanderlip ranch which enabled its managers to plant it scientifically.

"I spent last night in an orchard," he added, "rolling in my blanket for three hours' sleep on the ground so that I could give the owner of the orchard data concerning early morning conditions in his grove."

"And today you were lecturing at the university on medical climatology."

"Oh, yes, but the trips into the country and up into the sky are so much recreation. The diversity of the work makes it possible for me to go three days and nights at a time without removing my clothes and with only a few hours' sleep. The very

contact with the earth and the upper air keeps one healthy.

"In fact, medical climatology, linked up with aeronautics, is our next advance along curative lines. Instead of doctors ordering tubercular patients to distant sections, breaking up homes and causing untold misery and expense, they will simply send a bunch of them up in an airship to float at a certain altitude for so many hours a day. The effect will be marvelous. Our army aviators learned that their headaches vanished in their flights, that they could go up with a bad cold and come back without it. It will be my work and that of other practical meteorologists to determine the best strata for these patients. This must be done locally, since different sections are altogether different atmospherically."

Of aviation, Dr. Carpenter declared, "The airship—not the plane—with engine and all within the envelope, using the non-explosive, non-inflammable gas, will be safer and more comfortable than the present railway coach, so it will be a joyful as well as a beneficial trip into the blue.

"Did you know that Kipling is the prophet of aviation? Along in the '90's he originated the term 'air lanes' and all but visualized present aeronautics. The traffic is ready, waiting; we have only to build the ships and map out our ocean to ocean highways in air as on land. That is my chief job now."

Amid all his daily duties, with their interruptions, this citizen of the air has found time to write fifteen books on scientific themes, to lecture in biological universities and clubs, to arrange the gold medal exhibit in meteorology at the San Francisco Exposition, and to qualify as an international pilot of aircraft, ready for the license issued only to the favored few, which permits him to fly over all national boundaries.

Just after a thirteen-hour night balloon trip to gather data on air-currents, valuable to farmers as well as to aviators. Dr. Carpenter is at the extreme right in the picture

Mother Earth Tries to Swallow a Railroad

Thousands of car-loads of dirt have failed to satisfy this strange subterranean appetite

By Howard Egbert

THERE'S a hungry bit of land out in Darke County, Ohio, that is swallowing up a railroad track. Once in a while it bites off a freight engine or a string of cars trying to carry earth enough to stop its maw; but its regular diet is the railroad right of way, rails, ties and all.

More than three thousand car-loads of earth have been thrown into this "hole in the earth," as the engineers call it; but a child might have been at work with sand-pail and shovel, for all the good it did. The engineers won't admit yet that they are whipped, and the fight to save the railroad tracks goes on.

A Subterranean Lake?

A heavy rain always sends the cold chills running up and down their backs, because every rain makes the road-bed more unsteady. One of these days, it is feared, there won't be any road-bed, nor any tracks; and it is said that a whole railroad train might be swallowed up and engulfed.

several cars, endeavoring to stop this earth revolt, suddenly disappeared from view. The engine was never recovered, and is probably reposing a hundred feet below the surface at this point today. An effort was made to recover the cars, but they pulled away from their trucks and the trucks sank from sight.

Changes in Land Contour

Some pent-up force of tremendous pulling power has carried away every cubic foot of earth dumped into this stretch of five hundred feet, lowered the level of the railroad track, and, by some unusual underground operation, has made out of what was once level country a hill twenty-two feet high. Whole rows of trees have been swept into new locations, and the old Fort Recovery road—along which General Arthur St. Clair, pioneer Indian-fighter of America, marched with his little band more than a century ago—

Local tradition has it that an engine and several cars suddenly disappeared from view

is breaking up, crumbling, and disappearing entirely in some spots.

Indian Relics Come to Surface

Not long ago, in an effort to see how far a pole might be driven into the earth, George Kisiman, a railroad foreman, succeeded in driving a piling fifty feet into the earth, when the piling stuck. Pioneers in that section of Ohio, who have been studying the phenomenon for years, contend that the

This road once ran level across the railroad tracks. Now great fissures are forming in it, and while parts buckle others disappear. The large drainage-pipes at the left in the picture below were forced to the surface by an earth movement that is swallowing the road and a railroad

Although thousands of car-loads of earth have been dumped in, the railroad line is gradually sinking. The strange earth movement has moved trees ten feet from their original position

The great fissures which are constantly forming in the road-bed make traveling on the road impossible. Railroad men and geologists have been working for years to solve this problem

Geologists who have been enlisted in the fight say that probably a subterranean lake or river exists not far below the surface of the earth at this point. Their belief is borne out by the fact that the swampy lands on both sides of the track eat up hundreds of carloads of earth, and that pools of water are making their appearance on the surface.

Local tradition has it that some years ago a dummy engine and

piling ran into the forest of trees and logs that were dumped into the lake years ago in an effort to secure a firm foundation.

From time to time the shifting and changing of the earth at this point has brought to the surface relics of Indian days—cannon, trappings, chains, and equipment that General Arthur St. Clair must have used, and probably lost in the treacherous marshes as he marched against the red men.

How Gold Can Make Us Poor

Giant dredges for placer-mining may make the dollar shrink—Unheard-of fields of gold supply—What science is accomplishing

THE more gold there is in the country the poorer we all are. Does this sound startling and unreasonable? It may be startling, but let us see if it is unreasonable.

Gold the Standard

Gold is the standard, the yard-stick, by which the value of things in the United States—and in all countries that have adopted the gold standard—is measured. By it the monetary value, or purchasing power of the dollar, is measured.

It is easy to understand, then, that anything that lowers the intrinsic value of the standard—that shortens the yard-stick—diminishes the value and therefore the purchasing power of money; and it is also easy to see that, just as apples are less valuable when there is an over-supply in the market, so gold is less valuable at times when the quantity produced is greatly increased.

Gold Mining

Now, gold is mined in many parts of the United States, and the quantity of it available for coinage and other purposes is growing steadily. A certain quantity of the precious metal is normally absorbed in the arts and industries without causing a marked disturbance of our monetary system, which expands in harmony with the country's steadily growing wealth. But supposing that a new and prolific source of the yellow metal is found, or that some method is invented which will make it possible to extract the gold from its ores or from placer deposits easier and at less expense; the supply of gold may suddenly increase so that it will exceed the normal demand. The immediate effect would be a depreciation of gold; and as all other monetary values are measured by the value of

The man in the linen duster, standing on one of the dredging buckets, is Mr. N. Cleaveland, who built the giant dredge; and the other man is W. P. Hammon, dredging contractor

This gigantic dredge can dig from 12,000 to 15,000 cubic yards of earth and
gravel a day, and extract from it practically every bit of gold it carries

gold, their value would shrink in proportion. But it is clear that a dollar which is worth less than one hundred cents will buy less than a dollar of full value.

Why You Pay More

If today you pay from twenty-five to one hundred per cent more for your food, your clothing, and other commodities than you had to pay ten or more years ago, the over-production of gold is one of the principal causes. Of course during the war other causes contributed to the higher cost of living; but the upward tendency of the prices of practically all commodities was strongly marked long before the war.

In the years immediately preceding the war, improved methods of mining and the introduction of the cyanide process of extraction caused an increase in the production of the yellow metal, which led to its depreciation and the diminution of its purchasing power, hence to an increase in the cost of living.

In 1880 the production of gold in the United States represented a value of $36,000,000; in 1900 the output reached $79,171,000; and in 1918 $84,456,600.

The value of the metal absorbed for coinage purposes, which in 1850 was $31,981,739, reaching $62,308,279 in 1880 and $99,272,943 in 1900, dropped to $10,014,000 in 1918. This sudden drop was due to the contraction of our national coinage by reason of war conditions. The enormous surplus that thus remained in other than coined form emphasized the depreciation of the metal and helped to boost the cost of living.

In the light of these considerations, it becomes evident that the invention of a powerful dredge, which can dig up from 12,000 to 15,000 tons of gold-bearing alluvial deposits a day and extract practically every ounce of the yellow metal the gravel contains, can scarcely be expected to lower the high cost of living. Such an invention makes it possible and profitable to work over gold-bearing deposits so poor in gold that they could not be worked with profit by any other method of mining. The dredge shown in the pictures was built by N. Cleaveland, and is in successful operation along the Yuba River, near Marysville, California.

The dredge, together with the machinery for driving it and the apparatus for extracting the gold from the soil, is carried by a barge 165 feet long, 68 feet wide, and 11½ feet deep. The entire equipment weighs more than 2,000 tons. One hundred manganese steel buckets, each holding sixteen cubic feet of earth or gravel, are attached to an endless chain which runs over rollers at the top and bottom of

After the gold-bearing gravel and earth has been thoroughly
sifted and washed and relieved of the precious burden
of gold, the tailings are dumped in piles by conveyors

the digging ladder, which is raised and lowered by means of suspension tackle. The buckets have sharp cutting edges and can dig to a depth of 80 feet under water. The bucket chain is driven by electric motors. The dredge is kept at work day and night.

It Stops Once a Week

The earth and gravel dredged up by the buckets is dumped upon a revolving screen at the upper end of the digging ladder. The screen is made of perforated plates of high-carbon steel. It has a length of fifty feet and a diameter of nine feet. A powerful stream of water pumped by electricity washes the gravel and breaks up lumps. The gold is washed out by the water, and, together with other heavy matter in a fine state of division, passes through the screen and is washed into an adjustable distributor, which deposits the gold and other heavy materials on inclined tables indented by shallow grooves filled with mercury. As the gold is washed down it comes in contact with the mercury, forming an amalgam, which is removed from time to time to extract from it the gold. The gravel tailings are carried off by conveyor belts.

Once every week all dredging operations stop. The gold-mercury amalgam is collected, and fresh mercury is poured into the channels of the amalgamation tables. The amalgam is pressed into the form of bricks and sent to the distilling works. There the mercury is recovered to be used again, and the gold is molded into bricks and refined. The cost of operating one of these dredging equipments is about five cents a cubic yard.

This Chilean desert was once the world's only source of fixed nitrogen

CHEMISTRY

Nitrogen in War and in Peace

By Frank Parker Stockbridge

NITROGEN, the most democratic of all the elements, is the essential factor without which the war for democracy could not have been won. In the last analysis, war is an effort to discover which of two sides can liberate the most nitrogen where it will do the greatest damage. For all explosives are nitrogen compounds—their deadly effects the result of the ineradicable tendency of this liberty-loving gas to burst its bonds and hurl in every direction fragments of whatever has served to restrain it. From old-fashioned black powder to the most modern and powerful trinitrotoluol, or "T.N.T.," nitrogen is the basis of them all.

Even more essential, in peace, is the possession of nitrogen in usable form. Without its aid there could be no plant growth, nor could animal life upon this globe continue. Yet, while nitrogen is literally as common as air, since four fifths of the volume of the earth's atmosphere is free nitrogen (serving to dilute the essential oxygen and make it breathable, so that you would not be literally burned alive), the problem of obtaining sufficient nitrogen is one that holds the serious, even apprehensive attention of scientists, economists, and the Government itself.

Before the world became densely populated with people who live in cities, and who therefore depend upon the annual crops produced by others for their food, instead of living on the

fruits, nuts, and game that were the food of our ancestors, nobody worried about nitrogen. People went where food could be obtained; if they failed to arrive soon enough they starved.

When the World Faced Starvation

Up to less than a hundred years ago, the entire human race was constantly menaced by the possibility of famine and wholesale starvation, and nature's methods of supplying nitrogen to plants through the action of bacteria in the soil long ago became too slow to keep pace with the increasing demand of the human race for food.

For a great many years the world has been dependent for its supply of nitro-

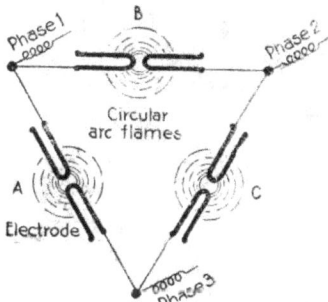

This shows the triangular arrangement of three Birkeland-Eyde single-phase furnaces with six adjustable electrodes

gen for fertilizer upon enormous deposits of sodium nitrate, or Chile saltpeter, found in the high, arid desert regions of Chile and Peru and nowhere else.

In late years there has been an important addition to this diminishing resource—the production of ammonium sulphate as a by-product of the coking of coal. But the total annual supply from both of these sources, about 2,500,000 tons from each, is still insufficient to meet the growing demand for agricultural purposes alone, while the war's demands created a situation little short of critical.

Crookes' Advice to Chemists

Twenty years ago Sir William Crookes, then president of the British Association, startled us by declaring that the population of the world was increasing so much faster than its food supply that the race would soon face starvation unless new means of increasing the earth's fertility were found. His words carried conviction, and his suggestion that chemists turn their attention to the development of practical artificial means of extracting the nitrogen of the air and "fixing" it in usable compounds stimulated experiment in that direction.

As a laboratory experiment, the fixation of atmospheric nitrogen was old. The main essential of all processes then known, tremendously high temperatures, running up even to 6,000°F., made the practical application of any of them doubtful. Chemists, however, set to work. The development of hydro-electric power at Niagara and elsewhere, which made it possible to produce high temperatures through the electric arc, turned attention to this means of accomplishing the result sought. Charles S. Bradley, an American engineer, almost at the time Sir William Crookes was pointing out the imminent need, began the first large-scale experiment at Niagara Falls. His process was not a commercial success, but a little later a plant was established at Notodden, Norway, where water-power costs only $3 a year per horsepower. There nitrogen products were successfully made.

Reducing Nitrogen from the Air

The process used at Notodden and later at several other plants in Norway, known as the Birkeland-Eyde, is not unlike that devised by Bradley. An electric arc is produced by leading a current of about 5,000 volts equatorially between the poles of an electromagnet. This produces what is practically a disk of flame six and a half feet in diameter and having a temperature of about 3,000° F. The disk really consists of a series of successive arcs which increase in size until they burst. Air is passed through this arc.

The first product of the reaction is nitric oxide, which, on cooling with the residual gases, produces nitrogen peroxide. The cooled gases are then led into towers, where they meet a stream of water coming in the opposite direction. Thus nitric acid is formed in the towers, in diminishing degrees of dilution. In the last tower the remaining gases are brought into contact with milk of lime, which combines with the gases to form calcium nitrate and nitrite. The nitric acid obtained in the other towers is combined with a base to form a commercial compound.

These Norwegian plants were financed by Germany, and their output of fixed nitrogen was almost entirely absorbed by that country. At the

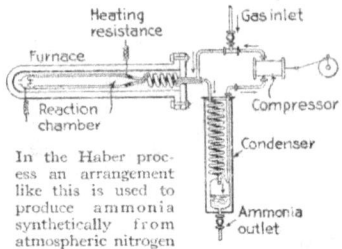

In the Haber process an arrangement like this is used to produce ammonia synthetically from atmospheric nitrogen

beginning of the war the annual production of the Norwegian plants was equal to about 10,000 tons of fixed nitrogen a year. At that time Germany imported annually 880,000 tons of nitrate of soda, equal to 137,000 tons of nitrogen. At the beginning of the war Germany had a stock of 1,000,000 tons of nitrate, equivalent to 156,000 tons of nitrogen, and a by-product ammonia capacity of 550,000 tons of sulphate of ammonia, equivalent to 113,000 tons of nitrogen.

It is known that the Norwegian plants have not been commercially successful. In Germany several other processes for the fixation of atmos-pheric nitrogen were developed, all of which helped to supply the enormous quantity of nitrogen products required in manufacturing explosives.

One of these processes, developed by two German scientists, Drs. Frank and Caro, who began their experiments in 1898, is known as the cyanamid process. It is based upon the fact that calcium carbide, largely produced as a source of acetylene gas, may be induced with comparative ease to absorb nitrogen, thus forming a combination of calcium, carbon, and nitrogen, known commercially as cyanamid. This is the only process that has been installed on the American continent, a plant in Canada, at Niagara Falls, having been in operation for several years, with an annual capacity of about 60,000 tons.

Production of Cyanamid

The carbide is placed in the furnace and the reaction is initiated by local resistance heating to a temperature of 1500°-2000° F., the conversion proceeding to completion without further heating. The nitrogen is obtained from liquid air, manufactured by compressing air to a density of 500 pounds to the square inch and cooling by expansion. When the liquid air begins to rise above its normal temperature of −313° F., pure nitrogen boils off. The compound of calcium carbide and nitrogen, known commercially as cyanamid, is itself valuable as a fertilizer; by treatment with superheated steam its nitrogen may be released to enter into combination with the steam, forming ammonia. In Germany about 600,000

In the electric furnace shown here calcium carbide is made by fusing lime and coke at a high temperature. The carbide combines with nitrogen at 2000° F. to form cyanamid

Making pure nitrogen gas by distilling liquid air formed under a pressure of 500 pounds to the square inch at a temperature of 313° below zero, one of the features of the cyanamid process

tons of cyanamid is being produced annually.

The other process of fixing atmospheric nitrogen, on which Germany mainly relies, is the Haber process. In this, nitrogen and hydrogen gases, under a pressure of 1,500 pounds to the square inch, are passed through a chamber electrically heated to a temperature of 1,170° F. As a result, the nitrogen combines with the hydrogen to form ammonia. Although this process has long been in operation in Germany, its technical details have been carefully guarded, and it took chemical and electrical experts employed by the United States Government nearly a year to discover its secret from patents obtained in this country.

The Haber Process

The Haber process involves the presence of what is known in chemistry as a "catalyst." It has been found that certain elements or compounds —the list is constantly being enlarged—have the remarkable property of causing other substances to combine chemically, often in entirely new formations, without themselves undergoing any change or entering into the new combination. A familiar example is the common device for lighting gas without matches, which con-

sists of a small bit of "spongy platinum," or asbestos fibers coated with platinum black. When this is held over an open unlighted gas-burner, the presence of the platinum causes the hydrogen in the gas to combine with the oxygen in the air with such speed and violence that great heat is generated by the reaction, the spongy platinum becomes incandescent, and in a second or two is so hot that the gas ignites.

In the Haber process the catalyst is spongy iron, although any one of several other substances probably would answer as well.

In the electric arc process the first product is nitric acid, which is directly usable for explosives. In the cyanamid and Haber processes the ammonia product is best adapted for use as fertilizer, but is readily convertible into nitric acid by passing a mixture of ammonia and air through a red-hot platinum screen acting as a catalyst.

The fact that nitrogen can be fixed directly in the form of sodium cyanide by the action of nitrogen gas on a mixture of soda and coke has been known for many years; but, while English, German, and American scientists have tried their hands on a commercial adaptation of this reaction, it is only recently that an American firm has been able to prepare sodium cyanide for the market by this process.

A Low-Temperature Process

Intensive study of the various methods of speeding up the reaction has led to the adoption of special apparatus and, at a temperature around 1800° F. with the assistance of a specially developed catalyzer, an unusually pure cyanide is formed. From the cyanide it is easy to prepare ammonia quantitatively.

One advantage of this sodium cyanide method of fixation, aside from the low temperatures used, is that when ammonia is made from the cyanide, another product of commercial value is also obtained—sodium formate. This latter material can be used as a starting-point for a number of artificial flavoring oils, for a whole line of useful solvents, or for making the formic and oxalic acids that are so necessary in our dyeing processes, for instance, and which were formerly imported chiefly from Germany,

The crude nitrate is shoveled into leaching-vats, where it is purified

The nitrate mining in Chile is nearly all surface work at little cost

Fighting Dust with Dust in Coal-Mines

It's in the non-gaseous mines that most of the dangerous explosions occur; and coal-dust is more deadly than gas

MIX any powder or dust that is composed of organic matter with the necessary amount of air, and you have a mixture as explosive as that which you use to drive an automobile. Flour explodes and sets fire to mills; threshing-machines burn up because crop-dust has exploded; and coal-mines that are quite free from gas are converted into death-traps because coal-dust has been allowed to form an explosive mixture with air.

Between four and eleven hundred American coal-miners lose their lives each year as the result of coal-dust explosions in mines that are considered safe because they are free from gas. In Belgium, which has the most gaseous mines in the world, not an explosion took place from 1892 up to the outbreak of war.

Causes of Coal-Mine Explosions

As the advertisement has it, "There's a reason." To find the reason, the Bureau of Mines has been making a very elaborate and painstaking series of investigations, as a result of which it has been able to make recommendations that, if followed, ought to make American mines as safe as those of Belgium.

Explosions in coal-mines are caused either by an open light igniting fire-damp or gas, or by an explosive igniting coal-dust. Without an open flame there would be no explosion. Matches, open lights, explosives which give off flames, mine fires, and arcing electrical machinery are all possible causes of explosions.

Miners Become Callous to Danger

Government engineers who have studied these conditions believe that if safety lamps or permissible electric lamps were used even in so-called non-gaseous mines, more than one third of all the explosions in the country would be prevented. Permissible explosives produce so small a flame and their flash is so brief that if they are used intelligently they will not ignite gas or dust. If dangerous explosives were excluded from the mines, fully one half of the present number of explosions would be prevented.

Even a very small percentage of fire-damp is dangerous, because it will render coal-dust explosive. Without the presence of the gas the coal-dust would not be explosive. The introduction of as little as one and one quarter per cent of natural gas is enough to ignite coal-dust.

Miners become so accustomed to the presence of dust that they overlook it. The fact that the coal-dust is generally scattered throughout the mines makes the danger of an explosion great. Once an explosion is started, it will sweep wherever coal-dust exists.

There are two methods by which coal-dust may be treated so as to render the mine safe from dust explosion. One of these is to wet the dust, and the other to cover it with rock-dust or some inert dust. Coal-dust is dangerous only when suspended in the air.

To Stop a Coal-Dust Explosion

As a means of stopping explosions that have started in spite of precautionary measures, a number of "barriers" have been tested.

The use of rock-dust barriers originated with J. Taffanel, of the French mine-testing station at Lievin. He found that by blockading the testing gallery with earth he could stop an explosion. This not being practicable, he worked out a method of placing shelves laden with rock-dust across and over the roadway.

The Taffanel shelves were tried in the tests made by the Bureau of Mines, and, while found successful, were not as sensitive as was desired, so six new types of barriers were designed. The principle on which the barriers are operated is to launch non-inflammable rock-dust into the air-blast that precedes a coal-dust explosion. The eddying currents mix the non-inflammable dust with the coal-dust; so, when the flame reaches this dense cloud, the rock-dust, being more plentiful than the coal-dust, absorbs the heat and extinguishes the flame.

To prevent coal-dust from exploding in a mine, barriers of rock-dust are set up. When the coal-dust explodes the rock-dust is shaken down. This prevents any further propagation of the coal-dust explosion

They are going down to drag men out of the coal-mine. It was not a mine filled with gas. Then why the smoke? The mine was full of coal-dust, which formed an explosive mixture with the air. Then the mixture was ignited by a flame

Keeping Meat Fresh a Hundred Years

Condensing the ox to preserve him forever: a new marvel of chemistry

By John Walker Harrington

These three articles are cod, sea-bass, and salmon cutlet, all dried in a vacuum oven. They represent the result of the newest scientific method of preserving fish, meats, and vegetables

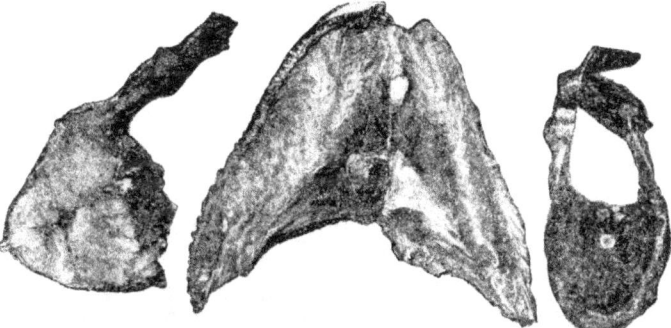

MEAT that can be shaken dried from a paper bag at the sound of the voice of the unexpected guest in the hall, and so served that the dear friend from the office will believe that its source was the family refrigerator, is the latest boon that science has evolved.

It means something to have on hand beef, sea-bass, and liver, as dry as the husks of the Prodigal Son, and yet capable of being converted by a little soaking into foods that have the flavor of cuts just from the butcher shop around the corner. Think what that means! You might be able to drop in for luncheon in your airplane at Mac Millan's headquarters at the North Pole next year, and eat what tastes like a perfectly new Delmonico steak.

The remarkable process by which the flavor is retained in flesh food was originated in the Harriman Laboratory, an institution founded by Mrs. E. H. Harriman in memory of her husband. As the late financier once told Dr. W. G. Lyle, the director of the Laboratory, he would have been a physician if his mind had not been turned to railroading.

The First Order

The laboratory devotes itself largely to discovering the errors commonly made in stoking the human furnace. When the United States entered the World War, the Government took over the establishment, which is situated on the grounds of Roosevelt Hospital, in New York city. The first instruction it received was from Lieutenant-Colonel J. R. Murlin, of the Division of Nutrition and Food of the Medical Department, United States Army. He wished to know

what could be done to prevent meat from spoiling without injuring its nutritive value. It was a hard riddle to tackle, but Doctors Falk and Frankel, of the laboratory staff, reached the goal. They were assisted in applying the process on a large scale by Professor Ralph H. McKee, head of the Department of Chemical Engineering at Columbia University.

This Meat Will Keep Forever

The thing that causes meat to become unfit for use, unless it is cured or refrigerated, is the breaking down of its chief constituent, protein, which belongs to the same family as the white of egg. Any egg antiquarian would be able to give us an idea of just how bad such substances become. The jerked beef of the Western plains—dried for days and sometimes weeks in the sun—is so tasteless that the Indians and the Mexicans depend on red pep-

pers to make their palates register food. The dried beef of the grocery store is treated with salt to keep it and with saltpeter to retain color. The new method does away with expensive and long-drawn-out preserving and smoking processes.

Its basic idea is the drying of the meat in a vacuum oven at so low a temperature that, after the water has been driven off, the protein and also the delicate ferments and aromas are not injured by the heat. The principle is the same as that of the vacuum kettle on which the candy-maker depends to keep his wares from scorching or growing tasteless in the making. For certain obvious reasons, the exact temperature used and other matters of the sort are for the present withheld; but the proof of the theory is in the eating. I have tasted all these products, and have seen them being dried.

First, the meat, cut either into inch cubes or in thin steaks, is laid on trays of wire netting. The trays are placed in an oven six feet square, which looks not unlike a bank safe, so heavy are its iron sides. Alongside is a pump that draws out the air and leaves the interior of the oven almost a vacuum. The moisture from the drying food is sucked into the big cylinder alongside of the oven, known as a condenser. The oven is kept warm by passing hot water through tubes with which it is lined.

It's Condensed Cow Now

The length of the process depends upon the size of the pieces. About ten hours are needed for such meats as beef and mutton; and from four to eight hours for fish. The meat is first freed from bones, gristle, and

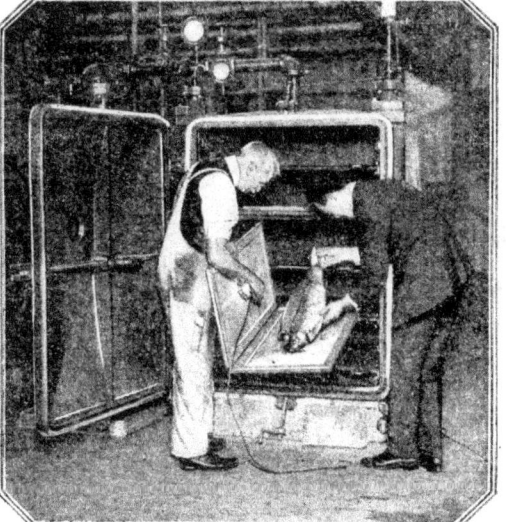

A dried carp is just being taken out of an oven from which all the air was pumped out before the drying process took place

superfluous fat, which at once reduces a carcass to half its bulk. The cubes come out one third of their original size and approximately one sixth of their former weight.

A condensed cow, allowing for the discarding of the bones, would therefore be eight or nine per cent as bulky as when she browsed the meadow. She has gone by the same vacuum route over which her milk may have gone to the condensery

Transportation Charges Reduced

Should this process come into general use, it would greatly reduce transportation charges on all meats. It is estimated that, commercially applied, it would cost as much as refrigeration. Vessels that cross the sea with chilled or refrigerated beef give up fully as much space to the necessary apparatus as they do to the brightly hued quarters themselves.

What a great saving in cargo space would result, then, from desiccating steers of the Argentine down on the pampas, sending them across the Atlantic, and steeping them in water to expand them into the choice cuts of the beef of old England!

A large proportion of the cost of meat in the United States is due to its transportation, first on the hoof to the railway, and then on the ice to the East. Heavy charges are those which pile up on the way from the stock-yards of the packers to the steel-yards of the butchers.

Question of Cost to the Consumer

Just what the prices of meats and fish would ultimately be to the small consumer depends upon the interest that the wholesale dealers take in the matter. Many thousands of tons of the flesh of the creatures of land and

sea are annually wasted which could be saved, were they not all eaten up in freight charges.

A Wafer of Liver

The Harriman laboratory is controlling the patent, with the idea of keeping it in trust for the public good.

The oven itself looks like a safe, so thick is the iron door. The moisture is sucked from the drying food into the big cylinder alongside of the oven

Food prepared in the new way is germproof. Even when exposed to moisture, only a slight coat of mold forms, and this is easily removed. Cheap containers of paper or pasteboard keep it very well, and large quantities can be packed in barrels, or even in burlap bags, without injury. For quick service meat in powder form is excellent.

Meat powder (*center*) that will some day be hash. On each side are pieces of meat before treatment and (*above*) after, when they are one sixth their original weight

The other day Dr. Frankel handed to an interested visitor a thin brown wafer decorated with a net work design.

"Very delicate," commented the favored one. "It's a new kind of chocolate, I should judge."

"Liver," was the reply.

Cut very thin and dried, liver can be preserved indefinitely. Fish having not a trace of salt, and retaining the fine, full flavor of sea food, are coming daily from the big airless oven at Columbia.

Also Dehydrated Vegetables

Vegetables also are being dried by this process, and the indications are that they will take a high place among dehydrated products. They retain the color and the taste of the garden, and experiments reveal that they also keep the invaluable vitamines—those little-understood forces that strengthen and nourish the body and keep away scurvy and wasting diseases.

A number of very fat rats, which are serving as test animals, are willing to testify to the worth of these supplies from the truck-patch.

Invitations to luncheon consisting of the vacuum dried foods are issued from time to time by the Harriman laboratory. The director has a French chef who has prepared some very delicious stews from the meats and vegetables sent from Columbia; but then, a French cook can concoct a dream out of an old soup-bone and a garlic suspicion.

Even Dried Hash Stood the Test

Hash made by American boarding-house experts has stood the best tests, however; and down at Camp Oglethorpe in Georgia, last summer, three hundred soldiers were served dried beef stew and never knew that they were not getting fresh cuts, although they had it for weeks at a time.

There will come a day, no doubt, when both the soldier on the march, and the housewife who can redeem herself as a bountiful hostess by just adding hot water and serving, will be singing the praises of the new processes.

Nature may abhor a vacuum, but human nature will sing no hymn of hate over delectable provender out of the heated void.

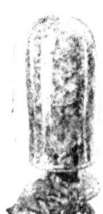

Vegetables also may be dried. Here are cabbage, carrots, potatoes, tomatoes, and cauliflower. They retain their color and taste and also the indispensable vitamines

Blood Will Tell

Is man descended from the monkey? Are you well or ill? Your blood crystals will tell

By Anna Heberton Ewing

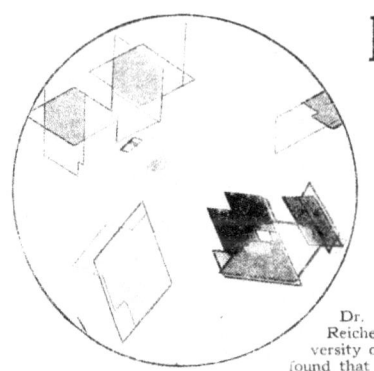

Dr. Edward Tyson Reichert, of the University of Pennsylvania, found that the blood crystals of a horse look like this

Blood crystals from another horse. The formation of crystals is the same, but the measurements and the groupings are different

THEY found the body of the dead man in his room. He was a Frenchman who had lived alone. It was clear that he had given up his life only after a terrible struggle. There was blood on the floor and on the walls—blood everywhere except upon the body itself. Nor were there any wounds. The man had been strangled to death. And the blood? The dead man must have wounded his murderer.

And so the detectives of the French town in which the crime had been committed looked about for a wounded man. They did not find him. There were finger-prints enough. They revealed nothing; for they did not correspond with any finger-print records at police headquarters.

At last it occurred to an official that perhaps the blood with which the room was so liberally bespattered should be analyzed. That was done. It was not the blood of a human being at all, but the blood of a bull!

Two Strange Murder Cases

The crime was more mysterious than ever. Here was a murder which had been committed by strangulation; the finger-marks on the throat were those of a strong man; yet the blood in the room was that of a bull. True, it would be easy to obtain blood from a slaughter-house; but why?

Someone remembered that one of the few persons who disliked the murdered man was one who worked in a slaughter-house not very far away. He was arrested. His finger-prints agreed with those upon the wall of the room where the crime had been committed. The man confessed; he was the murderer. Yes, he had spattered the blood of a bull around the room. Why? So that he might insist, should he be arrested, that he had fought and killed in self-defence. To bear out the story, he had even cut himself.

Another case:

The only evidence of a murder upon which the police could work was a pair of blood-stained trousers. The suspected murderer grieved, apparently sincerely, over the death. Indeed, he had evaded suspicion to a certain extent by taking an active interest in the investigation. When the trousers (his trousers) were discovered he assumed an air of outraged indignation. He had killed a goose shortly after the murder, and had splashed himself with its blood. The story was plausible; the man had kept poultry. The District Attorney ordered the blood examined. It was the blood of a human being. The man confessed.

Science to the Aid of Law

Thus science comes to the aid of the law. To Dr. Edward Tyson Reichert, the internationally famous physiologist and biologist of the University of Pennsylvania, belongs the credit of having built up the new science of blood crystallography, which has made it possible to bring criminals to book so surely. But that is, after all, only one phase of the wide application of Dr. Reichert's discoveries. There is hardly an aspect of plant and animal life which is not illuminated in some way by Dr. Reichert's work. Blood has always been held in a kind of superstitious regard by humanity. Hundreds of proverbs have blood for their theme. "Blood will tell" is one of them.

Just what it tells, Dr. Reichert's investigations begin to show us.

It all came about in a very curious way. One day a scientist in a laboratory was interrupted in the examination of a drop of blood. Impatiently complying with the demand upon his attention, he left his task for a few minutes. Returning, he resumed his work at the microscope. To his astonishment, he beheld upon the slide a totally transformed specimen. Hardly realizing the great

These small, bar-like crystals are found in the blood of the leopard

A tiger's blood crystals. The darker formations indicate thickness, not color

Because of the hybrid character of the mule, its blood presents an interesting study

One of the three forms of blood crystals in a human being; another form is like prismatic rods; another diamond-shaped

One of three forms of crystals in the blood of anthropoid apes. The diamond shape is similar to one of the human blood crystal formations

significance of the change that had taken place, he nevertheless recalled the way in which he had prepared the specimen. He experimented again. Once more the peculiarly formed crystals appeared. Scientists became interested and repeated the experiment, but made nothing of it. To Dr. Reichert and some other specialists the red crystals with their sharp edges and flat surfaces presented a scientific problem of irresistible interest. Did the crystals in blood really convey a message of which any practical use could ever be made?

Dr. Reichert decided to solve the problem. He secured blood of wild and domestic animals, the former with danger and difficulty. He made tedious and refined tests of human blood. An exhaustive study involving years of patient effort and highly specialized knowledge in biology, crystallography, and physiology began. At last he succeeded in disclosing scientific facts of inestimable value to every scientific man who studies living things.

Dr. Reichert's Discoveries

The blood is an extraordinarily complex fluid which consists of what is called the plasma, in which living cells, "corpuscles," are held in suspension. Most of us think of blood as red; yet not all blood is red. In the lower animals the blood corpuscles may be colorless or colored, and if colored they may be green, red, yellow, blue, violet, purple, madder, mahogany, brown, or lilac. Some blood has corpuscles of varied hues.

In all cases perhaps the principal function of the blood and in particular of the colored constituent of blood is the assimilation of oxygen from the

Bloodstains used as court evidence in blood crystal tests should be fresh, or only slightly clotted, to yield positive results for testimony

air. We breathe in order that our blood may breathe; for we care about oxygen only in so far as our blood corpuscles care for it.

Now, one of the discoveries recorded by Dr. Reichert was that the red coloring matter of our blood, which is called "hemoglobin," is closely related to the green coloring matter of higher plants, called "chlorophyl." Our blood is red merely because it contains iron; the blood of an octopus is blue merely because it contains copper.

The red blood corpuscles of the higher animals are inconceivably numerous. It has been estimated that the total number of cells in the human body is 26,500,000,000,000, and that of this number 22,500,000,000,000 are red corpuscles. Think of this vast crowd of corpuscles—numbering in the case of man more than 10,000 times the population of the earth—hurrying through the channels of our system at such a rate that the majority of them complete one entire circuit in the space

of less than a minute! The traffic of the New York subways is slight in comparison.

It is the crystals formed by blood which reveal so much to Dr. Reichert. Suppose he has a specimen of blood to be examined. Dr. Reichert adds oxalate of ammonium to prevent coagulation. Then he shakes the mixture with ether to free the hemoglobin from the corpuscles in which it is found. After that the ether is separated from the mixture, and some of the latter is placed on a microscope slide, protected with a glass cover, and sealed with Canada balsam.

Slowly the crystals become visible under the microscope. They can be identified by reference to the Reichert classification of blood crystals.

Soon after he began his investigations, Dr. Reichert found that the blood crystals of one species of animal can be distinguished from those of others and that blood crystals of the human being can be differentiated from those of the lower animals.

Blood of Apes and Human Beings

Striking is the likeness between the blood crystals of monkeys and human beings. Such close similarity does not exist between the crystals of the monkey or human being and those of any other living species. Blood crystals under the microscope shed a flood of light on Darwin's theory.

Dr. Reichert hopes to distinguish between various nationalities by blood tests, to fix race relationship more scientifically than is now possible, and even to trace hereditary traits. He has also directed his attention to the study of the cause and prevention of such phenomena as two-headed children, one-eyed calves, etc.

PLANT AND ANIMAL LIFE

Why I Know that Monkeys Talk

Dr. Garner's Life Work—His
Proofs of an Animal Language

By R. L. Garner

IT matters not to me whether you believe that monkeys talk or not, and it is not the purpose of this article to convince you either way, for convictions are no more a matter of choice than noses are.

The purpose of this article is to present the essential facts thus far tabulated on the subject of simian speech and allow the reader to draw his own conclusions.

The word *speech* is used throughout as a more exact term than *language*, which is often used in an ambiguous or figurative sense. Let us begin then, by asking: What is speech?

My reply is:

Any oral sound voluntarily uttered with the definite purpose of conveying a preconceived idea, concept, or impression from the mind of the speaker to that of another is speech.

From this plain and simple premise we proceed to collect the salient facts on the question of simian speech and briefly recount how those facts have been formed by many years of methodic research. The limits of space preclude many minor and incidental observations that corroborate the main facts.

All through my early life I observed instances of intercommunications between animals. For some years my studies were

In the picture at the right is Susie, one of Dr. Garner's subjects, learning to spell. She could assemble three letters of her name, but had no idea of what she was doing

Susie laughing and threatening to tickle Dr. Garner, who had just tickled her. Her laugh sounds very much like the chuckle of a human being

A lifetime of study and nearly a quarter of a century of self-imposed exile in the African jungles in a search for the truth about animal speech are summed up in this article. It was Dr. Garner's last word on the subject to which he devoted his life; for he died at Chattanooga, Tenn., shortly after preparing the manuscript for the article "Why I Know that Monkeys Talk"

only casual and the results incoherent; but my progress, though slow, was constantly in one direction, for I had faith in my own ability to solve the riddle of speech. In the meantime, I had sorted out certain sounds that appeared to qualify as elements of speech and others that did not. The former were voluntary, more or less modulated, and expressed a desire; while the latter were involuntary or accidental, and expressed no deliberate mental process. The one group I classed as speech sounds and the other as anomalous sounds.

One day I visited the Cincinnati zoological garden, where I saw a large mandrill caged with a lot of small monkeys of three or four different species. The cage was divided into two compartments with a small doorway between them. It was quite evident that the big mandrill was a source of terror to the small monkeys. I noticed that some of them were constantly watching his movements and from time to time uttering peculiar sounds. It was also clear that the sounds conveyed some idea to the small monkeys which inspired them with fear or quieted them, according to the conduct of the mandrill. I spent the whole day watching those animals until I was convinced that they could understand the meaning of the sounds well enough to be guided in their actions by the information conveyed. This incident opened a new avenue of study.

Among the great difficulties in determining the speech of animals, not the least is to distinguish the exact quality or intonation of sounds made by the same animal at different times, to remember the actions that attend them and the results that follow them. It took me a long time to learn that no two species of monkeys had the same vocabulary, and that strange monkeys of different kinds, when first brought together, could not understand each other, though they learned readily.

After countless difficulties, I went to Washington and sought the aid of Dr. Frank Baker. He let me have the use of two monkeys which were kept in a small annex of the Smithsonian Institution. Taking a gramophone to the building, I first placed the two monkeys in different rooms so that neither of them could see or hear the other. Then on the wax cylinder I made a record of the sounds uttered by the male monkey.

This was not difficult, for he was in a loquacious mood. Taking this record into the other room, it was reproduced to the female. She evinced great interest and anxiety. She rushed to the horn, looked into it and all around

it, thrust her arm into it, and chattered to it.

Then a record of her voice was made and repeated to the male, who became more excited and vociferous than ever. By repeating and varying these experiments I was convinced that these two monkeys absolutely understood

the sounds thus reproduced. Dr. Baker was likewise convinced.

After a cursory study of several specimens elsewhere, I selected the brown capuchin monkey because it was one of the most talkative. Incidentally I observed that there were certain sounds that they used more frequently than others. Upon one of these I focused my efforts, and by noting the

Dr. Garner in the cage where he sat motionless for long hours listening to and recording the talk of the jungle folk. The cage was made in twenty-four sections which when assembled made a cube six feet six inches square

actions of the monkeys when uttering or hearing that sound I soon began to make deductions as to its meaning.

The method by which I proceeded is so simple that any novice can follow it. I selected a young capuchin monkey in Central Park and made a clear record of its voice on a phonograph. In fact, I made several of these, each containing the sound that I regarded as most important. These cylinders were taken to Cincinnati and there reproduced to a specimen of the same kind whose conduct was carefully studied. A second machine recorded the sounds made in response and at the same time the conduct of the second monkey was noted.

Having made a score or so of such records and duly tab-

ulated the actions of the animals at the moment of uttering or of hearing the sounds, I was enabled to carry about with me and study those sounds at leisure, to compare them with others, and ultimately to make a tentative translation of some of them. With those records and data I went to Chicago, where the experiments were continued and amplified. After adding several new cylinders I returned to New York to resume and elaborate the experiments.

By certain manipulations of the phonograph, such as changing speed, reversing the cylinder, and other means, the sounds can be converted into divers forms, analyzed and studied in many aspects. Such experiments show the essential difference between musical notes and spoken sounds, which the keen ear of the monkey perceives more readily than our own ears do.

Now and again a new sound was added to the list and the experiments extended to four or five other species of monkeys. Finally, the chimpanzees in the Cincinnati Gardens confirmed my opinion that the higher types of animals had the higher types of speech, and this fact induced me to go where I could study the gorilla and chimpanzee in a state of nature.

Allotted space here precludes even a synopsis of my seven voyages to tropical Africa, where I have lived most of the time for twenty-seven years, during which I have owned and studied on my premises thirty-nine specimens of those apes, besides a greater number in a wild state.

Living alone in the depths of the great jungle, cut off from all social and intellectual intercourse with my own race, having no companion but an ape which was likewise isolated from his kind, it is surprising how quickly and how well we learned to

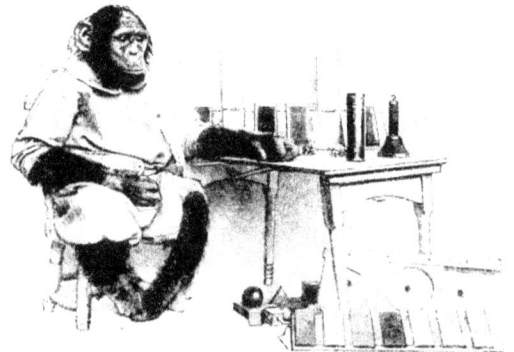

Susie in her own kindergarten, where she studied colors, geometrical forms, and numbers. With the bell on the table she would summon her keeper whenever she needed his attention

understand each other. In summing up the results of my researches in the African jungles I would cite the following cogent facts:

The phonograph shows that the higher types of simians have a greater vocal range and a greater number of phonetics, more clearly enunciated, more uniform in quality, and apparently more definite in meaning than have animals of inferior types. Those characters are more marked in the chimpanzee than in any other animal below man.

The next fact in the order of importance is that certain oral sounds of simians are essentially the same in contour and phonetic quality as certain sounds of human speech. Conspicuous among these are the basic sounds of deep "a," as in *war*; short "a," as in *hat*; long "u," as in *blue*; short "u," as in *hut*; short "o," as in *cot*; occasionally long "o," as in *more*; and the diphthong "eu," as in the French *peu*. Every simian does not utter all of these sounds; but the chimpanzee does, and there are other sounds more obscure.

While it is impossible to represent most animal sounds by letters of the alphabet, all of the sounds here cited are capable of being articulated to consonant elements, or vocalized, as it is technically called; and some of these, as uttered by the apes, actually carry in them incipient consonants, such as the initial and vanishing sounds of the semi-vowels "w" and "y," together with perceptible gutturals and labials. These features suggest a transition state in the evolution of speech and warrant the assertion that the phonetics of the ape are about as nearly like those of man as the physical type of the ape is to that of man.

The next item is that certain oral sounds of simians are recognized by other monkeys of the same kind, and their meaning is sufficiently definite to evoke a uniformity of response that justifies the assumption that those particular sounds have a meaning that serves the purpose of the animal, just as human speech serves that of

Susie had just had her photograph taken by flashlight. She didn't like it and was disinclined to face the machine again, but Dr. Garner told her in her own language that it was all right, and you can see by Susie's expression that she was going to take his word for it although still rather anxious

man; that the same sound usually produces the same effect upon those that hear it, and that certain other sounds uniformly produce certain other effects upon them.

Note also the fact that the sounds are habitually addressed to some particular individual or group, with the evident purpose of evoking a response from the object addressed, as must be inferred from the speaker repeating the

sound until a response is elicited; and it is apparent that the speaker is conscious of a definite meaning to the sound he utters, since observation proves that no simian habitually utters those sounds when alone.

The accuracy with which a monkey regulates the loudness, pitch, and quality of tone shows that he is aware of the values of speech sounds as a means of communication; and this fact implies that he possesses both the instinct and the faculty of speech.

It has been shown that all simians recognize and apparently understand the vocal sounds peculiar to their own race when those sounds are imitated by the phonograph and other mechanical agencies. These facts show that the sound alone is the medium of conveying the concept.

The vocabulary of every race of animals is measured by its actual, normal need. It consists of a few single sounds of categoric meaning, which are not qualified by any auxiliary terms or united into sentences. The paucity of words does not lessen their reality as speech. A word is the smallest unit of expression, but it is speech just as a single drop of water is as real water as a tubful.

All data focus upon the conclusion that every simian has the faculty of speech sufficiently developed to express any desire, need, or mental process as clearly as he is capable of conceiving it.

It is believed that man himself is evolved from a simian prototype. Why may not his speech likewise be evolved from the same source? If, as my research shows, the sounds uttered by simians perform the same functions in simian economy as human speech does for man, in what respect is it not speech?

Jim is aiding the research by making a phonograph record

This is Jim's expression after listening to his own record

Jim listening critically to the sound of his own voice

Watching the Heart-Beats of a Plant

Sir Jagadis Chunder Bose, a Hindu scientist, has discovered a secret of nature

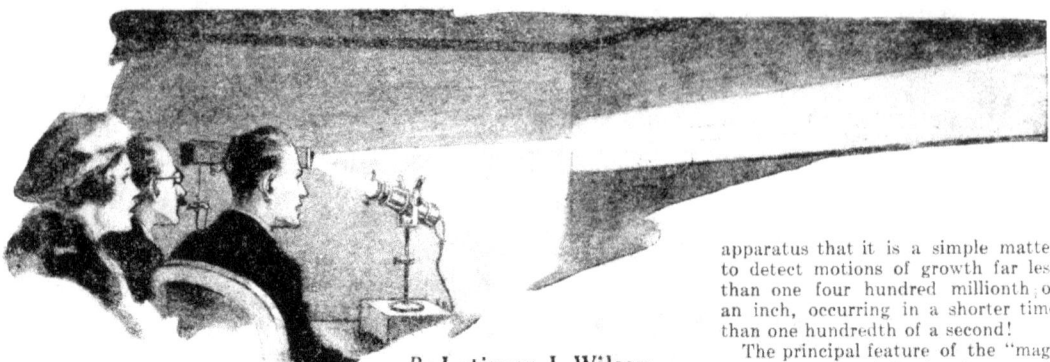

By Latimer J. Wilson

ATTACH a pen to a leaf and place a sheet of paper where the pen can move upon it freely. Then watch the leaf write in plain terms the most minute details of its passing sensations! Substitute a delicate magnetic apparatus for the mere crudity of a pen and the fact is accomplished: the dumb leaf will open its confidences to the eyes of man and indicate exactly how it responds to the influence of outside happenings.

He Reads the Thoughts of Plants

The man who has discovered the key to the innermost secrets of plant life, and who has exposed the very "life movements" of the trees, is Sir Jagadis Chunder Bose, founder of the Bose Research Institute of Calcutta, India.

The interested visitor who steps into the Institute at Calcutta will see a large tracing being automatically made in two curves. The first curve shows graphically the changes of temperature, the shifting effects of light and atmosphere that are taking place throughout the twenty-four hours of the day. The other curve shows the resulting effect in the response of a tree standing outside the building. The passing cloud, the temperature change, the fall of a raindrop or the shifting wind —all are reflected in the heart-beats of the leaf!

Only the larger movements of plants were recorded by the first instruments invented by the Indian scientist, but continual improvements brought about a means of greatly magnifying the small motions, until finally the curious discovery was made that every plant, like every living animal, has a "heart-beat," rhythmic and regular, which indicates its state of health.

So delicate is the recording apparatus that it is a simple matter to detect motions of growth far less than one four hundred millionth of an inch, occurring in a shorter time than one hundredth of a second!

The principal feature of the "magnetic crescograph," the instrument that makes visible the pulse-beats of the plant, is a long magnetic lever, the short arm of which is attached by a cocoon-thread to the leaf. The other end of the lever is arranged to move in front of a magnetic needle to which a small mirror is attached.

The very slightest motion of the

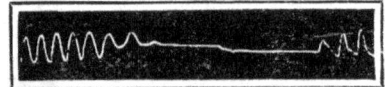

How ether affects a plant: first is seen the normal response before the ether is admitted, then is seen the stoppage of pulsation due to the effect of the ether. When the ether is blown off, the rapid recovery of the "heart-beats" may be noted

The magnetic crescograph, composed of a magnetic lever, a magnetic needle, and a small mirror which reflects a beam of light, forming a spot of light upon the screen. The motion of this spot shows at a glance the pulse-beats of the leaf attached to the arm of the lever by a cocoon-thread

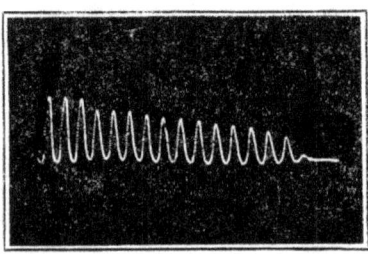

The heart-beats of a plant. One sees on the screen a graphic line of motion illustrating the effect of poison on a plant. The line-waves diminish until at death they form a straight line and the spot of light on the screen then ceases to move

thread-end of the magnetic lever meets with response from the needle with its attached mirror. When a beam of light is thrown on the mirror at an angle, it is reflected on a screen, and the slightest motion of the mirror shifts the spot of light. When the plant contracts under a shock, the movement is in a reverse direction; when the growth stops, the spot of light becomes stationary.

A magnification of ten million times is possible. If the human heart-beat were magnified to this extent at the rate of seventy-five beats a minute we would see about twelve million "palpitations" every second

Explaining the Mysterious Mimosa

Ordinarily the records in the study of plant growth are not necessarily magnified so greatly, and instead of being thrown on a screen they are marked with a point delicately adjusted near a plate of smoked glass. Sir Jagadis has observed various plants. The mimosa, well known as a "sensitive plant," is shown by the investigations to be merely more obviously sensitive than other plants.

The "nerve" of the mimosa leaflet, the *pulvinule*, corresponds in its up-and-down motion to the expansion and contraction of the animal heart. It is a "rhythmic tissue," the motion of which may be disclosed with light from the marvelous crescograph.

At the top of the mimosa may be found the youngest leaflets, while the mature leaves occupy the middle zone of the tree. At the bottom are the aged leaves. Investigations were made separately to find just how the age of leaves determines their response to stimulation and their recovery from shock. The effect was found to be very similar to what might be expected in animal life—the young and the very old differing greatly from the mature adults in their response to stimulation.

Did You Know that Plants Sleep?

What is the effect of cold on the heart-beats of plants? Sir Jagadis has found that intense cold paralyzes the action of the rhythmic tissues. In ordinary cold weather the trees that are acclimated merely "hibernate," when the leaves die. But the most astonishing result of his investigations is the disclosure that plants, like animals, actually sleep. During the period of rest there is a decided difference in the heart-beat of the leaflet from that of its waking hours. In its resting hours the plant is less capable of excitation, and, curiously enough, the resting time of the mimosa is from evening to morning. Its waking up time is about 8 A. M., and the maximum of activity is reached at from 1 to 3 P. M.

The Indian plant *Desmodium gyrans* has furnished many experiments. Its pulse-beats range from one- to two-minute intervals. A leaflet immersed in water was kept in a dim room for forty-eight hours before its heart-beats stopped. A cut leaf generally retains its rhythmic life-movements for a similar interval, at the end of which it can be revived by stimulants. The arrested growth of plants may also be revived by stimulation.

But what are the practical results to be obtained from the wonderful investigations conducted at the Research Institute? The more we know about plants the more we know about agriculture. It is no longer necessary to wait months, even years, to discover under what conditions plants thrive best. The crescograph answers our questions in a few minutes. Moreover, much light is shed on the life processes of higher animals. Highly

Courtesy of Hollinger

Sir Jagadis Chunder Bose is imbued with all the mysticism of the East, and yet he is an exact scientist in the fullest sense. His wonderful instrumental study of matter, living and non-living, has made him more of a Hindu than ever. "It was when I came upon the mute witness of these self-made records, and perceived in them one phase of a pervading unity that bears within it all things: the mote that quivers in ripples of light, the teeming life upon our earth, and the radiant sun—it was then I understood for the first time a little of that message proclaimed by my ancestors on the banks of the Ganges thirty centuries ago"

developed forms of life, such as are reached in the animal world, are so complex that much difficulty attends their isolation.

To study the effects of disease and the results of remedies in the simplest life-forms would be of great benefit when adapted to the higher forms. Life is ephemeral in animals, compared with that of plants. Man's fourscore years and ten is a puny age compared with the 5,000 year-old sequoias! Thus in the recorded pulse-beat of the plant can be observed the primitive problems of life. The effects of various drugs, the depression or the recovery

from shock, the effects of various stimulants, can all be seen in the high magnification provided by the magnetic crescograph.

When stimulated, the heart-beats of a plant become enhanced in size. Depressed with poison, they dwindle in size until the cause is removed, and if not removed the death spasm is seen in the quivering writing of the recording instrument, when the heart-beat stops. This physiological effect of drugs on the simple life-movements is regarded by physicians as leading to an important advancement in the science of medicine. It is possible to test the effect of drugs in a far simpler manner than that employed upon the higher forms of animal life.

Reading the Life of a Leaf

How interesting to sit in the darkened auditorium and to watch the screen upon which the crescograph is writing the autobiography of a leaf! In a vertical direction the spot of light oscillates. Gradually the audience becomes aware of a change that is taking place. There is a diminishing motion in the wavering image of the light reflected from the small mirror. Is the leaf about to expire?

The lecturer explains that the leaf is under the influence of an anesthetic. Ether has reduced its state of sensibility, and its normal heart-beats have temporarily stopped. When the fumes of ether are blown away, the pulsations begin. At first they are very small and weak, but gradually the spot of light begins to swing upward and downward in the rhythmic form similar to that of the normal motion.

One of the most interesting applications of Sir Jagadis' discovery was recently tried on a tree that was supposedly impossible to transplant. Previous attempts to transplant other trees of its kind had resulted in the death of the plant. Working on the theory that the shock had proved too great for the sensitive plant, the tree was subjected to the effects of ether during the operation. When it recovered from the ether after the replanting had occurred, the tree first shed its leaves, but it lived. Precisely as in the case of animals, an anesthesia proved beneficial in easing the shock of the operation.

Fumigating a Sick Tree

A balloon drops a tent over it and the deadly parasite-killing fumes are released

By P. Schwarzbach

WHEN you emerged from the measles your room was fumigated so that the rest of your family wouldn't catch 'em. Just so, when part of a tree becomes diseased, the other trees are also fumigated so that they won't catch diseases.

But fumigating rooms and fumigating trees are vastly different jobs. By closing the door you can shut off a room, but in order to fumigate a tree you must put a tent around it to shut it off from its neighbors.

What is the quickest, easiest, and best way of accomplishing this?

Mack Swain, of Los Angeles, will tell you to drop a tent over the tree by means of a captive balloon.

Perhaps you have heard of Mack Swain. He is a well known slap-stick moving-picture comedian. He has recently patented his idea of fumigating by balloon.

The tent is hooked to the balloon, which is moved until it is directly over the tree. Then the balloon is lowered. The tent opens like a parachute, and settles comfortably

The tent was fastened into place and then deadly fumes were released inside of it; the insects and their eggs were quickly killed, whereupon the balloon carried its tent away

over the tree. The balloon is then unhooked and starts back for another tent for another tree, while the tent it left behind is fastened to the ground.

Hydrocyanic acid does the fumigating. It is a deadly volatile poison, and as it fumes away it kills all the insects on the tree, and even the insects' eggs. This acid is also known as Prussic acid. It has the seductive odor of peach blossoms.

When the tent is securely fastened, the tanks of hydrocyanic acid are shoved underneath it and opened. After the fumes have done their deadly work the balloon is brought back to the scene of action, the tent is hooked on again, and away the balloon goes to another tree.

Mack Swain, the inventor of this device, is not the only moving-picture comedian who is interested in it; for Chester Conklin, who spends his working hours trying to dodge pies, dough, and soft tomatoes, has bought a forty per cent interest in the patent.

Insects were ruining a fruit tree, and the owner decided to fumigate it. A captive balloon with a tent hooked to it was brought directly over the tree and lowered. The tent opened up like a parachute and settled over the tree

King Silkworm
Thousands of Japanese spend their lives in the service of this autocrat

Silkworms ought to be beautiful, but look at them: they are just as ugly as any other worm. The silk industry originated in China; but the Japanese realized its importance to such an extent that in some of their villages every home is a silkworm nursery. The worms are here feeding on the mulberry leaf

Eggs, cocoons, and moths of the silkworm. In winter the eggs are carefully spread out on sheets of paper and placed on trays. When the larvae hatch out, their human attendants chop up mulberry leaves and scatter the pieces over the tiny worms. As they mature each worm receives a whole leaf a day

Japan takes in forty million dollars yearly in her silk industry. The government has established a free school near Kyoto for teaching the business thoroughly

But weaving at home is still carried on to a great extent, as witness this Japanese woman, who is employing a rather primitive method in the weaving of raw silk

Skeins of the beautiful shiny stuff as it goes to the "silk throwster." Its length is frequently a thousand yards

The thread consists of silk fibers wound together, the separate fiber having been actually spun by the worm

The silkworm is probably the only worm that gets any human help in the pursuit of its business. In spinning cocoons silkworms need plenty of space. Here the kindly Japanese are preparing aba, or supports for them to spin on

It is when the cocoon is completed that the ulterior motive of the silkworm-owners appears. The chrysalides in their cocoons are killed in boiling water, and the threads are separated. Then the filaments are unwound and spun for the weaver

This Farm-Hand Never Tires or Asks for Pay

Sit on the porch and watch the synmotor work

IT was not alive, apparently, and no human being seemed to be concealed about it, and yet the thing was seen cultivating a ten-acre farm in New Jersey. Down the rows of corn it went all alone, and never bruised a blade or chopped a root. It was uncanny to anyone who had never beheld such a sight before, and even to some who had.

It worked nights, too. Dimly outlined in the white moonlight, it could be seen threading its way with almost human intelligence and with mathematical precision, while the farmer slept peacefully in his near-by mansion and dreamed of waving corn-tassels. Around and around the field the thing moved, around a center which it continually approached. The corn had been planted in a spiral formation about a tall post capped with a circular drum or cask.

Close inspection reveals a thin wire extending to a central drum, around which the wire winds itself as the work of cultivating proceeds. That explains the spiral movement. The wire shortens itself by the same amount each trip around, and is used for steering the machine. Yes, it is a machine, after all. The wire, being perfectly tempered, cannot stretch. and an electro-coated surface protects it from rust. Its total weight is less than two pounds; yet a pull of six hundred pounds is required to break it.

The machine is a narrow tractor of special make, and it is called a "synmotor." The engine is a compact but very efficient gasoline type of about four horsepower. To the framework can be attached any of the usual

The furrows between the rows of dwarf corn are made by plows automatically steered

Showing the works of the synmotor

implements for cultivating the land. Plowing, hoeing, harrowing, and the many other operations are performed in the spiral path as well as in the straight course. A gang-bar for the attachment of the implements may be used so that several rows may be cultivated at the same time.

When the synmotor is utilized on a large scale, the farm is divided into convenient ten-acre circles, each section being planted and cultivated separately. Any vacant spaces between adjacent circles can be utilized for fruit trees, buildings, or the like. For that matter, the intervening spaces can also be cultivated by merely disengaging the steering wire and utilizing the tractor in the customary manner.

For intensive farming and overlapping seasons, the accuracy of the synmotor in following a given track is of great advantage. The machine does not disturb the small plants, and it can work very closely to the rows. The working tools are spaced the exact distance between rows and do not swerve from the spiral course. Strawberries, peas and other vegetables can be cultivated with the synmotor,

With such a machine as this, the laziest man on earth can sit in the shade and fan himself while gasoline does his work.

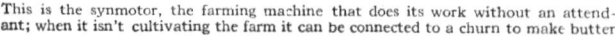

This is the synmotor, the farming machine that does its work without an attendant; when it isn't cultivating the farm it can be connected to a churn to make butter

On the big drum the steering wire is wound, thus drawing the cultivator ever closer to the center

Rubber on Its Way from Tree to Tire

He is tapping a tree by the "herring-bone" method. Latex, the rubber fluid, runs down the small cuts into the large central one and drops into a cup on the ground

The latex is poured from cup to can, and coolies carry it off. Of the many kinds of rubber-yielding trees, the heveas are most numerous; those shown below grow in Ceylon

The sticky, crude rubber leaves the forest by way of an ox-cart; later it goes through the "curing" process

In the nursery beds of a plantation the seeds of the hevea are planted in regular rows; but when the tree is allowed to reproduce naturally it drops a pod containing three seeds. The pod explodes

Then the seeds scatter, landing about fifty feet away. This is nature's dramatic way of planting seeds so that the future rubber trees will grow up outside of the spreading foliage of the mother tree

Nearly ready to leave Ceylon: the strips of rubber are being sorted before they are packed for shipping

Their last glimpse of their home land; it will not be long before they are rolling on State roads

Where the Pearl Button Gets Its Start

THE pearl buttons that you buy for your clothes are obtained from the shells of the fresh-water mussel. A lively mussel-fishing industry has sprung up along the banks of the Mississippi river and its tributaries. The mussel-fishermen go out, praying that they will find pearls worth, it may be, as much as three thousand dollars apiece. Once a year or so they find a pearl or a slug. But even the shells, which are thrown to one side and sold to pearl button makers, are worth twenty dollars a ton. What becomes of the dead mussels? Oh, they are thrown to the pigs or the chickens, or used as fish bait

The story of your pearl buttons begins with the infection of a fish with the larvae (glochidia, the scientists call them) of the mussel. The larvae, it seems, are expelled by the parent mussel, and attach themselves to the gills, fins, or other parts of fish unfortunate enough to be swimming around at the time. The men in the picture above are scientists who are sorting the fish that are to be infected with the larvae or glochidia of the fresh-water mussel

Since the larvae of mussels thrive on the gills of fish, we find that the Bureau of Fisheries has developed a method of collecting the fish in nets and transporting them to infection tanks, into which the larvae are introduced. It takes only a few minutes for the lively little larvae, or glochidia, to establish themselves in these free boarding-houses. The fish are liberated in ponds, all unconscious of the part they are playing in buttoning up mankind

A flat-bottomed, square-end boat, commonly called a John-boat, is employed. Two long iron bars are carried, to which are attached by short lengths of cord some one hundred hooks. The fisherman throws one of the iron bars overboard. As soon as the hook enters the open shells of a feeding mussel, the shell closes tightly

The black crappie carries the glochidia of several species of mussels and thus serves as an unwilling host

The glochidia of the "nigger-head" mussel, one of the most valuable commercial species, are the special uninvited guests of this river herring. The scientists also supply large-mouthed black bass and and other varieties for the glochidia

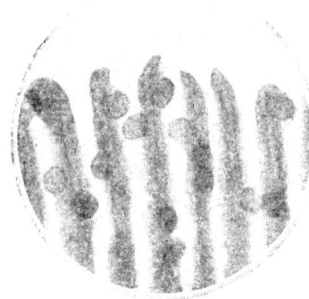

The little specks are the glochidia, or larvae, of pearl mussels; the straight finger-like projections are the gills of a fish. Behold the glochidia at home

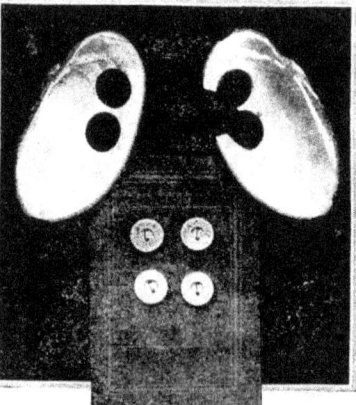

These buttons were cut from shells of mussels raised from the glochidia stage in the ponds of the Fairport Biological Station, of the Bureau of Fisheries. All of which shows what science is doing in the matter of your personal adornment

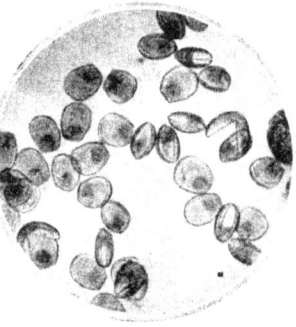

Here the glochidia—the beginnings of so many pearl mussels—are drifting around looking for a home. Eventually they will find it in the gills of fish

After a Coconut Leaves the Tree
Its varied career, ending at last in a soap factory

In the oil factory the coconuts are pressed or boiled. A thousand nuts yield twenty-five gallons of oil, which is used chiefly for making candles and soap

You can't blame the monkeys for dropping coconuts on people's heads, the nuts are so plentiful. Each tree turns out sixty nuts a year—starting when it's five years old

A sharp steel knife cuts the coconut in two. The husks are turned into fiber, called coir, which is used in making cordage, brushes, and matting; the nut is turned into oil

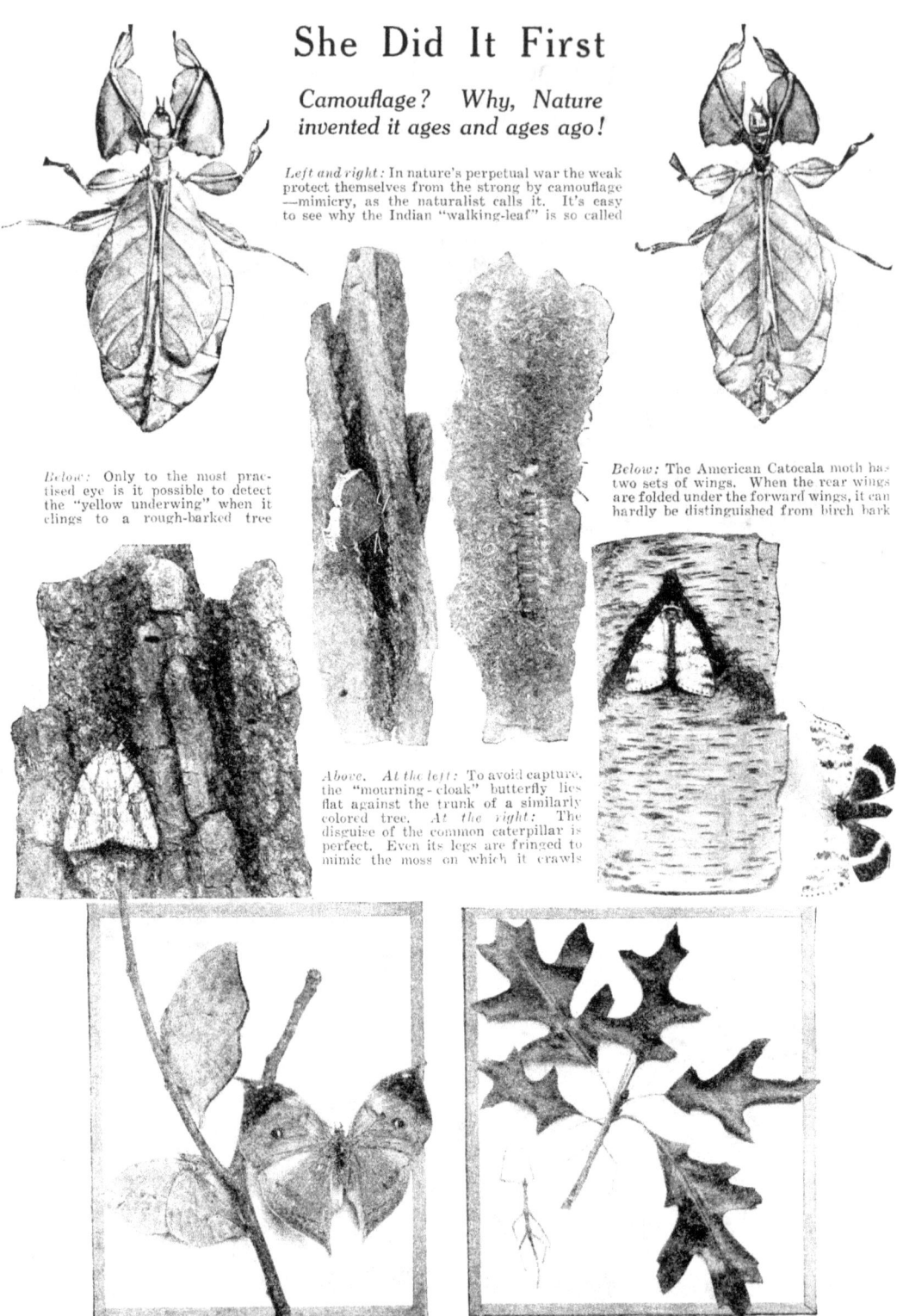

She Did It First

Camouflage? Why, Nature invented it ages and ages ago!

Left and right: In nature's perpetual war the weak protect themselves from the strong by camouflage —mimicry, as the naturalist calls it. It's easy to see why the Indian "walking-leaf" is so called

Below: Only to the most practised eye is it possible to detect the "yellow underwing" when it clings to a rough-barked tree

Below: The American Catocala moth has two sets of wings. When the rear wings are folded under the forward wings, it can hardly be distinguished from birch bark

Above. At the left: To avoid capture, the "mourning-cloak" butterfly lies flat against the trunk of a similarly colored tree. *At the right:* The disguise of the common caterpillar is perfect. Even its legs are fringed to mimic the moss on which it crawls

ELECTRICITY

What Next?

The Electromagnetic Gun?

A big electric power plant like this would have to go with each electromagnetic long-range gun to supply the power

IF the bombardment of Paris from a distance of more than seventy miles served any useful purpose at all, it was that of calling attention to the limitations of modern artillery. Through the popular press of the country, the public has learned that every gun—and, above all, the huge piece that shelled Paris—is short-lived.

Sir Robert A. Hadfield, England's greatest steel-maker, once stated that the actual firing life of a gun is less than three seconds. Why? Whenever the gun is fired (and it can be fired scarcely more than two hundred rounds), the hot powder gases wash away the steel lining of the gun-barrel as if it were so much wax.

Apart from the difficulties encountered because of erosion, as this washing away of the steel is called, ordnance officers would hesitate about building a great gun because of other difficulties encountered. A range of more than seventy miles can be obtained only by enlarging the powder-chamber so that enormous charges of explosives can be used, and by lengthening the barrel for the purpose of enabling the powder gases to act for a longer time on the projectile. If the barrel is inordinately lengthened, special machinery must be invented to turn it.

Theoretically, there is no limitation to increasing the size, caliber, and range of guns; but it requires more than mere theory to build guns that are efficient for the purposes of war. Guns must be made of some material, and, unfortunately, there are limitations to the physical strength of every material known to man. Once upon a time guns were made of leather or wood; but, as the range and power of the guns was increased, stronger and ever stronger materials had to be found to resist the increasing strain.

Medium-Sized Guns Most Effective

In the last fifty years the evolution of artillery has made rapid progress. Chemistry, physics, metallurgy, and engineering were pressed into service and contributed their share to the marvelous development of this particular and most scientific arm. But scientists are inclined to believe that the limits of possibility in the evolution of the present type of ordnance has nearly been reached.

The building of bigger guns is limited by considerations of a technical nature. The larger the gun, the slower the fire. This is a serious drawback. Practice teaches that medium-sized guns are the most effective.

The monster guns of the most recent type are as big as can be built with the steel presses and converters now in use. To exceed the present muzzle speeds of projectiles, explosives of greater power must be found. However, the known laws of the expansion of gases do not suggest the possibility of obtaining a gas pressure behind a projectile that will propel it at an initial speed much above 4,000 feet a second. The erosive power of high explosives must also be taken into account.

The Electromagnetic Gun

Because steel is so easily washed away by powder gases, because big guns are so difficult to make, attention has once more been directed to the electromagnetic guns proposed from time to time by engineers.

The electromagnetic gun is extremely simple in principle. Any magnet attracts a piece of steel. Suppose that electromagnets were arranged in a row, so that the first would attract a steel projectile, then the second, the third, the fourth, etc., until finally the projectile would be hurled into space. It sounds very simple, doesn't it?

Several guns of that type have been invented and patented. As they are based upon the same principle,—a principle that allows but little latitude in the method of its application,—the guns invented by Professor Kristian Birkeland, Levi M. Bowman, Eli M. Alderman, Paul T. Kenny, Foster, and a few others differ mainly in the arrangement of the solenoids and in the method employed to cut out the electric current from the electromagnets after the projectile has passed them.

Professor Birkeland's invention, probably the most important of its kind so far patented, contemplates a gun-barrel made of bronze or steel, which is surrounded by solenoids. These solenoids are supplied with an abnormally heavy current, that is to say, a current beyond the capacity of the solenoids as calculated by the methods now in vogue for determining the safe carrying capacity of a coil.

According to the plan of the inventor, the heavy current that energizes the first solenoid is shut off by the projectile as it reaches the coils toward which it is drawn by the magnetic force of the solenoid. At the same time, the next group of solenoids is energized. The magnetic force of the energized coils acts upon the iron projectile and adds to its speed; but again the current is cut off by the projectile before it has time to raise the temperature of the solenoid to a degree high enough to destroy it.

Is It Practicable?

The inventor calculates that for throwing an iron projectile weighing two tons and containing 1000 pounds of nitrogelatine at an initial speed of one thousand feet a second, a gun about ninety feet in length would be required. The projectile would be about nine feet long and nineteen inches in diameter. The gun solenoids may be made up of square wire, each solenoid containing 720 windings with a resistance of fifteen ohms. The length of each solenoid is made about three-eighths of an inch and the height about eight inches. With an electromotive force of 3000 volts, this will give a current of 200 amperes.

Major-General William M. Black, Chief of Engineers, U. S. A., does not believe in the feasibility of such a gun. At the thirty-fourth annual convention of the American Institute of Electrical Engineers, in June, 1918, he said:

"You can understand the impracticability of transporting the electrical plants required for any number of such guns. I think that you will agree that, until new discoveries give a much improved method of storing or generating electricity, smokeless powder will continue to be the most compact and convenient form of stored energy for guns."

Each of the dynamos would have to produce 3000 volts and 30,000 amperes, being the equivalent of about 120,640 horsepower

As the projectile passes through the barrel of the gun, drawn by the tremendous electromagnetic force of the solenoids, it breaks automatically the circuit behind it, and energizes the solenoids ahead of its course, thus gathering speed and momentum for its long flight through space

CONTACT WIRE

CIRCUIT BREAKER AND MAKER

SPRING

BRASS GUN TUBE OR BARREL

SOLENOID WINDING

CIRCUIT-CLOSING PROJECTILE

SOLENOID HEADS

Such would be the appearance and size of one of the electromagnetic guns of the Birkeland type if it were to be large enough to carry a large projectile a distance of one hundred miles or more

Making Safes Safe

Thereby causing an old art to become almost extinct

By P. Schwarzbach

Clang! A bell sounds and an indicator swings around: another safe is being cracked! As the man in charge reaches for his telephone, he looks at his indicator-board to see which safe it is. A complicated system of wiring connects his batteries with the room in which the safe is kept: the door, the windows, and the safe are all wired. What chance would Jimmy Valentine stand today?

REMEMBER Jimmy Valentine, the tender-hearted, nimble-fingered safe-cracker? Had he existed today, chances are he would never have become famous, but would have been nabbed on his first job. For invention has helped to make safes safe.

Let us suppose that Jimmy comes to life and decides to try his hand at a large safe of large content. He approaches stealthily, and discovers wires that indicate the presence of an alarm system; if he has had no experience with wires he may decide to cut or ground them.

In the first case he releases a relay by destroying the line current, and in the second he energizes another by increasing the current. In either case he unwittingly sounds the alarm.

"Ten years of hard labor," says the judge a few weeks later.

Suppose he ignores the wires and boldly forces the door of the room in which the safe is located? Again a relay is released and the alarm sounds in the central office. Or should he try the window, either by opening it or smashing the glass, the result will be the same. Raising the window trips a spring that breaks the circuit, and smashing the glass severs a tinfoil strip that is also an important part of the circuit.

The Central Burglar Alarm

This complicated system of circuits and switches is known as the "central burglar-alarm system"—nor does it stop at the windows and doors. Suppose Jimmy should miraculously get inside the room without disturbing the switches? He approaches the safe and kneels down before it.

Foiled again! There is another circuit-breaker under the rug that will quickly report to headquarters.

And the last circuit-breaker is in the safe itself: when the burglar starts to drill or melt a

hole in the safe, a resistance wire within is either cut or fused, and the alarm sounds once more.

A Safe that Whirls

And now for another safe safe, invented by Patrick Meehan, of Brooklyn, New York, that is known as the whirling safe. When the owner of the safe decides to close up for the night, he starts a motor located beneath his safe and below the floor. The safe, which stands in the corner of the room, starts whirling, and at each revolution a light flashes and a bell rings.

Jimmy decides to try his luck at it. He stands watching it for quite a while; there is no one passing by outside. The safe is whirling too fast for him to try his hand at the combination; it is located in the corner close to the wall and

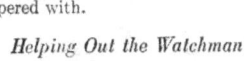

The whirling safe. At each revolution a bell rings and a light flashes. The burglar can't try to work the combination; he dares not stop the motor, since the light and bell will stop too. What is left for him? He might climb the latticework behind the safe and jump on to the top of it by breaking away the part that covers the safe—but the result? An alarm will sound throughout the building

there is no chance of his hanging on while it whirls—he would be knocked off and thrown down violently.

There are just two possibilities: he can stop the motor or climb to the top of the safe and drill through it. But, if he stops the motor, the light that flashes and the bell that rings at each revolution will stop too, and the neighbors will know that something is wrong.

Jimmy figures that his best plan is to drill through the top. He climbs up the frail latticework behind the safe and breaks away the part that covers the top of the safe.

Clang! A loud alarm rings throughout the building. Jimmy jumps down, and flees just in time to save himself from another ten years in prison.

The breaking of the latticework caused the alarm to sound. As Jimmy thinks it over he is sorry that he didn't risk stopping the motor, instead. But if he had the result would have been the same; for the whirling safe is equipped with a second alarm that automatically sounds when the supply of current is tampered with.

Helping Out the Watchman

One of the oldest, simplest, and best methods of guarding a safe is to have a watchman keep a constant eye on it. But suppose the watchman should be called away for a short while? Well, there has been invented a safe-protector that will guard the safe while he's gone. It was invented by George C. Smith, of Cambridge, Mass.

The watchman estimates the amount of time he will be gone and the hour at which he will return. He then sets an indicator at the time he expects to return, adjusts a weight attached to the clock, and goes out, quietly shutting the door after him.

What Happens When a Great City Suddenly

Photographs by courtesy of

"Looks like a storm coming," says the weather-wise business man, whose perch is the top floor of a skyscraper. Other people far down behind the windowed walls notice that it is getting dark, but they can see little of the storm that is approaching, because their view of the sky is limited. Then suddenly all the great city realizes that it has become dark in the middle of the afternoon. "Click! Click!" go the buttons turning on the electric lights, and the burden of instantly supplying about 175,000 additional horsepower is thrown upon Waterside, the generating station of the New York Edison Company. What would happen if preparation had not been made to meet such a huge demand? Either no lights could be furnished the darkened city or other power would have to be cut off. In other words, the coming of a severe storm would tie up the business of the city until conditions again became normal

When two million lights are suddenly switched on during the time when the greatest amount of power is being demanded by the city, the New York Edison Company is prepared. The approach of the dark clouds is announced by an ingenious device, the "wireless storm-announcer." A system of relays is connected with a coherer and a bell. When a storm is an hour and a half or two hours away the bell rings every fifteen or twenty minutes, and as the storm draws nearer the ringing becomes more frequent. Some storms pass away without ringing up the power station from closer range; but others advance so rapidly that quick work must be done

Notified by the wireless storm annunciator of the swift coming of lowering clouds, and knowing that the city in an hour or less will be plunged into darkness, the system operator at the central station has everything prepared for the increased load. The banked fires must be immediately ready, and the automatic stokers are set to work to generate quickly the steam-energy which, by mechanical magic, will soon be turned into electrical energy. While the city makes its demand without warning, the storm sends its warning ahead by the "static" condition of the atmosphere. The watchful operator in the power station becomes more than usually attentive, and soon the fires flare up, the boilers quickly respond, and the whirring generators begin to manufacture electricity

Turns on Two Million Lights to Meet a Storm

the New York Edison Company

Imagine a score or more of men stoking the fires of the huge electric power plant, to beat the approaching storm in the race toward darkness! That is what would have to be done were it not for the fact that the New York Edison Company at its Waterside station employs only automatic stokers. The whole operation of sending fuel into the fires under the boilers is automatically conducted under the guidance of a few intelligent workmen. The hard conditions and the strain of the storm are borne, not by human beings, but by the steel of machines. From the unloading of the coal from the barges, to the removal of the ashes, the process is carried on almost entirely by mechanical means

As the storm swoops down and the sky darkens, there is strict attention in the room where the indicators are arranged. Watchful eyes are turned toward the numbers of dials on the wall. Every district must be looked after and its requirements fully provided for in advance. Nothing is overdone, but enough electricity must be generated to meet the requirements. Both generators and storage batteries may be called to the rescue when a severe storm prevails, so the close watchfulness by the men before the indicators interprets the story told by the instruments, and the action of the plant follows

When the wheels of activity are whirring to generate the extra load of electricity needed to overcome the darkness of the thunder-storm, what does the effort really mean in the terms of figures? On a day (June 20, 1919) when the record storm of the year occurred, one sees at 5 A. M. about 75,000 horsepower being used by the city. The low ebb of Edison power was at hand. By the time the factories opened at 8 o'clock, about 165,000 horsepower were being drawn upon, while at 11 A. M. there were slightly more than 250,000 horsepower. The highest point reached in the afternoon would normally have been about 250,000 or 275,000 horsepower; but the extra load brought about by the storm caused the amount to jump suddenly up to about 425,000 horsepower! The huge increase, 175,000 horsepower, had to be prepared with scarcely an hour's notice

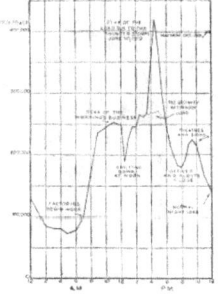

A Searchlight of More Than a Billion Candlepower

The new Sperry searchlight is so powerful that lead liquefies in its beams at a distance of twelve feet. Moving pictures can be taken in the beams of this searchlight with the speed required for genuine sunlight

By Latimer J. Wilson

HOW bright is the sun? In candlepower the sun's intrinsic brightness, as seen at sea-level, is equal to the light of 270,000 candles for each square inch of its surface. The crater of an ordinary electric arc is equal to 84,000 candlepower on the same scale. Now the sun has a closer rival in the dazzling brilliance of the new Sperry searchlight, which has a luminosity of 1,280,000,000 candlepower! In its powerful rays lead can be melted twelve feet from the crater.

The rays of light from the arc are projected in parallel lines by a sixty-inch parabolic mirror. Allowing for the loss of light in reflection, the searchlight carried out into space would be visible at practically the same distance from the earth as would an equal area of the sun. A piece of the dazzling solar disk clipped into a circle sixty inches in diameter would have a rival in this searchlight. What

a strange cosmic effect could be witnessed when the shaft of light sweeps across the sea to meet the rising sun!

Standing in the blinding path of rays emanating from opposite directions, the observer would notice that, for the first time in his life, he could stand in the direct sunlight without casting a shadow! The flame of the ordinary electric arc is itself cast as a shadow in sunlight when held a few feet from the ground. At a given distance the beam of the Sperry searchlight is about equal to the total light of the sun when the latter is well up in the sky above the horizon. Concentrated against the light of the sun at such a time, the searchlight actually equals the sun, and becomes such a formidable rival that it can counterbalance the shadow-casting power of the great orb of day.

This marvel of searchlights was in-

vented by Elmer A. Sperry. It consists of an automatically rotated positive cored carbon which burns to a crater-tip. The negative carbon is held by silver-tipped "fingers," while the positive is held in an airtight clamp tipped with quartz-crystal. The hot air is blown off by a motor-driven mechanism to keep the carbons and the great parabolic mirror at a safe temperature. The carbons are held at the proper distance apart and are automatically fed to keep a constant flame. The superior power of this new searchlight depends upon the intensity of the illumination of the gas which fills the crater of the positive arc. But for the device that makes possible the retention of a perfect crater whose image is projected in parallel rays from the huge sixty-inch mirror, the power of the arc would not have exceeded that of other searchlights.

A Typewriter that Writes in Electric Letters

A HUGE electric bulletin-board atop a newspaper building in a mid-Western city flashes the latest news. The device is operated by a typewriter. The bulletins are typed on the machine in the usual way, while contacts underneath the keys control the apparatus that flashes the corresponding letters to the roof. As each key is struck, the corresponding letter appears in electric lights in its proper position on the big board.

All the letters remain in position until the bulletin is completed. They are left standing a short while, and

then the operator puts them out by pressing one key.

The sign itself lights up on both sides. Each side is divided into sixty units, arranged in three rows. Each unit contains an arrangement of twenty-one electric lights. Various combinations of these lights form any letter or numeral, as well as other characters.

Selection of the units is done automatically as the writing proceeds. This is made possible by an extra platen on the typewriter. In this platen are sixty silver plugs, arranged to correspond with the sixty units on

the sign. A stylus rides over these plugs as the machine is operated, thus giving the desired connections.

Since the operator cannot see the result of his work from where he sits, a pilot-board is placed directly before him. This board has three rows of green-light bulbs, corresponding to the three rows of units. When he strikes a letter with the stylus resting on a certain plug on the platen, the corresponding letter appears on the display above, and automatically this causes the green light in the position corresponding to the unit and plug to burn.

Every Farmer Can Be His Own Chemist

ONE of the greatest aids in practical, scientific farming ever devised is the Truog Acidity Test for determining the acidity of soils quickly and accurately in the laboratory, farmhouse or field. The illustration shows the apparatus in detail.

The test depends upon the fact that zinc sulphide, when mixed with an acid, will liberate hydrogen sulphide gas, which discolors paper moistened with a solution of lead acetate, or sugar of lead. A small metal cup, having a capacity of 9 cubic centimeters, is filled with the soil to be tested. This is emptied into the boiling flask and 5 cubic centimeters of a mixture of zinc sulphide and a solution of calcium chloride, together

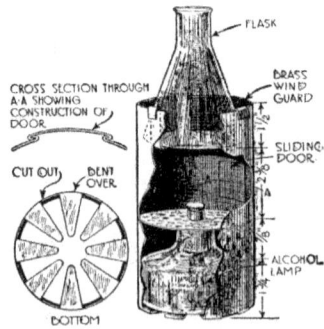

Here is the apparatus in detail. It is very simple to make

with 95 cubic centimeters of water are added. The alcohol lamp is lighted and after the contents have boiled for just one minute a strip of the dry lead acetate paper is placed over the mouth of the flask and the boiling continued for just two minutes more. The paper is then removed and its color noted. If the under side has been darkened the soil is slightly acid. If it is yellow to brown in color the soil is more or less acid and if brown to black very acid.

This test is much more positive and certain than the old litmus test. It indicates the degree of acidity of the soil, and from the data thus obtained the farmer can find the quantity of lime per acre needed.

PHYSICS

Dare We Use This Power?

Sir Oliver Lodge says atomic energy will supplant coal

By E. F. Richards

SIR OLIVER LODGE thinks that man is not yet civilized enough to use the energy hidden in ordinary matter.

"The time will come when atomic energy will take the place of coal as a source of power."

The man who spoke thus before the Royal Society of Arts in London was Sir Oliver Lodge —one of the towering figures in modern science, a man who has devoted the better part of his life to the study and interpretation of the atom. This new form of energy, which our great-grandchildren may utilize instead of oil and coal, has possibilities so appalling that Sir Oliver almost rejoices that we do not know how to release it. "I hope that the human race will not discover how to use this energy," he says, "until it has brains and morality enough to use it properly, *because if the discovery is made by the wrong people this planet would be unsafe.* A force utterly disproportionate to the present sources of power would be placed at the disposal of the world."

Sir J. J. Thomson, England's great authority on the atom, gives a picture of this terrible form energy that wins one over to Sir Oliver Lodge's view. He tells us that the atomic energy stored in an ounce of chlorine "is about the amount of work required to keep the *Mauretania* going at full speed for a week," and that the splitting up of the atoms in any substance would involve enormous transformations of energy. "In fact, the explosion of the atoms in a few pounds of material might be sufficient to shatter a continent."

Everything Is Made Up of Atoms

These eminent men of science have no particular atom in mind when they speak thus of the fearful possibilities of atomic energy. They mean the atoms of any familiar thing—radium,

iron, copper, wood, or stone. The food we eat is made of atoms, and so are the tables and chairs in our houses. Every one of us, then, locks up within himself immense stores of

"I hope the human race will not discover how to use this energy until it has brains and morality enough to use it properly," says Sir Oliver Lodge, in explaining the terrible possibilities that lie not only in radium but in a piece of wood or iron

energy. In a little finger there is enough energy to run all the trains in the United States for a few minutes, if we could but release it. The high explosives detonated to hurl millions of shells during a campaign are not so terrific in their possibilities as the atoms of which our bodies are composed.

The first intimation of the fundamental facts upon which the modern conception of the atom is based came to us through the study of X-ray tubes. In these Sir J. J. Thomson discovered particles weighing two thousand times less than the lightest atom then known, the hydrogen atom. These particles, which we now speak

of as beta particles or electrons, were subsequently found to be given off also by radium, as one of the products of the breaking up of its atoms. For, though the atom ordinarily remains *undivided*, we know today that it is not *indivisible*. In certain cases the atom breaks up of its own accord, as in the case of radium, shooting off the fragments at speeds which make a rifle bullet appear like a snail in comparison.

Active Radium Particles

There are, indeed, good grounds for suspecting that all matter, not only radium, is thus shooting off particles and giving out energy. But we become conscious of the fact only in the case of radium and a few other radioactive substances. Radium is giving up its atomic energy more rapidly and more violently than limestone, for example, which explains why it seems more explosive than other elements. A radium atom is like a two-ton gun firing a hundred-pound shot. Just like the gun, the rest of the atom recoils after having been fired. This is not merely a speculation, a picturesque guess. *The recoil has actually been observed.* After five such projectiles have been fired, radium settles down into another existence—a quieter existence, such as lead or something chemically like it. A uranium atom fires off four such projectiles in order to become radium.

All this is definitely known. The old alchemists were not so far wrong when they dreamed of turning iron into gold. Nature is constantly changing one element into another before our very eyes. When we know how to control the electron in an atom we may be able to make gold out of base metal at will.

Sir Oliver Lodge concludes: "Our descendants, instead of burning one thousand tons of coal, will make energy out of an ounce or two of matter."

Now the X-Ray Has Gone into

THE *uncanny property that X-rays have of penetrating substances, whether they are transparent or not, is now turned to practical account. If the X-rays can discover just where the safety pin lodged that baby swallowed, why shouldn't it also explore the hidden interior of a casting or a piece of timber for defects? So British scientists have reasoned, with results shown in the collection of pictures on these two pages.*

The X-ray has been used to discover whether there is a flaw or crack in a metal weld. The steel manufacturer has radiographed his castings and forgings in order to localize the blow-holes and other concealed imperfections. The airplane inspector, with the aid of the ray, has searched for hidden faults in workmanship that outwardly satisfied the eye.

Here was a casting for an airplane engine that would have passed any inspector. Major J. Hall Edwards, of the Royal Air Force, turned on the X-ray. And what did the X-ray show? A large blow-hole. Much work had been done on the casting which would have been saved had the X-ray photograph been made earlier. But the experiment proved that the time has come when the careful manufacturer must add the X-ray to his inspecting equipment

In the box resting on the floor is an X-ray tube. Upon it rests a wooden member that is to form part of an airplane. Upon the wooden member a fluorescent screen is placed. The man who is bending down is looking at the fluorescent screen for defects in the wood. This is the quickest way of examining wooden aircraft parts with the aid of the X-ray for defects

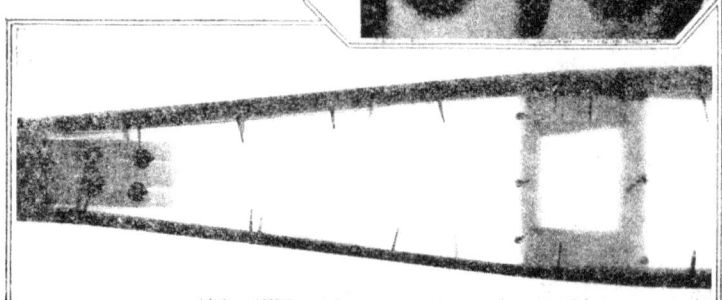

A brave man's life was saved when this X-ray photograph was taken. The internal strength of this block was badly fitted; each of the screws had split the wood; the work was altogether discreditable. And so the piece was rejected. If it had not been for the X-ray it might have been a part of an airplane which would have collapsed at a critical moment

The X-rays showed a defect in the gasoline-feed canal of this carburetor, thus saving hours of effort in locating the trouble

the Business of Manufacturing

the armoring of a cable, if you have
concealing mistakes, turn on the X-ray

If you wonder whether there is a hidden corrosion in a gas-cylinder, if you suspect that gutta-percha has been adulterated, if you have reason to believe that a workman is cleverly concealing his mistakes or his carelessness—turn on the X-ray.

When a strut or spar of an airplane is completely covered with fabric, veneer, or plywood, it is easy enough to fool the inspector. But you can't fool the X-ray. The detection of one such fault with the X-ray resulted in the rejection of dozens of completely finished main wing-planes in a British factory. In one fuel-tank the rivets were found to have heads on the outside only. Cracks in the airplane timber are sometimes cleverly hidden by gluing a shaving over the sand-papered surface. The eye overlooks them, but not the X-ray.

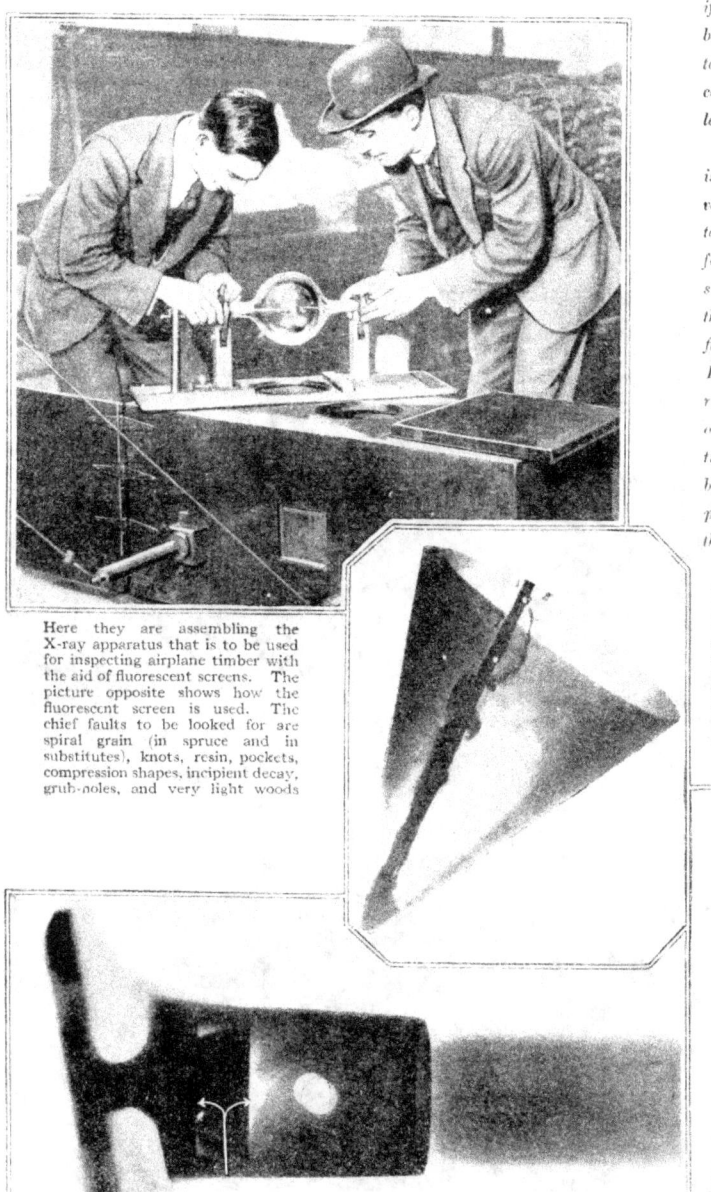

Here they are assembling the X-ray apparatus that is to be used for inspecting airplane timber with the aid of fluorescent screens. The picture opposite shows how the fluorescent screen is used. The chief faults to be looked for are spiral grain (in spruce and in substitutes), knots, resin, pockets, compression shapes, incipient decay, grub-holes, and very light woods

In every well conducted aircraft factory you will find a notice that reads: "A concealed mistake may cost a brave man his life." In spite of that, mistakes have been concealed. Now comes the X-ray to protect the flyer. Here we have an X-ray photograph of a fuel-tank for an airplane. Its defective riveting and soldering stand out glaringly

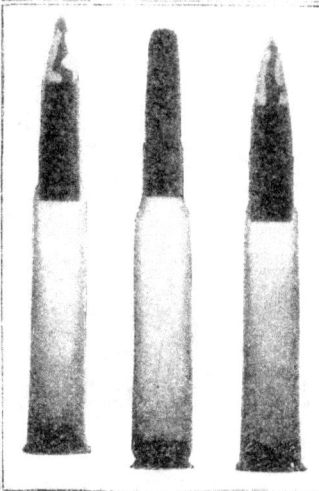

The wooden skid of an airplane was cut off too short to fit into its socket. In order to make up the length, a piece of packing (here marked with an arrow) was introduced into the space below. The workman responsible thought he could "get away with it"; but the X-rays found him out

Major J. Hall-Edwards of the Royal Air Force photographed loaded rifle cartridges to show how the distribution of the lead would influence their flight. In the center is the old round-nosed bullet

Is Sir Oliver Lodge Right About Atomic Energy?

Sir Oliver Lodge says that if the atomic energy in an ounce of matter could be utilized it would be sufficient to raise the German ships sunk in the Scapa Flow and pile them on top of the Scottish mountains. "I think," he adds, "we are on the brink of a great discovery"

PSYCHOLOGY
Is Your Child Left-Handed?

Why, according to psychological tests,
left-handed people ought to remain so

PARENTS, teachers, and educators have long been
puzzled by the left-handed child. Some have argued
that the "left-hander" should be taught the use of the
right arm; others believe in the saying, "Let well enough
alone." Recently a group of psychologists, headed by Dr.
W. Franklin Jones of the University of South Dakota, have
got on the trail of the left-handed.

Their first move was to study arms in general. They
measured the wrists, muscles, palms, bones, etc., of about
20,000 left- and right-handed men, women, and children.
They tabulated the results, noting any accidents that had
happened to the arms, as well as other useful data.

Which Arm Is Your Larger One?

These investigators found that in every person one arm
was larger than the other. Most of those whose right
arms were larger were right-handed, and those whose left
arms were larger left-handed, which seemed to indicate that
we are all born "handed."

But several exceptions to the rule
had to be accounted for persons with
longer left arms who were right-handed,
and vice versa. It was found that
the "handedness" of these persons had
been acquired. They had been forced
to use the wrong arm, either through
some accident or through misguided
efforts of their parents.

Now that our "handedness" had
been established, the next step was to
discover whether harm was apt to
result from these "transfers."

On questioning the "transfers," the
investigators found that more than
half of them had stammered or stut-
tered at some time in their lives. Was
this merely coincidence, or was there
some real reason for it?

Brain psychologists advanced the
following theory: The brain centers
involved in speech are located in one
hemisphere of the brain in the left
hemisphere for the right-handed, and

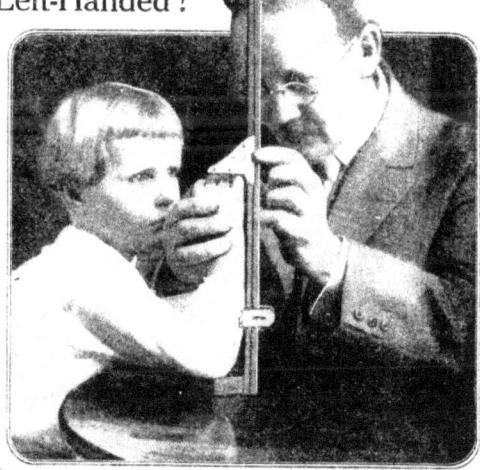

Measuring the child's arm to discover
whether he was born right or left-handed

Testing the skill of both hands by determining
how long it takes to drop steel balls into a tube

in the right hemisphere for the left-
handed. If a left-handed child is
forced to write with his right hand,
his writing center will be developed in
the wrong hemisphere. This may
result in speech-hesitation.

Why Some People Stammer

To establish more definitely the proof
of this, a stammerer was experimented
on. The victim was an eight-year-old
boy, a left-to-right "transfer," who
had just begun to stammer. The boy
was set to writing with his left major
arm, and in a short time the stammer
disappeared. Other similar experiments had the same results.

Taking it all in all, this investigation seems proof conclusive that left-
handed children should not be forced
to use the right hand.

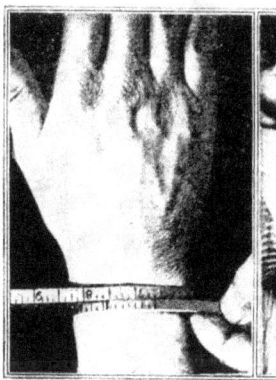

The arm muscles should
be relaxed and the fingers
spread in measuring the
circumference of the wrist

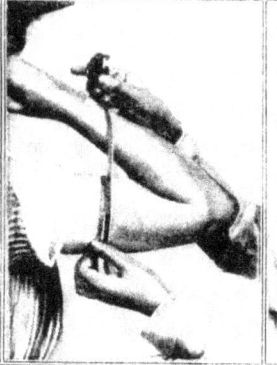

Before the muscle swell of
the arm is measured, the
subject clenches his fist
and shakes it vigorously

The muscle swell of the
forearm also helps to de-
termine whether one was
born right- or left-handed

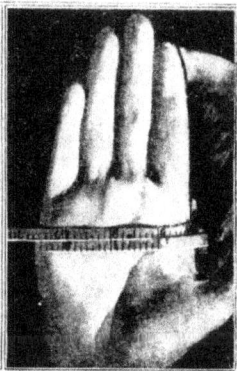

After flattening out the
palm of the hand, with the
fingers touching, a tape-
measure is drawn round it

The Measure of a Man

By Hinton Gilmore

In the sphygmograph, the pulse-beat drives up a button on the device, and this motion is transmitted to a needle, which makes a tracing on a camphor-smoked film

MODERN advance in the science of diagnosis has made it possible to weigh and measure the vital organs and functions of the body with accuracy. The day of pulse-feeling, temperature-taking diagonosis is almost a thing of the past. Instead of submitting himself to a cursory checking over, the patient nowadays —especially if he goes to an up-to-date hospital or sanatorium— will be studied, calipered, and charted according to an elaborate system in which mechanics play an important part. Here is a list of alarmingly named instruments that are ready to lend their aid in the measuring of a man:

Microscope, chemical balance, polariscope, stethoscope, sphygmograph, sphygmomanometer, dynamometer, polygraph, electrocardiograph, respiration apparatus, hemo-stoscope, X-ray, gastroscope, proctoscope, cystoscope, laryngoscope, ophthalmoscope, thermometer, and esthesiometer.

And here are some of the distinct laboratories that collaborate in the

Taking moving pictures of the stomach while it is at work. The patient stands against the cylindrical plate-holder and the X-ray is applied from the rear

study of a single case: chemical, bacteriological, urinary, blood, metabolism, fecal, X-ray, anthropometric.

In the laboratories every important structure and function of the body is studied and compared with normal standards. Such an examination requires team-work. Half a score of experts, with their assistants, must contribute their special expert knowledge of fact and technique in working out the data necessary for a complete comparison of a sick man with a normal healthy man.

Hours of Expert Work

Such an examination requires, in the aggregate, many hours of careful concentrated attention and must necessarily occupy several days for its completion; but the information gained is invaluable.

The dynamic capacity of a man is measured by the dynamometer, a machine for testing every one of a set of muscles. Plotted on a strength graph, the results show the relation of the patient's strength to that of a normal man of the same height and weight.

If further data are desired, the actual horsepower of the patient may be determined by the eurostometer, or horsepower machine. This is a simple device in the operation of which the

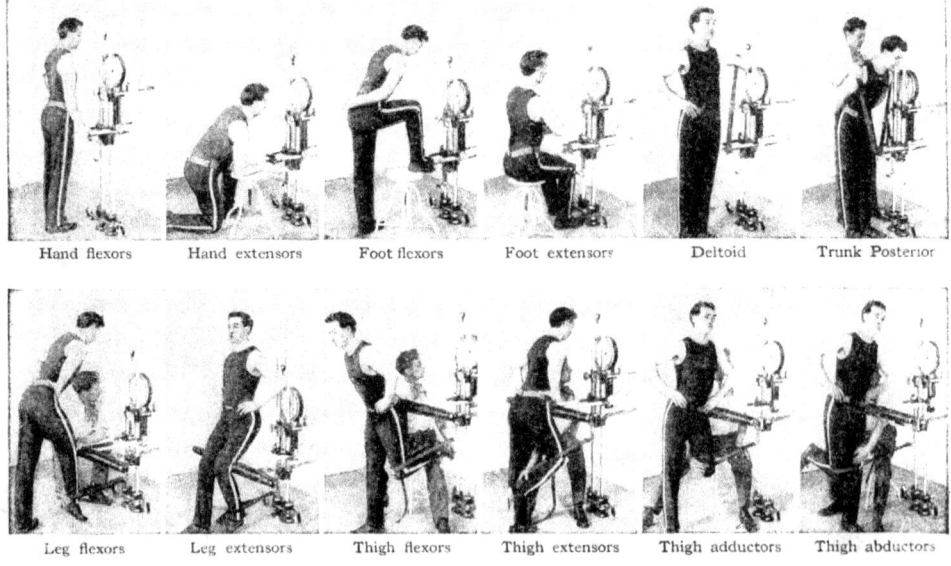

Hand flexors   Hand extensors   Foot flexors   Foot extensors   Deltoid   Trunk Posterior

Leg flexors   Leg extensors   Thigh flexors   Thigh extensors   Thigh adductors   Thigh abductors

The dynamometer is a device by which it is possible to determine the strength of every important muscle or set of muscles. The machine is capable of making forty-five separate tests. To de-termine the strength of the individual undergoing the test, the capacity of the various sets of muscles is totaled in foot-pounds. The average man has a total strength of 4,500 pounds and the patient exerts his strength against a resisting brake; the result is registered on a scale-beam marked in horsepower instead of pounds avoirdupois.

Lung power is readily demonstrated by the spirometer—simply a cylinder within a cylinder, the upper one being lifted by the air-pressure generated in vigorous exhalation. The lung capacity is registered on the beam as the cylinder is lifted.

Abnormal blood-pressure—one of the most significant functional changes in the body—is determined by a comparatively simple device. A rubber tubing is wrapped around the bared arm and inflated, so that the pressure of the blood is exerted against a tube of mercury. This registers the pressure of the blood with absolute precision. Such serious disorders as arteriosclerosis (hardening of the arteries), Bright's disease, and anemia are usually detected by the simple test.

The Electro-cardiograph

Abnormalities in the action of the pulse are made apparent by the sphygmograph—a mechanical apparatus through the operation of which exact tracings of the pulse are obtained.

The discovery that contractions of the heart are accompanied by the production of electrical cur-

rents has finally led to the perfection of apparatus by means of which information can be obtained respecting the morbid conditions of the heart, both organic and functional, which cannot be obtained in any other way.

This apparatus is known as the electro-cardiograph, and is not available except in the larger medical institutions throughout the country. Deviations of the heart action from the normal are registered in graphic form, and provide an accurate index to diseases of the heart.

The X-ray, of course, is invaluable

The respiratory calorimeter, for determining the ability of the body to assimilate food. The patient is connected to the machine by breathing-tubes. His supply of air comes from the machine, and his exhalations go back into it, passing into a container of soda-lime, which absorbs the carbon dioxide. By a comparison of the oxygen inhaled with the carbon dioxid exhaled it is possible to determine the progress of metabolism

in the study of pathologic conditions. It is especially important in observing conditions in the stomach and intestines. The patient first eats what is known as a test meal, in which bismuth, a harmless metal, has been mixed. This admixture renders the meal opaque to X-rays as it passes through the stomach and into the intestines. Ulcers of the stomach, intestinal kinks, adhesions, and the like may thus be definitely located.

Another important apparatus in diagnosis is the respiratory calorimeter. This is specially valuable in cases of suspected diabetes.

Other Apparatus

By means of the X-ray the presence or absence of stones in the kidney and of other diseased conditions of this organ may be determined. By means of the cystoscope the interior of the bladder may be inspected.

The gastroscope permits a visual examination of the interior of the stomach. This recently perfected instrument may render material service when other methods of examination prove ineffectual.

The bronchoscope not only permits an examination of the larynx, trachea, and lungs, but also facilitates the removal of foreign bodies.

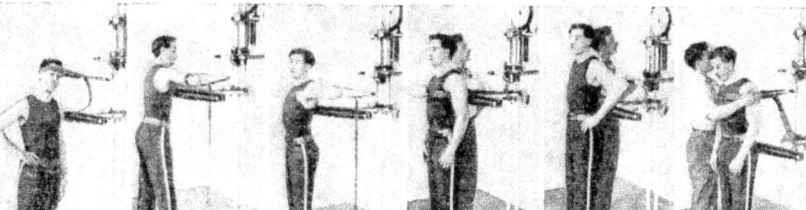

Arm flexors   Arm extensors   Latissimus dorsi   Pronators   Supinators   Neck anterior   Neck posterior

Neck lateral   Shoulder retractors   Pectorals   Inspiration chest   Inspiration waist   Trunk anterior   Trunk lateral

average woman can develop 2,500 pounds. Under normal conditions the trunk-anterior test should register greater strength than any other muscles; but this is not always the case. The

foot-extensor muscles, as shown by a series of tests made with the dynamometer, are usually the best developed in the average individual, owing to their constant use in walking

What's Your Speed?

Not that of your car, but of your brain?

By Ernest Welleck

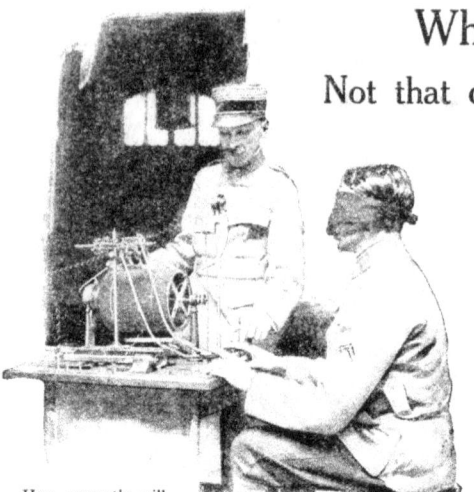

How promptly will the blindfolded soldier respond to a touch? It depends upon his brain speed

HOW promptly does your brain respond to impressions by muscular reaction? If you don't know, and the chances are you don't, go to Dr. Amar, in Paris, and let him find out for you. He will submit you to a few psychographic tests, and at their conclusion will tell you your speed to the hundredth part of a second. Incidentally, he will tell you whether you are fitted for the profession or occupation in which you are engaged or intend to become engaged.

His Own Invention

The apparatus that Dr. Amar uses is a psychograph of his own invention, a device for registering the promptness, intensity, duration, etc., of muscular responses to impressions received by eye, ear, or sense of touch of the person tested. The apparatus has a cylinder covered with paper coated with lampblack. This cylinder is revolved by clockwork at the rate of one revolution a second. A vibrating needle, which makes one hundred double vibrations a second, marks a wavy line that serves as time measure for minute fractions of a second upon the blackened paper. The muscular reactions of the subject tested are transmitted by air-pressure to two needles, which mark a record of these reactions on the cylinder.

Dr. Amar will seat you in front of a table equipped with the testing apparatus. Directly before you are two little pneumatic drums. When you have received your instructions, you place a finger upon the membrane of one of the little drums, your eyes focused on something that resembles a small camera.

A few minutes later a touch of the Doctor's finger upon the shutter releases allows a flash of the electric light in the box behind the lens to reach your eye. At the same moment— so it seems to you—you press your finger upon the drum-head. The air-pressure in the drum simultaneously causes a needle to mark a line, more or less curved, on the lampblack-covered paper.

You Read Your Record

Then the paper is taken off the cylinder. Here is the mark of the Doctor's signal and there is the record of your reaction. The Doctor counts the number of waves of the vibrating needle on the paper, and informs you that 20/100 of a second elapsed between the signal and your response. And you imagined your pressure to have been simultaneous with the signal!

You are assured that your brain functions normally—that the time for simple visual reactions in normal subjects averages between 0.195 and 0.21 seconds.

Tests that Involve Deliberation

In the tests for reactions involving deliberation, the same device is used. You place one finger of the left hand on one of the little drums, one finger of the right hand on the other drum. You are informed that the left drum means blue, the right drum red. The Doctor flashes a red or a blue light through the lens of the camera-like device, and you signal back the impression by pressing the right or the left drum-head. On examining the record on the cylinder, you find that it took you more than twice as long to react in this visual test as in the simple visual test in which you were not called upon to decide whether the light was red or blue.

From hundreds of observations like these Professor Amar has drawn interesting and valuable conclusions which enabled him to determine the aptitude of the individual tested for certain vocations, a problem of importance in finding employment for the thousands of soldiers returning from the war.

The statistical material so far collected shows that the age of the subject, between the limits of eighteen and forty-five years, does not materially affect the time of simple reactions. Subjects whose occupation demands alertness—for instance, designers, typists, and mechanics—react more promptly than farmers, who are invariably slower by 0.02 seconds or more. The records of persons who have sustained injuries of the brain, or who have been operated on because of such injuries, show much higher figures: 0.32 seconds for visual signals, 0.24 for sound signals, and 0.21 seconds for touch signals represent the average for that class of invalids.

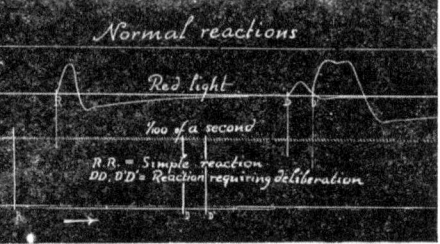

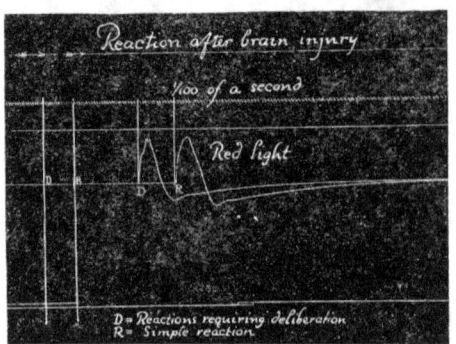

Above, a record of the simple reaction to a visual signal and of reactions requiring deliberation by a normal person; and, below, a record of similar tests on a young coppersmith whose brain has been injured

Games Blind Men Play

Helping them to see with the ends of their fingers

Photographs © Publishers' Photo Service

Raised numbers on dominoes notched so as to fit together have added the game to the list of amusements for the blind

By turning T's and L's and their respective triangular and square backs in different positions to represent different numbers, problems in mathematics are easily solved

After dominoes, checkers are easy. White checkers are round and black square. The black spaces on the board are cut out

Chessmen have pegs that fit into holes in the board. The white men have sharp points on their tops; the black tops are rounded

Best of all, blind men can now play cards. At first glance, their deck looks just like ours, but letters from their alphabet are embossed in the corners of the cards to represent suit and character—the outside one for suit, the inside one for character. Three dots above and one below, on the inside, mean Jack. Three of these mean a good hand in poker

71

How the Cipher Expert Works

The message that reaches him may be a wild jumble of letters and figures, but it soon gives up its meaning

By Charles A. Collman

THE military cipher expert gains only a small share of war's honors or glory; but in the quiet of his office, his swift, subtle, and scientific efforts in deciphering intercepted dispatches and orders many times bring defeat and disaster to the enemy.

Other branches of the military service, engaged upon more active duty, gather the material, which is laid later before the man whose task is the solution of ciphers abstracted from enemy communications. The scout patrol-boat, silently plowing the waters off our Atlantic coast in its quest for submarines, suddenly picks up from the ether the following message, which is recorded by the receiving instrument of the wireless operator on board:

3 S I U O W S E S I L D R Z L A L A
N A B D E n L Y D a N L F C O E T I
U O R G H F K A I R L M T E M T

The wireless operator instantly perceives that this is only too evidently a cipher dispatch of the enemy; and, since there are no other enemy craft suspected of being in these waters, it must be a communication sent by one U-boat to another.

No Cipher Indecipherable

The message is at once rushed by the quickest available means to the nearest office of the Intelligence Department, where it is submitted to the scrutiny of the cipher expert. It is accepted by him as an axiom that no practicable military cipher is mathematically indecipherable. But it requires a remarkably long time, as a rule, to decipher even the simplest kind of message. Therefore speed is essential, so that advantage can be taken of the news contained in this secret communication before it is too late.

So the cipher expert starts to work immediately. His office is amply supplied with all the known data and material by means of which the most up-to-date cipher communication may be solved. He is provided with tables of frequency of the language of the enemy country, covering single letters and digraphs of duals or pairs. He knows that ciphers are divided into two general

classes — the transposition cipher and the substitution cipher; and his first endeavor is to learn in which class this dispatch belongs.

How the Human Mind Works

Substitution ciphers may be made up of substituted letters, numerals, conventional signs, or combinations of all three. The human mind works along the same lines always. The inventive man creates, but his supposed invention is, in principle, many hundreds of years old. For instance, the basic system invented by the Abbot Tritheim (1462-1516) represents the foundation of the modern substitution

The pattern cipher cannot be decoded without the aid of this ground card

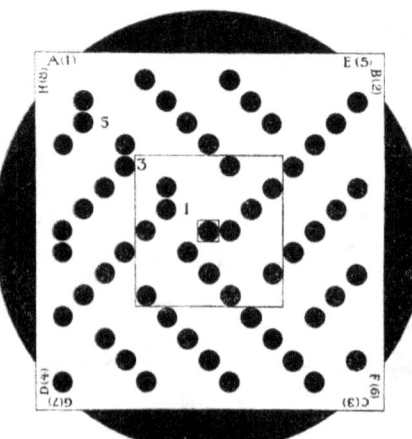

The perforated pattern, which, when placed on the ground card, discloses only the letters to be read

ciphers which are most widely used. Transposition ciphers are limited to the characters of the original text, which are rearranged singly according to some predetermined method or key. This method of secret writing is used to obviate the difficulty of deciphering substitution ciphers, since it is essential that the recipient of the message solve it quickly.

Fortified with his knowledge of the various known cipher systems, the cipher expert closely examines the message that has been submitted to his scrutiny. Much depends upon this initial scrutiny, in which all of his natural powers and talents are called into play. He soon makes up his mind that his frequency tables are of no use in this emergency, since this is not a substitution but a transposition cipher. Furthermore, he recognizes it as a special type of the latter class, from certain plain indications, such as the use of the figure "3" and the junction of the letters "Da," which represent a key. In fact, he suspects that he has before him a form of transposition cipher known as the "net," "pattern," "stencil," or "lattice" cipher.

As part of his office equipment, the expert is provided with most of the known forms of "nets" or "patterns," which consist of perforated squares of tin or cardboard. In this instance he discerns the use of the medium known as Fleissner's improved pattern cipher, in which the pattern or stencil called into play has fifty-seven small round hole cuts out of or punched into the carton, each of which has room for one letter or number.

The Fleissner "Pattern"

The pattern is a perfect square, and the perforations have been made according to certain mathematical calculations, so that, when the pattern is placed upon a sheet of paper, it may be moved in four different positions, and, letters having been written each time in the openings, the entire pattern surface will be covered with a cipher of fifteen lines having fifteen letters each, or 225 letters in all.

The reader need only refer to the accompanying chart to fashion for himself a pattern out of a sheet of common writing paper, when the problem will present itself to him in the most simplified form. The entire pattern surface is divided into squares, running from No. 1 to No. 7, and the size of the communication will depend upon the squares selected, as follows:

The entire pattern surface, that is, to the seventh square, presents 15 x 15 openings, or 225 letters

	No. of Openings	Letters
Sixth square	13 x 13	169
Fifth square	11 x 11	121
Fourth square	9 x 9	81
Third square	7 x 7	49
Second square	5 x 5	25
First square	3 x 3	9

How to Use the Pattern

In the corners of the pattern are indicated the four different positions in which it should be moved: namely, A, the first position; B, the second; C, the third; and D, the fourth. It is possible, also, by reversing the surface of the pattern, to obtain four additional positions (E, F, G, H): but, since no communication of such extended length is under consideration, this matter may be left to the further experimentation of the reader. The pattern should be moved always from left to right.

From the above sum of the letter openings should be deducted the central perforation, which is not intended for the substance of the cipher message. This opening has a special significance, since here is placed the capital letter (A, B, C, or D) that indicates the position of the pattern in which the enciphering of the message was begun.

The cipher expert, bearing these facts in mind, recognizes from the use of the number "3" that the third square of the pattern has been adopted

in writing the cipher, and verifies this fact in that exactly forty-nine letter openings have been used, although in two instances two letters each have been placed in the openings.

Shifting the card about according to well known rules, he solves the "pattern" cipher

Now, in beginning his work of deciphering, the expert draws on a sheet of paper a square that corresponds in size with the third square of the pattern, and fills it in with 7 x 7 small circles in positions corresponding with the pattern perforations.

The next step that the expert takes is to write the letters and combinations of letters of the hostile radio message, which results in a table of seven lines, of seven letters each, as follows:

```
S   I   U   O   W   S   E
S   I   L   D   R   Z   L
A   L   A   N   A   B   D
En  L   Y   Da  N   L   F
C   O   E   T   I   U   O
R   G   H   F   K   A   I
R   L   M   T   E   M   T
```

Now the decipherer places his pattern upon this table of letters, and discovers that the letters "Da" fit into the central hole. So it is evident to him that the writer of the message began his communication with "D," the fourth position. He therefore arranges the pattern so that "D(4)" appears in the upper left-hand corner, and the openings in the pattern disclose to him the following series of letters: S O R Z A L C T O K L M. This does not make sense, and the expert realizes that the writer adopted position "D" for purposes of mystification. Further observation inclines him to the belief that the use of "Da" as a key indicates that position "D" was utilized merely as a "stall," and that the real message was begun with position "A."

The Message at Last

Turning the pattern into the next position, with "A (1)" in the upper left-hand corner, he reads as follows: "Will Be N N E Fire   " He perceives that he has hit upon the proper solution of the cipher, and, turning the pattern to the left again, to position "B," then to position "C," he obtains the remainder of the message, which reads: "Will be N.N.E. Fire Island Light, Tuesday, four A.M."

All is now known. The proper naval authorities are notified, the machinery of the coast defense is set in motion, and on the day, the hour, and in the position indicated, two U-boats are captured or sunk. And perhaps the cipher expert gains at least part of the credit of the achievement.

It may be found interesting to construct a pattern such as the one presented out of tin, cardboard, or even writing paper. The perforations can readily be cut out with a penknife.

Hello, Central! Give Me Berlin

A MODERN army without telephone service is in about the same fix as a big city would be if all the wires should go dead. It follows that every effort must be made to keep the telephone service in the field at top-notch efficiency, and therefore daring linemen and operators follow close at the heels of the troops.

On the fighting line "central" is housed far underground or in carefully camouflaged shelters; but the larger stations back of the front are usually safe from all but aerial attack, and they set up their switchboards in the comparative comfort of a large tent or, better still, an abandoned house.

At least, an army "central" isn't asked foolish questions, nor does he have to "listen in" to his gossipy neighbors

In addition to seeing that messages go to the right places, the telephone squad must watch out for breaks and make quick repairs. Formerly the squad had to be on the alert for an enemy "listening in"; but in the latter part of the war the lines were so laid that the use of spy microphones by which messages were stolen from the wire was almost impossible.

The army central's task, while made less hazardous by peace, is none the less strenuous. The title of this article may be literally lived up to, for the army centrals will have a large part in aiding the work of food distribution among the starving enemy.

Can Music Cure as Well as Charm?

How a concert pianist turned her music to the business of healing

By W. T. Perry

Working on a deaf man. The man places his forehead on the lid and his hands on the sides of the instrument, while the pianist plays major chords very loudly

HAS music a new and hitherto little recognized curative value? That different emotions are excited by different sorts of musical strains has been appreciated since first Pan played his reeds. The easy cadence of waltz time invited our fathers and mothers to dance, just as the sound of the so-called "jazz" prompts the present generation to step and glide or slide about. The snap of military march time quickens the pulse. The soft crooning of some Southern mammy lulls to sleep a tired, fretful pickaninny.

Miss Anderton's Theory

Miss Margaret Anderton, a concert pianist, and the originator of what she terms the *piano causerie*, which consists of prefacing each program number with a few cleverly chosen words with the idea of creating a better understanding of the numbers played, has studied the sedative and, in what she believes to be an astonishing number of cases, the curative properties of musical sound vibrations. Miss Anderton had been making an analytical study of the effects of musical sound vibrations for some time when the war came. She was in Canada at the time, and she became strongly impressed with the belief that she could make use of her ideas in helping soldiers who were wounded or victims of shell shock.

Perhaps her own words describe the work best. She says:

"The concert artist should use her skill for the scientific application of

music as a medicine under a doctor's prescription, a different matter from simply playing music to soothe or divert.

"It is high time that art became altruistic. Musico-therapeutics should not be a commercial asset. The artist, giving of her highly trained abilities to the service of science and the medical profession after years of private research work, feels profoundly the use of music as a curative means. The procedure is psychological."

The purpose of musico-therapeutics,

A man who had lost the use of three fingers was taught to play with two fingers; as a result the paralyzed fingers became almost normal

Teaching a man suffering from shell shock to play the xylophone. At first he is almost afraid to strike the instrument, but gradually becoming more and more confident, recovers his mental poise

therefore, is curative. It is not an entertainment. It is not to play to invalided soldiers just to pass the time away, merely to amuse and please them. Music is to be administered with a definite, scientific purpose, and only after careful study of the medical reports of each individual case and of the patient's physical history.

Different sound vibrations produced by different musical instruments, different keys, and different time are selected for each individual case.

Thus a man who had been a structural carpenter, and who was a victim of shell shock, was patiently taught to play the xylophone. At first he hesitated to strike the bars even gently. Step by step, he was persuaded lightly to tap the keys and later to pick out "God Save the King," a tune he remembered Soon he was hammering away lustily, and, incidentally, recovered his poise Another interesting case was that of a soldier who, after a high-powered shell had exploded near him, although he was untouched, believed himself deaf.

Distinctly Not for Enjoyment

This man could hear absolutely nothing. After inducing him to place his forehead on the piano-top and to press his hands against the sides of the instrument while loud major chords were played, his hearing became acute.

As Miss Anderton points out, music so played is not supposed merely to be enjoyed in the way that normal, well folk enjoy music. It is rather that the tone vibrations, varying in quality rhythm, and intensity, may, when properly applied, alleviate and often cure neurotic conditions.

Minds that Work While Bodies Sleep

Some authenticated cases of a much doubted phenomenon

By Edwin F. Bowers, M.D.

THE impossible is the thing we don't believe; the incredible the thing we have not yet seen. Both are states of mind common to the average human being. That is why any account dealing with manifestations of the subliminal mind—that mind which works while we sleep and at all other times, and which so few understand must always be bolstered up with affidavits and attestations if it is to carry conviction. That is why the phenomena of unconscious cerebral and bodily activity have been considered, even by scientists, as old wives' tales.

And yet, among medical reports of abnormal mental conditions, and in the Proceedings of the Psychical Research Society, hundreds, if not thousands, of well authenticated cases of most extraordinary activities of sleeping persons have been recorded. In some instances the mental feats accomplished far transcended the normal capabilities of the individual.

Turning Sleep into Money

Such a case is the intuition—or perhaps it was the clear subconscious grasp of business detail—of a Russian banker who was addicted to the habit of getting up at night and looking over his papers while asleep. The banker had been examining the prospectus of an oil company about to be formed, in which he had planned to buy an interest. After mature deliberation with his objective mind (the mind we use while awake) he decided not to "take a chance."

However, a few days later his agents told him that, following his instructions, they had bought heavily for his account, at the same time showing him a letter in his handwriting authorizing this purchase—a letter he had written while somnambulistic, and of which he had not the slightest conscious recollection. Within two years the banker had added two and a half millions of dollars to his account—which puts him in the championship class of sleep-walking money-makers.

But, while one person does something constructive during a somnambulistic attack, a dozen do some destructive or absurdly foolish thing.

Such, for instance, is the case of an English nobleman who, missing his shirts almost at the rate of one a night, accused his valet of stealing them. The very next night the erratic nobleman was seen stealing out in his pajamas, carrying a soft-bosom dress shirt under his arm.

Proceeding cautiously to the back of the barn, he secured a spade, dug a shallow grave, and buried his shirt. Investigation showed that all the remaining shirts had been interred in the same way.

She Hid Her Ring from Herself

Dr. John D. Quackenbos, of New York, tells me of an instance in which a patient, a young woman well known as a successful writer of short stories, was taking treatments by hypnotic suggestion which materially increased her powers of concentration and quickened her imaginative faculties. Through an oversight on the part of one of the nurses, she was permitted to leave for home while in a somnambulistic condition.

When she "came to" the next morn-

Sleep-walking made millions for a Russian whose wits were keenest when he was asleep

He might be called a "sleep-writer," since most of his writing is done while he is sound asleep

ing, she found that a valuable diamond ring was missing. Calling on the doctor, she made known her loss, adding that she had the impression that she had given the ring to a beggar.

Dr. Quackenbos immediately put her into the hypnotic condition, and gave her strong suggestions to the effect that when she returned to her home she would remember where she had hidden her ring.

Within an hour she called up, exclaiming: "Doctor, I've found my ring in the lining of an old muff—a muff I was going to give away today."

A Sleeping Novelist

Dr. Quackenbos himself is a rare and remarkable example of constructive somnambulistic activities, for some of his rather voluminous writing is done while he is objectively sound asleep.

It is his practice to arm himself with pad and pencil on retiring. On awaking in the morning he frequently finds that he has covered sheet after sheet of paper with a perfectly coherent essay. It was in this manner that Dr. Quackenbos wrote his fascinating study called "Body and Spirit."

In this connection it may be interesting to know that Dr. Quackenbos holds firmly that anyone who will take the trouble to cultivate the faculty can develop a psychic mental stream which will inevitably sweep ideas, impressions, and memorizations into objective consciousness, there to be utilized in solving business problems, and in assisting in the conception, construction,

Burying his shirts was the by-night occupation of an English sleep-walker

and completion of all work of a creative nature.

He advocates that one should comfortably compose himself and go to sleep for an hour or more with the problem, the story-germ, or what not, firmly fixed in mind. When the objective mind relinquishes the burden of thought—in other words, when the subject goes to sleep—the subconscious mind takes the matter up and carries it forward, together with the memory impressions of the subject.

The Sleep-Swimmer

A most interesting case, showing the unique coordination between muscle and mind in a somnambule, is recorded in which a young man, totally unable to swim in his normal waking condition, was accustomed to getting up at night two or three times a week and swimming across a river two miles wide.

Psychologists insist that, were this young man to be awakened while swimming, he would inevitably drown, for his objective mind could not transmit to his motor nervous system the impulses toward actions with which it was not itself acquainted.

This is supposedly the same form of mental and muscular correlation that enables the somnambule to walk fearlessly, and usually with safety, upon some precarious ledge or dizzying height.

While the erratic antics of the sleep-walker usually have a "happy ending," the uniformly accepted belief that no

accident ever befalls him unless he be suddenly awakened is fallacious. Numerous deaths from accidents to sleep-walkers testify to this.

A most remarkable case of somnambulism, combined with "externalization of faculties" and other psychic powers, was reported by a famous New York alienist and neurologist only a few months ago.

The subject was a Bavarian peasant girl, simple, good-hearted, and very ignorant. The gentleman in whose home the girl worked as a domestic was a student of psychic phenomena and a hypnotist of considerable skill.

Drunkards and sleep-walkers are said to be providentially protected

He had, it seems, developed a wonderful telepathic rapport with this girl, and had brought her to his friend, the neurologist, for experimental work.

Thrown into the cataleptic state by hypnosis, the girl would inhale deep draughts from a bottle of the strongest ammonia, under the suggestion that it was perfume of roses. She chewed a strychnia tablet, perhaps the bitterest thing in the world, with gusto and relish, under the belief that it was sugar. Blank cartridges were fired off behind her head, without producing a single quaver of shock appreciation. Tested as to her accuracy in telepathy, it was found that she could read her employer's every thought.

Her Mind Saw Another Room

The alienist, to extend the scope of the experiment, directed that the girl—or her subliminal mind—proceed upstairs, enter a certain room (his daughter's bedroom), and describe what she saw there.

After an interval the girl announced that she was in a bedroom, and described in detail the physical characteristics of the doctor's daughter, a little girl of eight years.

Asked to count the number of chairs in the room, she announced that there were nine—two more than the number usually kept there.

Thus far the results might have been due merely to the ability of this girl to read the doctor's mind and to describe what he well knew. But, on being asked what was on the mantel, the girl replied, "A picture of a horse."

Now, the doctor was certain that there was no picture of a horse in his daughter's room. So, leaving the subject lying on the couch, the experimenters proceeded to the little girls' room.

The mantel contained only the usual school-girl trifles.

"Just a clever mind-reader, after all," said the doctor. "She can't externalize her seeing faculties. She sees only what you and I have in our minds."

"Hold on," said the doctor's friend, stepping over to the mantel and picking up the photograph of a horse that was lying flat on the shelf. "What's this?"

It was a snap-shot of one of the doctor's horses, taken in the country by his little girl.

To explain these things is more difficult than to describe them. It is generally believed, however, that sleep-walking is only a form of auto-hypnosis. Somnambulism is generally confined to children, or to the youthful—in other words, to those happy people who still preserve illusions. Frequently, however, it accompanies a neurotic disposition; or it may result from great mental agitation. Lady Macbeth furnishes a classic example of this.

Do Spirits Talk through the Ouija Board?

Perhaps it is that subconscious ego whose memory is better than yours

By A. J. Lorràine

ABOUT you hovers the unseen. You place the ouija board upon your knees or on a table, and let your hand rest lightly on the planchette, the little three-legged carriage that rides over the ouija board. See that you are comfortably seated, and your arms not cramped, but free to move. Stop your thinking machine; try to make your mind, as far as possible, a blank. If you want a successful séance, see that the circle of friends about you consists of persons who take a serious interest in the proceedings. Scoffers are not helpful; a quiet, serious atmosphere is most favorable to good results.

A time passes in silence. The right conditions must be allowed to establish themselves.

Look — now the planchette stirs. A hidden power seems to move it. At first a few straggling, random movements. This may continue for a while. Presently, pointing, one by one, to the letters of an alphabet printed on the face of the board, the planchette begins to spell out a message:

```
M * * * A * * N * * * Y M * * * O * *
O * * * N * * * S A * * * G * * * * O
I L * * * I * * V * * * E * * * D A * *
G * * * A * * I * * N I C * * * O * * *
M * * * E . P * * * A * * * T * * * I * *
E * * * N * * C * * E W * * * O * * *
R * * * T * * * H M * * * Y N * * * A * *
M * * * E .
```

With these words one of the most remarkable personalities that ever spoke through the ouija board announced herself to the world. It is sometimes said, in criticism of those who inquire into occult phenomena, that nothing but trifles and dribble ever comes to us from the agencies, whatever they may be, that speak to us through mediums and other unusual channels.

Anyone who says this has either not read the communications of Patience Worth, through the hand of Mrs. Curran of St. Louis, or else is wholly devoid of literary sense. Read the book and judge for yourself. Selected excerpts from the records have been collected into a volume by Casper S. Yost (Henry Holt and Co., publishers). There you will find romantic tales of times long since past, and a series of poems of singular charm, written in a quaint old English dialect. And the remarkable thing is that Mrs. Curran, through whom the communications came, is herself not a writer, and the very language in which they are couched is strange to her.

Patience Worth claims to be the discarnate spirit of a person that once walked this earth, even as you and I. What does psychic science have to say

Skeptics are disappointed in the character of the messages received with the aid of the ouija board. They forget that even the greatest of us are but human beings. Sir Isaac Newton did not always discuss mathematics at the breakfast-table, nor Napoleon the best ways to win a battle. It is much more to the point if the spirit of Newton refers to a special dish of which he was particularly fond in his lifetime, and if on investigation we find his statement to be true, than if he were to dwell long and learnedly on gravitation

about it? To understand the situation, you must begin by understanding yourself.

Ordinarily your hand obeys your will; its motion fulfills some conscious purpose of your mind. When, therefore, you find the planchette under your hand performing movements and spelling out words that were not in your mind, and which you had in no wise purposed, you are naturally surprised.

One explanation that presents itself for this strange course of events is that some unknown outside influence, perhaps a spirit, guides your hand, irrespective of any knowledge or purpose on your part.

Are You Absent-Minded?

And this is a possible explanation. But, before we finally adopt it, let us pause for a moment. Think: in the ordinary course of life, do you not sometimes perform actions that are not controlled by any conscious act of the will — actions, perhaps, that are even contrary to what you intended? Are you never absent-minded? Have you never handed the conductor on a street-car a transfer and held out your hand for the change? Have you never dipped your pen in the mucilage bottle? Or, if not that, then some equally absurd lapse you surely have committed sometime, as sure as you are human, though you may not have gone so far as the young man who started to take a girl to an ice-cream parlor and found himself escorting her to the undertaker's.

Such lapses remind us that many of our actions are automatic. You exercise your will to decide the main issues, and then leave the execution of the details to — to what? Well, psychologists call it your subconscious

THE war has done so much to revive interest in the ouija board that the writer of this article made the decision to approach the whole subject of spiritualism with an open mind. It will not do to accuse those who manipulate the ouija board of being frauds; there are too many of them. Besides, it is the business of science to investigate and not to prejudge. In this article the author sets forth the results of investigations that have been conducted by psychic researchers for years. It will be followed by articles on other phases of spiritualism.

self. This other self is a very remarkable person, even though once in a while he plays curious pranks with us by doing the right thing at the wrong time. I am not going to say that your other self is in all things your superior: in some of life's affairs your dominant conscious ego has to hold the helm. But in some things the subconscious or subliminal self is indeed your superior. His memory is phenomenal. It seems almost as if he never forgot anything. Ordinarily he takes a modest place in the background, so that, though you have been living with him all these years, you may hardly have noticed him at all. But there are ways of drawing him out into the light, where you can observe him. He notes many things that you pass by unobserved. The clock has just struck the hour. You did not count the strokes. But it is not too late. Though you were paying no attention, he was on the job; he kept a memory record of the sound. And in your mind you can go back and count the strokes.

A revived form of an old pastime: A wedding ring suspended from a thread tied to the fingers performs various automatic motions. The more usual method is to put your elbows on the table, hold your hands against your temples, with one end of the silk thread at each temple, the ring hanging in the loop of the thread under your chin and swinging to and fro in a wine glass. It counts out answers to your questions by striking the sides of the glass, so it is believed

Subconscious Memory

There is not a second in your waking hours when you are not constantly receiving impressions through your senses. Your eyes are open and are seeing *something or other* as long as you are awake. How much do you remember of what you see? Far more than you ordinarily suppose—at least, if "you" means both of you, your ordinary self and your subconscious self.

Perhaps you say, how do I know that the subconscious self remembers so much? As long as I am not conscious of what he remembers, how can I test his memory?

There are several ways of digging down into the memory of the subconscious self—that is to say, into your subconscious memory. The first method is very simple. Select any little incident in your life—something quite trivial, say. To start with something easy, you may pick out some fairly recent occurrence. Perhaps you went to see a friend off at the railway station, or you met an acquaintance in the street and stopped

Mrs. John H. Curran, of St. Louis, through whom "Patience Worth" has communicated a number of romantic prose stories and poems

to exchange a few words. Whatever it may be that you select for the experiment, fix your thought on that event in a dreamy sort of way, and let your thoughts drift freely around the various associations that come to your mind in connection with the incident. Do not make any effort to recall things; just let your thoughts come floating in as they will. You will be surprised at the amount of detail—trivial detail—that you will thus find streaming back to your mind. The faculty of thus resuscitating memories can be improved by practice, and is not without its use. It may help you sometimes to recover articles mislaid or lost.

But what, you may ask, has this to do with the ouija board?

One of the several methods of tapping the subconscious is with the aid of the ouija board, or, more generally, of automatic writing. For the planchette is not necessary: it is merely an adjunct to facilitate the process. Certain persons possess the natural faculty of automatic writing. If a pencil is placed in the hand of such a person and is allowed to rest on a piece of paper, and if this person takes his attention off the action of his hand,—for instance, by reading a book or by engaging in conversation,—presently the pencil between his fingers starts to write, tracing words and sentences having no relation whatever to the book or the conversation. It is just as if some other person were writing. The operator himself (or herself) does not know what has been written, any more than do the other persons present, until the writing is inspected.

The Case of Miss C—

Then, it is often found to contain references to experiences that had been long forgotten by the subject, and that frequently cannot be recalled even after the writing is deciphered, though other evidence may establish their truth. So, for example, Doctor Morton Prince, Professor of Nervous Diseases at Tufts College Medical School, quotes the case of a certain Miss C——, who, in a hypnotic trance, narrated a strange tale, ascribing it to the intermediation of the spirit of a person purporting to have lived in the reign of Richard II. She gave many personal and historical details which, on examination, were found to be correct, although many of them could not have been ascertained without extended research. Miss C——, on coming out of the trance, was unable to say how the intimate historical knowledge displayed in her hypnotic utterances could have originated. She could not recall reading any book that would have furnished her with the information. But a script produced by her in automatic writing disclosed the fact that a book called "The Countess Maud" had been read to her by her aunt fourteen years earlier, when she was eleven years old. Both ladies had so completely forgotten the contents of the book that they could not recall even the period with which it dealt.

We see, then, that we must be cautious in ascribing the motion of the ouija board, and the messages spelled out by it, to the action of extraneous influences, or perhaps of discarnate spirits. For, from what has been said above, it is clear that much of what appears to the subject as the creation of another mind may be merely the resuscitation of impressions wholly forgotten by the ordinary self, but indelibly fixed in the phenomenal memory of the subconscious self.

Hélène Smith from Mars

But how about the remarkable literary quality of the productions of Patience Worth? Mrs. Curran, through her ordinary consciousness, has shown no signs of any remarkable genius in that direction. And how about the quaint old English dialect consistently followed throughout her planchette writings accredited to the discarnate Patience Worth?

Well, in the matter of literary quality it is admitted that Patience Worth's productions are unsurpassed by any auto-

matic writing on record. But as to the dialect, the performances of a certain Swiss medium known under the pseudonym of Mademoiselle Hélène Smith totally eclipse her feat.

Hélène Smith has completely kicked over the traces, so to speak. *She* writes not merely in an archaic tongue, not in a mere outlandish dialect. Her speech is not of this earth at all. It is written in characters hitherto unknown, in words found in no earthly dictionary and for a good reason: Hélène Smith claims to be the spirit of a departed inhabitant of the planet Mars.

But there is something very suspicious about Hélène's Martian language. Its syntax, its grammar, the order of the words in the

A dream drawing. It makes us suspect that some modern paintings may be produced by automatic methods

sentence, are precisely the same as in French, Hélène's native tongue. More than that, the individual words are found, on inspection, to show a strange agreement in the number of syllables with the corresponding French words.

Subconsciously Evolved Dialect?

It would be difficult to find an example that would illustrate more tellingly the extraordinary faculties of the subconscious self than this performance of Mademoiselle Hélène Smith, who must have evolved, unknown to herself, a complete language and a cipher code, and have then proceeded to use them consistently, without stumbling, without hitch, in writing lengthy descriptions of alleged doings upon the planet Mars. As compared with this, it would have been a relatively easy task for Mrs. Curran to evolve, subconsciously, an "archaic dialect," which, by the way, has never been shown to have been actually spoken at any time or place.

As for the undeniably inspired character of Patience Worth's poetry and prose, this will not greatly surprise anyone conversant with the psychology of creative mental activity. Numerous men great in literature and other fields have told us that it seemed to them that not they, but some other person or persons for them, created their works—they merely wrote down what was revealed to them. Dickens, in his letters, speaks of the characters in his books much as he would of real personalities,

Psychographic signature of Archdeacon Colley, alleged to have been received after his demise by the Rev. G. Henslow. Above is shown the natural signature of Archdeacon Colley, for comparison with his alleged spirit signature

Characters used by Hélène Smith, and purporting to be the letters of the Martian alphabet

and he tells us that the famous Mrs. Gamp, perhaps his greatest creation, often spoke to him "with an inward monitory voice." Robert Louis Stevenson ascribes his inspirations to Brownies, who "tell him a story piece by piece, like a serial, and keep him all the while in ignorance of where they aim." Instances of this kind might be multiplied. Briefly, we may say that genius seems to consist in an unusually close contact of the conscious self with the subconscious, an unusually developed power to summon up before the view of the fully awake mind the dim phantasms with which the subconscious realm is peopled.

Psychology Explains Much

Are the revelation of the ouija board and other modes of automatic writing, then, to be wholly explained in the matter-of-fact terms of psychology? Is it all just a freak of the intricate workings of the human mind and body?

This does not follow. Psychology sheds light on the situation, and perhaps explains much. But there is seemingly also much that remains unexplained. So, for example, in a séance reported upon by F. W. Myers in his work on "Human Personality and Its Survival After Bodily Death," the operators of the ouija board were placed out of sight of the company, who in silence selected a photograph from an album and fixed their attention on it. Presently the ouija board began to spell out either the name or the initials of the person whose photograph had been selected, or else it wrote out some characteristic feature descriptive of the picture. Out of five cases three were correctly identified.

In another case a series of questions were put to the board: "How many shillings has Miss X in her purse?" "Four" (correct). "How many coins in my purse?" "Five" (correct, though the interrogator thought he had many more). A card was picked at random out of a pack, looked at by one of the party, and then laid face down on the table. "What is the color of the card?" ouija was asked. "Red" (correct). "What number?" "Seven" (correct). "What suit?" "Hearts" (correct).

Not only handwriting, but also drawings, have been produced by the automatic motion of the artist's hands. Such drawings, it is remarked, have a certain value for the artist which they do not possess for the ordinary individual

Another instance cited by Myers has an amusing side. The pencil in the hand of a certain lady gifted with automatic writing persisted, on one occasion, in writing time after time the word "Goat." She was then informed that the gentleman standing beside her was named Nanney.

In instances such as these, which are well authenticated, the ouija board and automatic writing generally seem to give marked evidence of telepathy or thought-reading. But there are cases in which the hypothesis of telepathy must be strained almost to

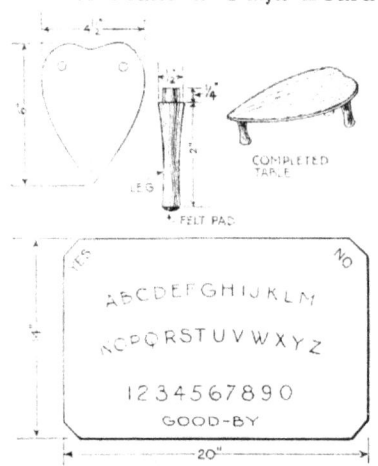

How to Make a Ouija Board

A homemade ouija board that "answers" questions fluently can be made from a clear perfectly straight board about 20 inches long, 14 inches wide, and ³⁄₈ inches thick. Round off the corners and sandpaper it until smooth. Trace on it the letters of the alphabet with a pencil, and black them with a brush; also trace the words "yes" and "no," and the numbers, and at the bottom the word "Good-by." Coat the board with two coats of varnish, and rub smooth. While the board is drying the table can be made. Cut from a board of ³⁄₈-inch material a heart-shaped piece 4½ inches wide at the widest point and 6 inches long. Next make the three wooden legs 2 inches long and glue them into holes in the under side of the table. Cut three disks of felt from an old hat and glue one to each foot of the little table

the breaking-point to account for messages received. Dr. Hyslop, in his book "Contact with Other Worlds," cites a number of such examples.

On January 28, 1902, Dr. Hodgson suggested to the alleged spirit controlling the famous medium Mrs. Piper that he should impress the daughter of Mrs. Verrall, in England, with the view of a person holding a *spear* in his hand. The medium replied: "Why a *sphere?*" Dr. Hodgson corrected her: "A spear." The control agreed to try. Both Mrs. Verrall and her daughter are noted mediums. On February 4 of the same year Mrs. Piper's control reported to Dr. Hodgson (this being the first séance since the giving of the message) that he had succeeded in transmitting the message about the "sphear" (so Mrs. Piper spelled it). Meanwhile Mrs. Verrall had, on January 31, received through her automatic script a message partly in Greek and partly in Latin, which translated read as follows:

"A universal seeing of the *sphere* fosters the mystic joint reception. Why did you not see it? The flying iron [Latin idiom for *spear*] will hit."

Mrs. Verrall is a classical scholar, so the receipt of a message in Greek and Latin is not in itself very surprising. Still more remarkable, therefore, is the case of a message given by Mrs. Verrall to her control in Greek, and repeated together with the translation and the reply by Mrs. Piper in America, though this lady is ignorant of the ancient languages.

Another Automatic Script Case

In other cases automatic script has been found to bear a marked resemblance to the handwriting of a deceased person purporting to be the author of the communication. Material of this kind has been obtained particularly by Mr. B. F. Underwood through the writing of his wife. Says he:

"Fully aware that, under certain conditions, from the submerged self may be sent up memories which cannot be distinguished from newly acquired knowledge, still, I am confident that Mrs. Underwood's hand has written names and statements of facts, not only once but several times, which were not part of her conscious self."

Faking the Supernatural

How spiritualistic frauds are perpetrated

By Hereward Carrington

He places the white robe in a black bag. Then he puts on a pair of black gloves, invisible in the darkness

The "medium's" properties are hidden in various secret places —a hollow chair-seat, the curtains, in the inevitable mandolin

Gradually he shakes the white stuff loose, and it appears to grow. When it is well up he darts out and throws it robe-like around his own shoulders

Suddenly a form appears clad in snow-white robes, a turban on its head, with flowing beard and hair

A DOZEN awed spectators are gathered around the cabinet of the medium. The medium and cabinet are searched. Nothing suspicious is found. The room is darkened. On the carpet, two feet or more in front of the cabinet curtains, a white spot appears. It increases in size, turns and twists and writhes as if it were a thing in agony. It turns and tumbles, and grows larger and larger, before the eyes of the spectators.

Finally the thing assumes the size of a person. A head and face now appear. "Lights" are called for. There, standing before the bewildered spectators, is a form—clad from head to foot in snow-white robes, a turban on its head, with flowing black hair.

The lights are turned down; the "spirit" begins to melt: it turns and twists again, dwindles and dwindles, until at last nothing but a white spot is seen upon the floor. The lights are turned up. The medium is found in a deep trance within the cabinet. No sign of the spirit is seen. No explanation of the miracle is forthcoming!

Have we seen a genuine "spirit materialization"?

I am not now attacking spiritualism in any way. My sole object here is to expose an old form of fraud in the interest of honest spiritualists as well as of the public. Most materialistic manifestations of the kind here described are faked. The forms are so crudely produced that it is hard to believe that even willing believers are fooled.

The hair, turban, and white material are hidden in various secret places—a hollow seat in the chair, the cabinet curtains, a hollow boot-heel, a dummy watch, in the mandolin inevitably used for the "physical manifestations," etc.

Under cover of the darkness, the medium extracts these various articles, dons the beard, hair, and turban, and places the long white robe in a black bag, which he has also hidden. He also puts on a pair of black gloves.

Lying on the floor well behind the cabinet curtains, he stretches out his hand. The gloved hand holds the bag containing the white stuff. Gradually he works and shakes this loose. It appears to build itself up and grow. Then he works it upward. When it is well up, he darts out from behind the cabinet curtains, throws it about his own shoulders, and calls for "lights."

The reverse process is similarly executed. The medium creeps back into the cabinet, lets the robe gradually sink to the floor, tucks it into the bag, and whips it behind the curtains. He then replaces the various articles in their hiding-places, and calls for lights and an investigation. Nothing is found except the medium.

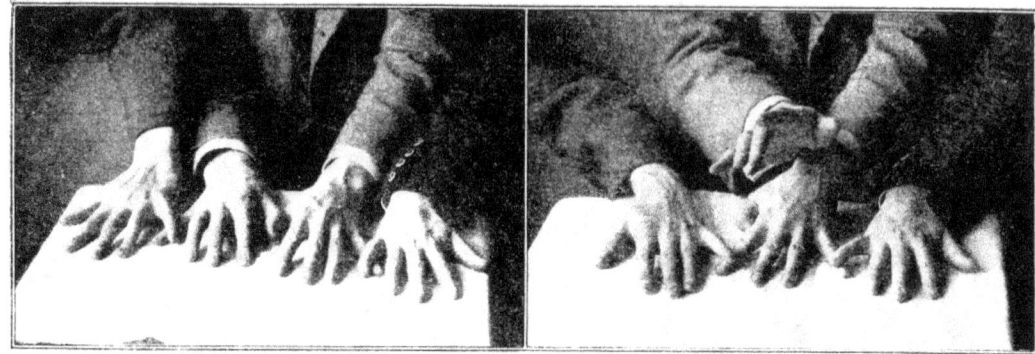

The medium's hands are in the center, his little fingers being held down by the little fingers of the persons on either side

By substituting one hand for two under cover of darkness and excitement, he gets a hand free to carry on his "manifestations"

Taking the Spirits
How mediums produce some

By **Hereward Carrington**, Ph. D.,

YOU sit around a table with ten other believers in spiritualism, the hands of all eleven of you spread out flat and your little fingers touching. Next to you is the medium a big, likable fellow with a kind, frank face.

"If the spirit moves," says he, fixing his eyes on you, "I shall produce some startling phenomena with the very hand that your little finger now touches."

It sounds incredible. "Not if I know it," you say to yourself, pressing your little finger harder against his. You are determined that he shall not remove that hand not for the fraction of a second.

Then the lights are lowered. Under cover of the darkness, the medium is seized with a series of convulsive twitches and spasms. His hands shake, and the table trembles.

However, you are not to be caught napping. You exert still more pressure on your little finger, and his hand is still there or was. For lo! a dinner-bell is suddenly rung in your ears, a whistle is blown, a gong is struck, and a flashlight shines in your face.

You are not only amazed and bewildered you are really scared. And when the lights are turned on, there, just as you anticipated, is his hand with your little-finger resting on it. You could swear away your reputation that he had not moved it. How, then, did it all happen?

A Very Simple Trick

Do you recall the instant when he was seized with convulsive twitches? At that time he was drawing his hands nearer and nearer together until, in one violent spasm, he had withdrawn one hand altogether the very hand under your little finger!

Oh, yes, you regained control of it again instantly, and you forgot all about the loss of it a moment later.

The fact is that the medium substituted the outstretched first finger of his other hand for the little finger of the hand you held so that then you and the man on the other side of him were both controlling the same hand, leaving the remaining hand free to perform the manifestations.

Can you imagine anything which would be more simple than this?

A More Elaborate One

But there are more elaborate demonstrations of the same character. The whole hand of the medium may be apparently controlled; yet he manages to release it all the same! He sits in a chair and places his two hands on his knees, while an investigator sits on either side of him. The man on his right side grasps his right wrist, while the medium himself grasps the wrist of the man on his left side. Thus a circle is formed in which the medium is holding the wrist of one man, and is, in turn, having his other wrist held by the other man. Obviously, the wrist in the grasp of the other man cannot possibly be used to produce the manifestations. But they do take place, nevertheless.

After the lights are turned out, the medium requests the man on his right to remove his hand for a moment.

"I want to use my handkerchief," he says. In a moment he returns his hand.

The medium's hands placed on his knees appear to be securely held, but he gets one hand free without your knowledge. *Above*) You think you have his *right* hand firmly in your grasp, but in reality it is his *left* wrist you are holding

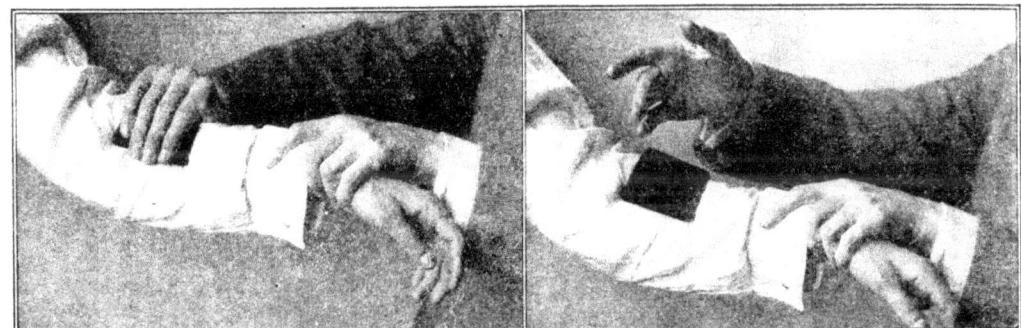

When he grasps your arm, the medium presses around it a piece of lead concealed in the palm of his hand

The medium has removed his hand, but the lead clinging to your arm gives the impression that the grasp continues

Out of Spiritualism

of their many "manifestations"

Author of "True Ghost Stories," etc.

Returns it? What he does, in reality, is this. He slips both feet forward and crosses his knees, the left knee being on top. Then, when he requests the man on his right to hold his hand again, he allows him to hold his left wrist—the one that is holding the wrist of the man on his left. The right hand is now free. The man on the left cannot tell that the hand has been removed, as, indeed, it has not; and the man on the right cannot tell that any change has taken place in the position of the hands, since he grasps a wrist which the medium tells him is his right hand. Since there is only one knee, the trick cannot be discovered if an investigator feels the other knee.

But some people will tell you that they have received "spirit touches" while they knew they held both hands of the medium securely all the time the manifestations were going on. In this case, the medium blindfolded the sitter, if the séance was held in the light, and with his teeth extracted a long feather from under his vest, and with it produced soft "spirit touches" upon the head, face, and hands of the gullible person.

When the medium's hands are held securely he sometimes produces "spirit touches" by means of a long feather drawn from beneath his waistcoat with his teeth

A Famous Feat of the Eddy Brothers

An ingenious "holding" test, made famous by the Eddy Brothers in their so-called "light seances," was carried out as follows:

Three chairs were placed in a row in one corner of the room. The medium sat in one—the right-hand one, as viewed by the spectators. The other two were occupied by investigators. The medium and the third man (in the opposite end chair) then grasped the arms of the man in the middle, one hand on the wrist and one hand high up on the arm. In this way the man in the middle was able to tell instantly if any hand was removed. A curtain was then pinned over the bodies of the medium and of his investigators. Presently manifestations took place. How?

Under cover of the cloth, the medium abstracted from his pocket a small piece of lead about the size of his hand. He bent this tightly around the arm of the man next to him, giving him the impression of being held by that hand.

Thus the hand could be removed and the impression remain—the left hand still grasping the arm lower down. Both hands were thus felt, while, as we have seen, only one was actually employed in holding the man—the other being free to play musical instruments and produce other "phenomena."

Of course, various devices have been resorted to in an endeavor to prevent mediums from producing "phenomena" fraudulently, particularly in dark circles.

Some of the Devices

Among these may be mentioned rope-ties of all kinds, chains, padlocks, handcuffs, etc., and especially various ways of holding the medium so that he cannot escape. These are known as holding tests. The aim of the medium is to evade these and, by releasing one hand or foot, to produce "phenomena" with the free member.

Sometimes "phenomena" are produced a long way from the medium—so far, indeed, that the sitter feels sure he could not have reached that spot, even had his hands been free. In such cases, the medium has produced from his pocket a long jointed rod known as a "reaching rod," and after opening it is enabled to reach objects four or five feet away—to ring bells, shake the tambourine, etc.

The above are some of the simple methods by which fraudulent "physical mediums" bamboozle their sitters. I do not wish it to be understood, however, that I do not believe in any genuine phenomena of this kind. On the contrary, I am quite convinced that they sometimes occur. And it is for this reason that we "psychical researchers" are so anxious to eliminate the fraud. Spurious money does not prove that no genuine money exists; it is the same with spiritistic phenomena. I merely wish to warn would-be investigators of possible disappointments.

Out of the flaming electric arc there came a human voice; there was part of a broken sentence, then the flame snapped out and the voice with it

A Voice Spoke from the Storm

But the startled hearer found a scientific explanation of his uncanny experience

A VOICE out of the night and the storm.

The voice fell on ears that knew there could be no one to utter the call. It was a voice apart from all human relationship. The hearer confesses that his flesh seemed to creep and thrill.

And yet, it was the day of the material present and not the eery time of ghost and gobl'n. The place was East Lansing, Michigan, and not the domain of witches.

It is Professor Herman Vedder, of the engineering department of Michigan Agricultural College, who relates the experience. Those who remember their Jules Verne will remember the thrill with which they read of the message that came over the telegraph line to which it was supposed no one could have access except its makers. Professor Vedder's experience was like that, only the chance of human agency seemed even more remote.

The circumstance was related by the professor to a group that had remained following a meeting of the Lansing Engineers' Club at the college, and was discussing some phases of a lecture that had just been delivered on advanced theories of electrical science. The group had been chatting informally, and the conversation led up to the experience told by Professor Vedder. Someone had just said that the physical seemed to taper off into the superphysical. •

The circumstance in question occurred back in the days when wireless telegraphy was under the close observation of students and investigators, before antennae were strung from the house-tops of experimentally inclined boys. Professor Vedder was a student of the new wonder.

On the night in question Professor Vedder was sitting over his instruments, in a crashing thunderstorm. How wireless would act, with the heavens surcharged to the limit with electricity, was the matter under observation. Presently, however, prudence dictated withdrawal.

"I drew back hurriedly from my apparatus," said the Professor. "A flash of lightning ripped into my station, and across one of my instruments there blazed a flaming electric arc. Out of the flame came a human voice—I heard part of a broken sentence. It lasted for an instant. Then the blaze snapped out and the voice ceased.

"It was some time afterward that the explanation was worked out. You perhaps know that an arc light responds to the resonant effect of the human voice. What is known as the 'singing arc' is well known to physicists. That is, a telephone circuit is introduced into a current supplying an arc light, and out of the arc word vibrations can be made to come.

"Now, on the occasion of which I speak, a sudden flash of lightning had formed an arc across part of my apparatus. This circumstance befell just at the time the telephone wire leading from the home of one of the other professors had been blown by the storm across my wireless aerial. The voice was that of the professor's wife, who was attempting to telephone the grocer. Later she told me what she had said, and the words I had heard fitted in with her sentence. But the first unexplained effect was most uncanny."

ENGINEERING

Changing the Course of a River

A powerful dredge that sucks up the bottom of the sea

UNDERCURRENTS, the rise and fall of the tides, the flow of a river as it follows its course,—all battling together beneath the surface,—constantly disturb a river-bed. Its soil is picked up, swept away, and dropped where the waters see fit. Whereupon man comes along with his dredge, gathers up the soil, and puts it back again. But sometimes changes are desired, as, for instance, when the channel must be deepened.

When the Soil Is Soft

When the soil that is to be dug up is soft, a suction dredge is generally employed to do the work. A pipe is lowered until it touches bottom. Then a pump, which is attached to the upper end of the pipe, is started. The soft soil is sucked up and dumped into a well within the ship. Later it is carted away.

Suction Dredge

The dredge on this page is a suction dredge, but it is constructed to suck up hard, stiff clay—a very unusual combination. The pinwheel-shaped cutter at the bow of the ship makes this possible. The cutter is attached to a shaft that runs up through the suction-pipe. When the pipe is lowered into the water, the whirling cutter first plows up the clay, which the pipe then swallows.

Eight Revolutions a Minute

The cutter-head is composed of a set of hard cast-steel knives, and is turned at the rate of eight revolutions a minute by a 175-horsepower compound engine. It is

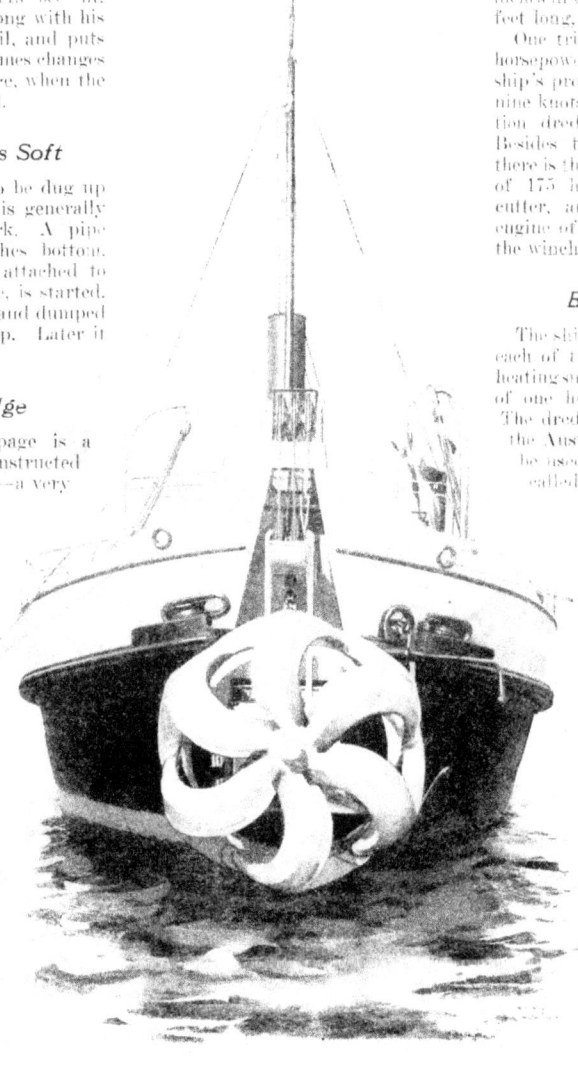

possible for the pipe to project itself to a depth of forty-three feet, and to raise one thousand cubic yards of clay an hour. The clay is not dumped into a well in the ship, but is carried off through a second pipe, twenty-eight inches in diameter and fifteen hundred feet long, to some point on shore.

One triple-expansion engine of 725 horsepower drives the pump and the ship's propellers. She has a speed of nine knots, and is one of the few suction dredges that is self-propelling. Besides the triple-expansion engine, there is the auxiliary compound engine of 175 horsepower for turning the cutter, and an auxiliary compound engine of 80 horsepower for working the winches.

Built in Holland

The ship has two cylindrical boilers, each of thirteen hundred square feet heating surface and a working pressure of one hundred and eighty pounds. The dredge was built in Holland for the Australian government, and will be used on Australian rivers; it is called the *South Australian*. The length of the dredge is one hundred and thirty-seven feet, six inches; the breadth, twenty-nine feet, six inches; depth, eleven feet, six inches.

Dredge Tested

When the dredge was tested she exceeded all expectations. Instead of raising one thousand cubic yards of stiff clay in an hour, she raised eleven hundred and twenty-nine cubic yards. And when she was put to work where there was a sandy bottom she sucked up twenty-six hundred tons an hour.

The pinwheel-shaped cutter at the dredge's bow chops up the hard clay of the ocean bed, which the suction pipe then swallows with ease

Making the Tides Work for Us

When the world's coal supply is finally exhausted we may turn to the tide motors which commerce now scorns

By Joseph Brinker

HARNESSING the heat of the sun and the power of the tides and making them do useful mechanical work for mankind have always been two great dreams of inventors. Perley Hale, of San Diego, California, has invented the method for taming the tides shown in the accompanying illustration.

Trapping the Tide

Two series of reservoirs are located one above the other between high- and low-water marks. Each of these is open on the sea side by means of automatic gates working by gravity, so that on the rising tide water is trapped in the upper reservoir. If this reservoir is of sufficient volume, enough water can be stored in it to operate a water turbine in the bottom. The water passes through the turbine and down into the lower or waste reservoir, where it is retained until the tide has fallen sufficiently to allow it to drain off into the surface of the ocean.

The working of such a tide-water plant may be compared to that of a reciprocating engine having a gigantic piston which moves up and down twice a day (once for each tide). Mr. Hale's plant is very different in nature from the motors which are driven by the intermittent action of the waves and are much more complicated machines.

The rapidly rotating turbine, mounted on a vertical shaft, could be made to drive a generator to provide current which could be used in manufactories built on top of the reservoirs or transmitted inland by wires for other power purposes.

While a tide-water plant like that shown might be located at any point along a coast, more power could be secured if it were situated where the rise of tide is abnormal—as in the Bay of Fundy, where the rise and fall is seventy feet. The head of water in the top reservoir above the turbine would then be greatest and similarly the power developed from that head.

While both the harnessing of the heat of the sun and the power of the tide are feasible, engineers condemn both enterprises as commercially impracticable. It is the tremendous cost of the construction of the reservoirs and the small power output due to the low head of water, except in extreme cases, which make the cost of tide power so great.

Tide Power is Far Off

Some day, when all the available river dam sites are utilized, when the supply of coal is as short as that of gasoline today, and when the heavy fuel oils have been exhausted or reserved for the navies of the world, if there are any at that time, then perhaps the tide motor will succeed.

The water which flows into the upper reservoir during high tide is allowed, when the tide recedes, to flow through a turbine wheel into the lower basin and thence into the sea. The power may be used for any desired purpose

By Tunnel from London to Paris

After a hundred and sixteen years of discussion the English Channel tunnel may be built at last

By Waldemar Kaempffert

"THIS is one of the great things that we could do together."

Napoleon was the speaker. The man to whom he addressed himself was the English statesman, Charles Fox, who visited the First Consul in France after the treaty of Amiens had been signed in 1802. What was the great thing "that we could do together"? Build a tunnel under the Channel to connect England and France — the proposal of Mathieu, one of the foremost French engineers of his time.

For one hundred and sixteen years the fear of war has thwarted the men who had the plan at heart. Napoleon was fighting England again soon after his conversation with Charles Fox, and thought no more of the Channel tunnel. His nephew, the Third Napoleon, tried to revive interest in the project, but the Franco-Prussian war quenched his enthusiasm.

England steadfastly opposed the tunnel. For centuries she had been an island. She had developed political liberty after her fashion partly because she was cut off from the Continent; she was safe from invasion because she was surrounded by stormy waters. Direct physical connection with the Continent was a military menace.

Yet there were broad-minded men in the English government who saw that England had much to gain by the building of a tunnel. In 1875 England and France signed a treaty which defined the tunnel rights of the two countries, provided for the flooding of the tunnel in time of war, and empowered a British and a French company to begin the work of excavation. Shafts were sunk on both the English and French sides seven years later, and tunnels were driven from these shafts out under the sea for a distance of six thousand feet. Then Joseph Chamberlain, Secretary of the Home Department, stepped in and, with the assistance of the courts, stopped the work. Both the French and British companies moved heaven and earth to recommence operations. Wolseley bombarded Alexandria in 1882, thereby hardly improving the feeling of England and France for each other. Wolseley, a popular hero after his Egyptian triumph, branded the whole Channel tunnel enterprise as insane, and voiced the opinion of conservative

One of the early plans for the Channel tunnel provided an entrance at Dover in the form of a winding railway to climb to the top of the cliffs. What might have happened to this exposed entrance in time of war is suggested in the picture

England when he argued that the tunnel would destroy the military isolation that had saved England from invasion for centuries.

England's Dread of Invasion

England invaded and conquered! The idea alarmed even such cool-headed scientists as Thomas Huxley and Herbert Spencer, with the result that they carried in person to the House of Commons an enormous petition, signed by tens of thousands, protesting against the resumption of work on the tunnel. A Parliamentary committee decided against the tunnel companies in 1883. The tunnel was dead. Nearly every great English engineering project for improving the means of communication with the Continent has met with similar absurd opposition.

The Channel Tunnel, Limited, and the Compagnie Continentale du Chemin de Fer Sous-marin, the respective English and French companies, must have been composed of extraordinarily cheerful financial optimists. If the tunnel was dead after 1883, they at least were alive. They stayed alive by paying taxes so as to keep their charters in force, and engaged engineers and geologists to make further studies of the technical problems that would have to be surmounted. Year after year, application was made to the British government for permission to resume work. France had always been in favor of the project. In 1913 Mr. Asquith gave some hope that the Channel tunnel might be considered anew. Then came the invasion of Belgium by Germany in 1914. Tunnel schemes were thrust into the background again.

How the War Changed England's View

Strategists who are now fighting in France realize what a stupid mistake the military advisers of the British government made in objecting to the tunnel. England must henceforth be able to reach Liège or Antwerp as quickly as a rail-borne German army. Besides, England is no longer isolated, in the old sense. The submarines and the airship have destroyed her insularity. To be sure, no invading troops have been landed on English soil; on the other hand, the sea, England's mightiest bulwark, has not been able to prevent attacks on her shipping by submarines or the bombardment of her towns by aircraft.

According to Albert Sartiaux, engineer for the French tunnel company, 20,000,000 passengers have crossed the Channel since the outbreak of the war, and millions of tons of munitions and supplies. A tunnel would have released for Atlantic service 1,500,000 tons of shipping and an army of dock-laborers. He estimates that 30,000 troops and 30,000 tons of supplies a day could have been transported by a Channel tunnel, on the basis of six trains an hour for twenty hours. Think what this would have meant in the early days of the war, when hours were precious! The tunnel can be built for $80,000,000. It has cost

VIEW OF CHANNEL, LOOKING EAST, SHOWING THE COURSE OF THE TUNNEL.

CROSS SECTION AT MIDDLE.

TRANSVERSE SECTION OF CHANNEL, LOOKING S WEST

For a hundred years the fear of war thwarted plans for a tunnel under the English Channel. Strangely enough, the greatest of wars has revived the project, which will probably be undertaken after the conclusion of peace.

England more than that for the lack of a tunnel.

Thomé de Gamond, who devoted the best part of his life to the problem of the Channel tunnel, made about fifteen hundred experimental borings in France and England, and went down three times in a diving-bell in order to bring up specimens of the Channel bed. Although Mathieu first proposed the tunnel, Gamond is rightfully its father.

The latest plans, for which Sir Douglas Fox of England and Albert Sartiaux are responsible, and which will in all probability be carried out after the end of the war, provide for two tunnels, each eighteen and one

half feet in diameter, to be driven under the Channel from Shakespeare Cliff, near Dover, to Sangatte, between Calais and Boulogne.

The distance would be about thirty-seven miles, twenty-four of which would lie under water. At their lowest point the rails are to lie 325 feet below water-level. For a short distance the maximum grade is ninety-six feet a mile; the prevailing grade is twenty-six feet a mile. At about every two or three hundred yards there are to be connecting passages between the tunnels.

The digging of the tunnel would be marvelously easy compared with the

driving of the tubes under the Hudson and East rivers or tunneling under New York to provide an aqueduct for Catskill drinking water. England and France were at one time connected. The evidence of that connection is to be found in the similarity of the geological strata in southern England and northern France.

Machines will burrow into the bed and discharge the material excavated on endless traveling belts that discharge their load directly into cars. There will be no manual labor—no shoveling.

First, a trial tube of about eleven feet diameter is to be run from Dover to Sangatte. It will carefully test the

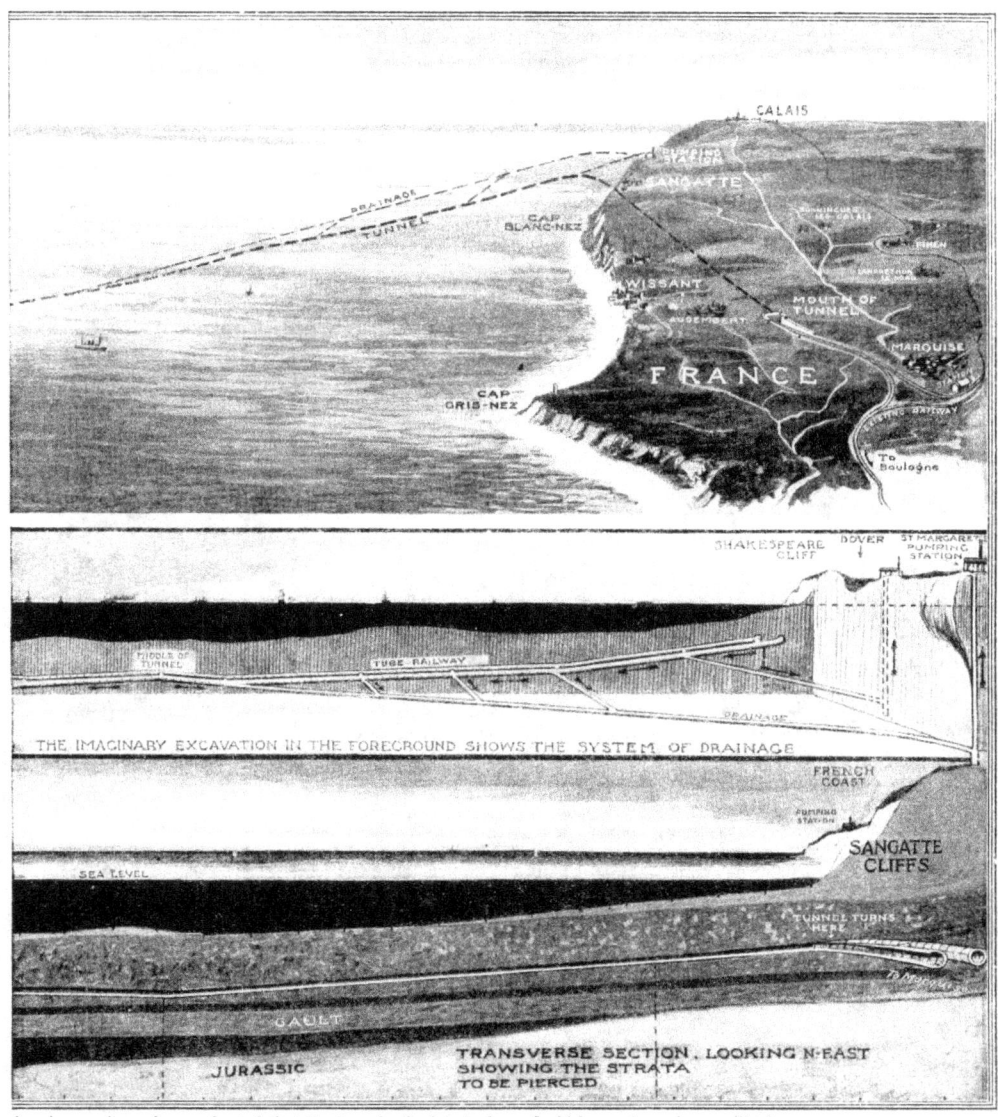

American engineers have estimated that the tunnel—the latest plans of which are shown here—will take less than five years to build. The cost is placed somewhere between $80,000,000 and $90,000,000

ground to ascertain the precise location of any fissures or faults. It will be used as a drainage tube, and will rise up to the center of the Channel, so that water will flow down in each direction and be pumped up at Dover and Sangatte.

It will take four years to construct this tube, but it will reduce the time required for the entire work. By its means chambers will be excavated in the middle of the Channel, and from these chambers it will be possible to drive the tunnels both from the shore ends and backwards from the center, and to carry off the excavated material through the tube.

The French have consented that the power-house shall be stationed at Dover under the complete control of the English. The mere pulling of a switch handle would cut off the electrical power in time of war.

There is to be a dip in the tunnel which is to form a water-lock. An officer at Dover has only to open a sluice-gate in order to flood the tunnel from rails to roof for a mile. It is an ingenious method of blocking communication with the Continent, and it ought to commend itself to investors who might worry about the cost of restoring the tunnel. The water could easily be pumped out.

Thirty-five years ago, when the tunnel was actually in course of construction, it was thought that the work could be completed in six and a half years. American Engineers now say that this time could be reduced to a little more than four years. As for the cost, that would remain at the original figure of $80,000,000 possibly $90,000,000.

From the interest that the British government has been forced to take in the Channel tunnel as a result of the war, it may be inferred that its construction will be the first great engineering feat to be undertaken after the war.

60,000,000 Horsepower Ready to be Harnessed for Work

When these giants are set in action, the real age of electricity will begin and our dreams will be realities

By William H. Easton

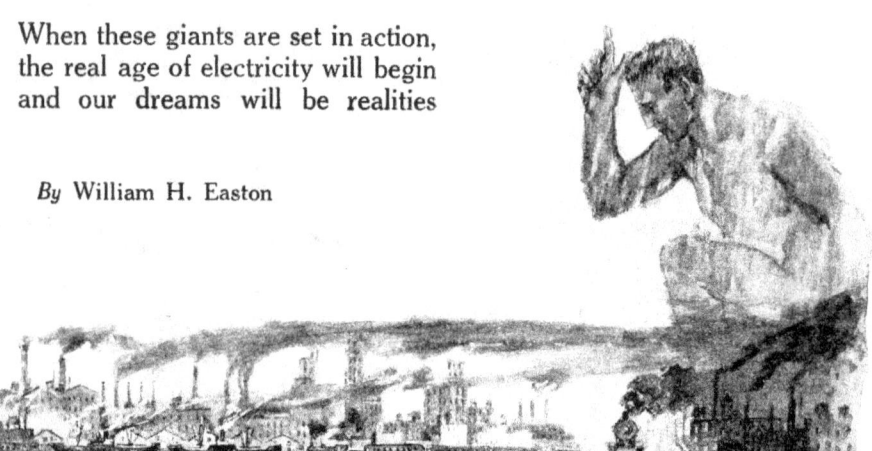

THE Age of Electricity is about to dawn in America. To some people, especially dwellers in large cities, this may seem an odd statement. Everywhere are to be found applications of electricity,— telegraphs, telephones, electric lights, trolley cars, factories operated by electricity,—so that the Age of Electricity appears to be already at high noon.

Electricity for Everybody

But appearances are deceptive. In spite of the vast amount of electric energy generated today, we have, thus far, passed through merely the preliminary stages of the electrification of the United States. Only a small portion of our territory is reached by electric power lines, and those of us who are fortunate enough to obtain electricity at all are not able to use it freely for all of the many purposes to which it is suited. Even a millionaire can hardly afford to heat his home electrically—the ideal way.

But a new era is approaching—a time when not only every town, but every farm, mine, and woodsman's camp will be able to obtain electric power, and to use it with almost the same disregard for its cost as the city dweller uses water. This will indeed be the Age of Electricity. Many have unleashed their imaginations and dreamed about this wonderful time, but

today we are past the necessity of dreaming. The coming of the Age of Electricity is inevitable.

Though this era would have come in any event, the Great War swept away all obstacles to its advance and brought it much nearer. We know now that the nation's welfare depends upon power and transportation, and that our present methods are inadequate.

It is plain to everyone that the fuel famine of last winter would not have occurred if we had had available an ample supply of electric power, transported from mine and waterfall through cables, thus relieving the railroads from the necessity of hauling the vast amount of coal now needed in every part of the country.

But what reason have we to believe that the supply of electric power will be much greater and its cost less than at present?

The cost of electricity, like that of any manufactured product, depends very largely on:

The 3½-ton electric truck leading the horse-drawn wagon typifies the present and past in farm power

(1) The efficiency of the machinery for producing it.
(2) The cost of the raw material.
(3) The volume of production.

The more efficient the apparatus, the cheaper the raw material, and the greater the volume of production, the lower the cost.

The apparatus for producing electricity can now be considered practically perfected. There is room for improvement, of course, and new discoveries may revolutionize present methods, but machines as they stand are satisfactory. We can generate electricity in any desired amount, transmit it any distance, and utilize it in a thousand different ways.

Unused Raw Material

One of the most important developments is the great increase in the size of steam turbines, which have leaped from 30,000 to 100,000 horsepower in capacity within the last four years, while their efficiency has come close to the theoretical maximum.

The raw material for electricity is power. The power now generally used is steam power, but we also have at our disposal another kind of power. From thirty to sixty million horsepower of waterpower exists in the United States, and of this vast reservoir only about six million is in use. The rest has been closed to development largely by Federal laws, which were designed to prevent its monopolization. Congress is now, however, endeavoring to frame laws that will permit the proper utilization of this vast power supply; and this important step toward cheaper electricity will unquestionably be taken in the near future.

The third element affecting the cost of electricity is the volume of production. In the early days of the industry, when the production was small, the cost was very high. But, as the demand for current grew, larger and more economical plants were built. Thus the cost was reduced. Reduction in cost gave rise, in turn, to a greater demand, which again caused greater production, with a further reduction in cost. And so the process has been cumulative, until today electricity costs a fraction of what it did twenty-five years ago. At present we can look forward to the probability of using water-power extensively on one hand, and huge, economical steam units on the other.

The inevitable result is a reduction in cost even with the present demand. But it so happens that two great industries stand ready and waiting to use electric power just as soon as a marked decrease in cost and increase in production become possible. These are the railroads and the electro-chemical industries. Soon a large proportion of our railway mileage will be electrically driven; and soon nitrogen will be electrically reduced from the air, metals will be electrically refined, and chemicals will be electrically made. Only a little improvement in the supply of electric power is needed to electrify these industries.

The railways and the chemical factories will use more power than is now being generated for every purpose. This extraordinary increase in the demand will obviously have its effect on costs, and prices will fall to a point undreamed of now.

Distribution Is Important

Another factor in reducing the cost of electricity is the method by which it is generated and distributed. Let us suppose that there are five neighboring towns, and that the maximum amount of current demanded by each at any one time in the course of a year is as given in the following table:

TOWN	MAXIMUM H. P. DEMAND
A	5,000
B	10,000
C	15,000
D	20,000
E	25,000

This list totals 75,000 horsepower. If each town is supplied by a separate plant (now generally the case), the combined capacity of the plants must

Motor-driven ensilage-cutters will be common farm machinery when the age of electricity really gets here

be at least 75,000 horsepower. But if all five plants were joined together into one system, probably 65,000 horsepower would be enough, because the maximum demands of all five towns would not occur at the same instant. Furthermore, only the largest and most economical of our five plants would be operated continuously, the others being held for "peak" and emergency service. Small and uneconomical plants would be scrapped.

Difference in Waterpowers

To give a specific example of the economy of this arrangement, a careful census shows that the Chicago district

The raw material for electricity is power, and we have at our disposal from 30,000,000 to 60,000,000 horsepower such as this

is operating power plants totaling 1,050,000 horsepower in capacity, but if all power were supplied from a single system, a capacity of only 825,000 horsepower would be required. The saving in fuel effected thereby would amount to 6,000,000 tons of coal a year, while the saving in labor, land, buildings, equipment, and many other items would be equally great.

With waterpower plants the benefits of interconnection are especially marked, because waterpowers vary so much in their characteristics. Some, like Niagara, can be depended upon for a continuous supply of water at all times; others have high and low seasons, and of these some can be readily supplied with storage reservoirs and some can not. When such powers are interconnected, each can be used to the best advantage with minimum waste of water and maximum economy and reliability.

The interconnection of near-by plants, being therefore the most economical and satisfactory policy, is bound to take place rapidly, especially when waterpowers are brought into service. This process has already been begun in the United States, and England has used it successfully for some years.

Part the Railroads Will Play

A reduction in cost will take place, but mark this point: Independent plants supply only the area immediately around them, but several plants brought together into one system are interconnected by a network of wire, and these will be so laid out that they will cover effectively the whole of the inhabited territory lying between the plants. With interconnected systems, whole districts will be supplied.

What applies to single plants also applies to small systems. These will in turn be wired together, and the larger systems thus formed will also be interconnected. The process will continue until there is hardly an inhabited part of the country that is not reached by electric power.

The backbone of the systems will be the railroads, which, being electrified, will carry the main power lines through the densest population, and from these lines branches will run off in every direction.

Compressed Air Faces Its Greatest Battle

Building a two-story tunnel of concrete under the Hudson river

By Robert G. Skerrett

THE mason's master task awaits him half a hundred feet below the tide of the Hudson river. General George W. Goethals is ready to stake his reputation in building a great vehicular tunnel, fashioned of concrete blocks, to span the water gap between New York city and New Jersey. The choice of this material is in itself an amazing novelty for such an undertaking, and the project otherwise has warrant enough for wonderment. A comparison will help to a fuller realization of the unusualness of the scheme sponsored by the General.

London's Vehicular Tunnel

The Rotherhithe single-level vehicular tunnel under the Thames at London has a maximum external diameter of 30 feet, runs under the water for 467 yards, and from portal to portal measures 4,863 feet. In driving the way for that structure, built up of cast-steel segments, the excavating "shield" cut its path through a firm bottom composed of clay intermixed with loam, pebbles, and shells. This earth formed a "blanket" which greatly helped to hold in place the needful air at the face of the shield, so that the workers could stand out under the sheltering apron and do their digging. So firm was the substance of the river's bed that an advance of twelve feet was considered a good day's work.

Even with the advantage of a compact bottom through which to burrow, the shield called for an hourly supply of 1,000,000 cubic feet of compressed air to hold the mud and water at bay. Now see what the double-level Hudson tunnel will involve in its building. The outside diameter will be 42 feet, the shield having a frontal area nearly twice that of a 30-foot apparatus, and its lower edge will lie not less than 92 feet below mean low tide. More than 2,000,000 feet of compressed air per hour will be needed in this undertaking.

For every foot of submergence the water-pressure increases at the rate of .43 of a pound; and between the top and bottom of the Hudson tunnel shield there will be a difference of 18

pounds pressure. This will invite grave risk if the bottom is not stable. The mud in the Hudson is known to be soft, and may therefore be easily displaced by the pent-up air seeking to escape upward from the front of the shield. A "blow-out" of this sort might mean certain death to the men laboring under the apron and possible harm to those within the shield. To

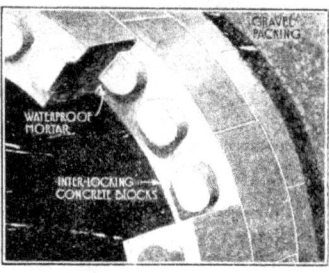

General Goethals' plan is to build the Hudson vehicular tunnel of interlocking blocks of concrete three feet thick

juggle with the contending forces of air and water, it may be found necessary to lay a blanket of sail-cloth surmounted by clay and sand over the river-bed above the line of the tunnel.

Interlocking Blocks of Concrete

As the shield moves forward, the masons will build the tunnel at the same pace, using for their structural units interlocking segmental blocks of concrete three feet thick. As rapidly as these are set in place the joints will be made tight from the outside by means of a special waterproof mixture. The inner face of each seam will be caulked, and the whole interior of the joint, between the blocks and their interlocking recesses, will be filled with mortar forced in under pressure. Finally, the inner face of the tunnel will be coated with white cement of a hydrolithic type.

Completely to fill the space between the cut made by the shield and the

exterior of the great cement sheath, gravel will be passed through the tail of the shield and driven into place around the outside of the tunnel structure, thus binding the tube to the river-bed and unifying the compressive force from without in a manner tending to make the circular sections stronger and more firmly set. The partitions, roadways, walks, ducts, etc., will be fashioned of steel, reinforced concrete, etc. Certain parts of these features will be fastened to the main body by means of sturdy anchor-bolts.

The internal clearance of the tunnel will have a diameter of 36 feet. It will be divided horizontally into an upper and lower level, each having a driveway 22 feet 6 inches from curb to curb. It will carry six lines of traffic, three moving eastward and three westward.

Ventilating the Passageways

By means of electrically driven blowers the two levels will be thoroughly ventilated, so that the carbon monoxide exhausted by motor vehicles can be drawn off promptly, and other blowers will force in a steady stream of fresh air. The under-water section of the project will have a reach from shore to shore of 5,300 feet, and from portal to portal the total span will be 8,050 feet, which is 3,187 feet longer than the Rotherhithe structure. It is estimated that the entire project can be finished well inside of four years, at a total outlay of $12,600,000. The approaches will have a 3½ per cent gradient, and from end to end the vehicular tube will be lighted by electricity. On the upper level there will be walks for pedestrians.

The tunnel will easily accommodate a daily movement of 100,000 vehicles and exceed a 24-hour carrying capacity of 75 big ferry-boats running continuously. Where fog and ice would interfere with the water traffic, the subaqueous highway will be available without interruption. There are the best of reasons for believing that this tunnel, charging less than the ferries, would pay for itself in twenty years.

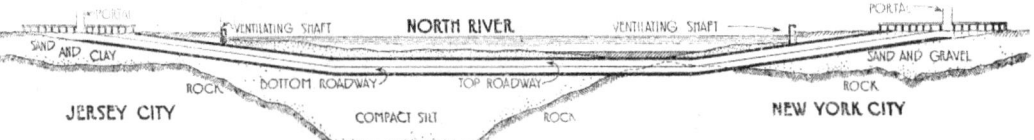

Showing the character of the material through which the tunnel will be built. One of the dangers to be overcome is the soft mud of the river-bottom, which invites "blow-outs"

New York Digs a New Way Out

With a large portion of the food supply of 7,000,000 people having to be carried across the Hudson, the necessity of adding to the means of getting over the river has long been evident. The new vehicular tunnel will be a two-story one, allowing the streams of traffic to flow in both directions uninterrupted. There will be special passageways for foot passengers. The tunnel will be 8,050 feet long, with an under-water section of 5,300 feet, being longer by 3,187 feet than the Rotherhithe single-level vehicular tunnel under the Thames at London. It is estimated that 2,000,000 feet of compressed air per hour will be needed to provide safety for the builders. More than 100,000 vehicles can pass through it in a day. The cost is estimated at $12,600,000

Building a Bridge Under Water

Neither a tunnel nor a bridge, yet both in one, paradoxical as it sounds

By F. W. Fitzpatrick

FOR years the commercial need of a railway crossing over, under, or through the British Channel has been recognized, for quick transit between London and the great cities of Europe is necessary. Always the bugaboo of an invasion of England has risen to throw cold Channel water on these projects. Today a railway link between the two Allies is deemed absolutely essential. Had there been such a crossing it would have undoubtedly facilitated the handling of troops and supplies during the war.

In England there is a Commission already at work studying plans, finances, etc., for the project. So far, a tunnel bored under the bottom of the Channel is the one plan receiving most serious attention, chiefly because there is a fear that any other form of structure may be blown up by a mine or depth bomb.

Years ago twenty-odd schemes were submitted. One much talked of was a trackway laid on the bottom, on which track a sort of platform could be hauled back and forth across the Channel. In turn, a train, run from the ground track upon this moving platform, could be whisked across on the surface of the water, without interfering in any way with the rest of navigation.

This year I have submitted to the Commission a project for effecting the crossing by means of a "subaqueous bridge"—neither a tunnel nor a bridge, and yet both in one, paradoxical as that may seem.

It Is No Longer in the Experimental Stage

There are several examples of river crossings in this country, one in France, one in England, and one in Russia, and plans have been drawn up for many such crossings. The scheme is no longer in the experimental stage.

Where there is very deep water, such as an ocean inlet, to cross, a body 250 or more feet deep, of course the difficulties multiply. Long spindle-legged concrete or stone anchors or "supporting piers" have to be built unless the tube is to rest upon the bottom. In the latter case the approach grades are bored under the bottom, and in the

"My subaqueous bridge crossing consists merely in lowering a previously built tube into place, either right on the bottom of the waterway or raised above the bottom on piers," says Mr. Fitzpatrick the inventor of this system

former case the construction of such piers a hundred feet or so high and all under water is an undertaking of considerable magnitude. But in shallow water this subaqueous bridge is ideal.

One of the First Projects Was for Duluth

One of the first projects of this kind seriously considered was the narrow crossing of the canal at Duluth.

The canal at this point cuts a long strip of sandy shoal that would have been an admirable and much needed freight terminal. But the authorities were frightened by its novelty.

There is a crossing there now, an aerial bridge that took every bit of ingenuity I possessed to get approved by the War Department, and that affords only a crossing for vehicles and pedestrians.

The scheme is ideal for the crossing from San Francisco to Oakland, where in the seven miles of water there is no greater depth than seventy-five feet.

San Francisco's Needs

There has been a good deal of talk of a bridge, then of a tunnel. It is generally recognized that something must be done to relieve the railroad isolation that now grips San Francisco and handicaps its terminal railway business. About seven years ago the newspapers and local engineers were quite enthusiastic about this subaqueous bridge crossing, but politics and lack of funds and the war prevented the plan from being actually realized.

Just now there is a movement afoot to get the authorities to doing something in that crossing matter. It is important, necessary, and would be of tremendous benefit to the city of San Francisco.

At intervals there will be lighthouse-like vent-towers projecting out of the water to the proper place, and sunk into position on the piers. With the ends of the tube

Divested of all technical details and terms, my "subaqueous bridge" crossing consists merely in lowering a previously built tube into place, either right on the bottom of the waterway or raised above the bottom on piers. It is the simplest, the most effective, the most cheaply maintained and least expensive water crossing.

Built Like Concrete Ships

My plans are for a concrete tube, made in sections ashore, as we build concrete ships. Sections about three hundred feet long are the more easily handled. They are square, large enough for a double track, provision being made for drain and ventilating pipes, electric wires, etc. Each section has a temporary bulkhead at the ends and is launched as you would a ship, and towed to the point where it is to be sunk in a trench previously dredged or on piers built in place.

To sink the tube section the bulkheads are knocked out. The water fills the tube, and it sinks, directed by divers, into its proper place. The ends are fastened to the other sections. Another and another section is lowered. When all are in place the water is pumped out, the connections perfected, tracks laid, lights, ventilation, and drainage installed. Your "subaqueous" bridge is complete and ready for operation—the cheapest and best water crossing ever devised, one not subject to winds and storms like a bridge, nor disturbed by currents or tides, nor painfully bored underground. If the traffic

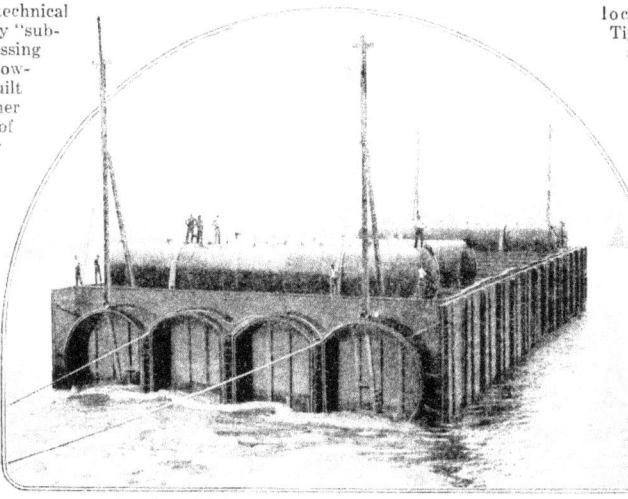

That Mr. Fitzpatrick's scheme is not impracticable is shown by this picture of the process carried out in building the portion of the New York subway that goes underneath the Harlem River. In this case the sections were built of steel instead of concrete; but they were sunk, as Mr. Fitzpatrick advocates, in a carefully prepared trench. This was in the year 1913

becomes too great for two tracks, another double-track tube is laid alongside the first, and another and another later on, as needed. Thus the "subaqueous bridge" can develop without in any way disturbing the first tube or its traffic.

How the Sections Are Joined Together

The joints in these tubes are so devised that when the two sections are butted together the easing into exact

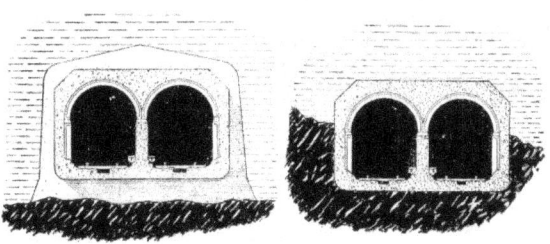

The really troublesome problem would be to keep the tube from floating off from the anchor piers. In these two drawings Mr. Fitzpatrick shows how to prevent such a possibility

aid in ventilation. The concrete tubes are to be built in sections ashore, towed to sections temporarily closed for transit, the tubes could be towed like large scows

location is automatic. Tighten one line of bolts and the juncture is as solid and water-tight as any part of the structure.

At the ends of the tunnel or subaqueous bridge the approaches lead through open cuts or troughs exactly as one would expect for the regular bored tunnel under a waterway bottom.

Some twenty years ago definite plans and estimates were made for four crossings, railway, street-car, and street traffic, near the Cortlandt Street ferry line into New York. The crossing would have required nine anchorages, and was estimated to cost $3,000,000.

Careful plans are being made at present for the Dover-Calais Channel crossing. What will be the cost? It is too early to state definitely, but we have gone far enough to be sure that the cost would not be in excess of forty per cent of what a regular bored tunnel under the Channel would be.

The tube idea was at first opposed by engineers. Now it is regarded as fundamentally correct. It was thus that the New York subway was built under the Harlem River.

At a conference of railroad men before whom I was advocating this tube crossing for a certain river here, the chief engineer of one of our greatest systems was loud in his opposition. Finally, as a clincher, he suggested that the piers necessary to *support* the tube would have to be wonderfully strong, and that the load in the tube would be so great as to produce a sag in the middle of each section. It was only after much bantering on the part of his colleagues, and the illustration of trying to keep a closed glass tube down in a glass of water, that it dawned upon his expert mind that the piers were not for support, but for anchorage. The really troublesome problem would be to keep the tube down in place and prevent it from *floating up* off its anchor piers.

Here is the Chinese idea of a handsome memorial arch. An emperor of the Ming dynasty placed it at Peking centuries ago

Beneath These Arches Fame's Highroad Passes

It was the Romans who set the fashion of spanning public thoroughfares with huge ornamental arches, usually built to commemorate victory. This one, the Arch of Constantine, built in the year 315 A. D., is one of the best preserved of all ancient Roman monuments

Most famous of all modern triumphal arches, the great Arc de Triomphe de l'Etoile in Paris. Napoleon began it in 1806, and it was completed a decade after his death. The total cost was more than twenty million dollars

When the Yanks come marching home. Sketch by Thomas Hastings of an arch of welcome on Fifth Avenue

The Dewey Arch was built over Fifth Avenue, New York, to welcome the Admiral from his triumphs. To be ready in time, work was rushed day and night on the structure and on the plaster-of-paris statues that adorned it. In spite of the hurry, critics agreed that the arch possessed impressive dignity and beauty. It stood 100 feet high and the flimsy materials weighed more than ten tons

Were things done now as in the old days, Germany might be suffering from falling triumphal arches and, as it is, the Brandenburg Gate in Berlin—under which Bismarck marched his captive Danes, Austrians and Frenchmen—has gone out of the triumph business. In the time when the "Me und Gott" partnership flourished, only Wilhelm could pass through the central opening

A Box-Car Filled Every Ten Minutes

And it takes only two men to do the work that was formerly accomplished by a large force

FOR the sake of greater efficiency in handling coal at Baltimore, one of the greatest coal-distributing ports in the United States, the Baltimore & Ohio Railroad Company has adopted a system of loading and unloading, storing and handling coal, invented and patented by Francis Lee Stuart, formerly the chief engineer of the railroad.

The system comprises many novel features in the arrangement of the docking, storing, and distributing facilities, and the use of machinery for loading and unloading, which formerly was accomplished by a large force of shovelers, stevedores, wheelers, and other laborers, in addition to numerous locomotives for spotting and switching the coal-cars in the yards.

An Electric Loading Machine

One of the most important time- and labor-saving machines used in connection with this system is shown in the pictures. It is a loading machine driven by electric power, requiring the services of only two men—an engineer, who is stationed in the cab, and a loader, who directs the operations of the loading arm of the machine. One of these machines, with its crew of two men, can load a box-car with coal in ten minutes and an open or gondola car in less time.

The machine is used for loading the coal from vessels into railroad cars. The coal is taken from the hold of vessels moored to a long pier, and deposited, in hoppers which feed it to the thirty-inch main conveyor belt. This belt runs the entire length of the pier, is located between the rails of the main track, and is driven at uniform speed by electric power. The loading machine is mounted on trucks, and runs on the center track. It is driven by a motor, and can travel the entire length of the pier, so as to be brought in line with any car standing on the adjoining tracks on either side of the center track.

Getting the Coal on the Car

The conveyor belt runs through the loading machine, which is equipped with a twenty-inch intermediate conveyor operated by the motor of the loading machine. This intermediate belt carries the coal up an incline, and drops it into a hopper, which

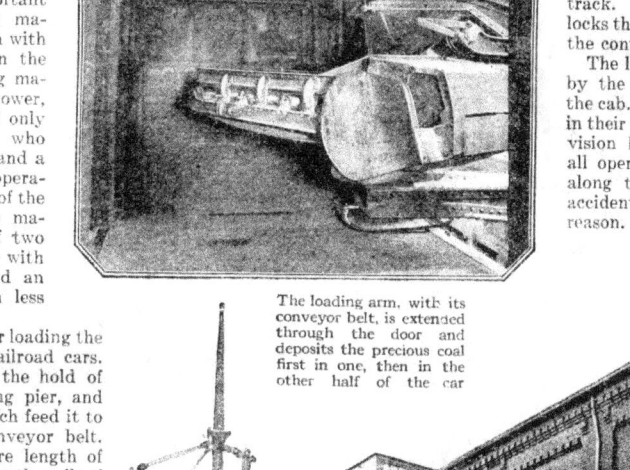

The loading arm, with its conveyor belt, is extended through the door and deposits the precious coal first in one, then in the other half of the car

deposits it on the twenty-inch delivering belt of the loading arm.

The coal is automatically weighed as it passes through the loading machine. Before loading, the tonnage to be delivered to each car is determined. The register is set for one half the quantity, the loading arm is swung into position for loading one half of the car, and, when the predetermined quantity of coal has been delivered, a bell gives the signal to stop the loading. Then the arm is swung to the other side of the car and the loading is completed.

Control of the Mechanism

The control of the entire mechanism for driving the belts and for running the loading machine is located in the cab. The power is supplied by three wires strung along timber bulkheads between the rails of the center track. A fourth wire interlocks the main belt control and the control on the hoppers.

The loader is put in motion by the closing of a switch in the cab. The belts are started in their proper sequence. Provision has been made to stop all operations from any point along the pier in case of an accident or for any other reason.

RECEIVING HOPPER

CAR LOADING AND AUTOMATIC WEIGHING MACHINE

LOADING ARM

CONVEYOR BELT

The man in the glass cage on top of the loading machine has the powers of a wizard. By a few levers he controls the entire mechanism of the conveyor belts, directs the movements of the machine, and operates the loading arm

It would take the lifting power of four thousand men to do the work of this lofty steel giant, which was built at the League Island Navy Yard, and is said to be the world's largest crane

Juggling Heavy Loads in Mid-Air

IMAGINE a giant having the lifting power of four thousand men, whose head is 230 feet above the ground and whose arms could reach outward 190 feet, and a very clear idea of the size and strength of the massive crane that was built at the League Island Navy Yard can be obtained.

Four locomotives, each weighing 100,000 pounds, can be juggled in the air at one time by the huge crane. Its great mechanical arm reaches outward and picks up one or more of the locomotives at a distance of 190 feet, dangles it in the air 141 feet above the ground, and sets it down at a distance of 41 feet from the central supporting arch.

With the power of this mechanical giant the Egyptians could have built the pyramids in short order. The temple of Karnak, with its massive blocks of solid rock, could have been piled up with ease if this giant crane had been at the beck and call of the builders.

What great achievements may not be expected in the future when it is employed lifting weights to the height of a city skyscraper?

With empty hands the steel crane can operate at a speed of ten feet a minute; but when it carries a heavy load, the rate of about two and a half feet a minute is required for its operation.

One of the 100,000-pound locomotives being juggled in the air from the strong cables of the giant crane. What a thrilling ride to take in mid-air!

In the picture at the right. Getting ready for a swing in the air. Compare the size of the tackle with the size of the men

Compare this giant with the man. It is a giant crane that can dangle four locomotives in mid-air, whose brawny arms of steel can reach outward from 41 to 190 feet, and can raise a heavy weight to a height of 141 feet. The giant stands 230 feet high. It is capable of lifting 400,000 pounds to the height of a ten-story building

Using Up the Coal Crumbs

One way of staving off the exhaustion of the coal supply

By Ernest Welleck

MANY years ago a French writer and philosopher was asked what he considered the most striking difference between human beings and animals. "Animals," was his terse reply, "are always wasteful; human beings only in times of plenty."

The American coal industry furnishes a typical illustration of the truth of this epigram. With almost incredible wastefulness surface outcroppings were first depleted; later the coal underground was attacked. Only the largest and richest veins were worked, while the thinner and less easily accessible veins were neglected and buried under masses of blasted rock.

For many years coal was principally marketed in large blocks. The small sizes, usually mixed with low-grade culm, the tailings of coal mines, were piled up in enormous heaps which were often destroyed by spontaneous ignition. In those days of abundance, when it seemed impossible that the supply could ever become exhausted, nobody even thought of utilizing the low-grade tailings of the mines. Millions of tons were dumped into rivers and lakes or used for filling low ground in place of earth or rock.

The intensive development of our industries after the Civil War created an ever-increasing demand for coal. Coal production developed by leaps and bounds and soon reached enormous figures. It became clear that the available coal deposits were by no means unlimited.

And so, at last, the method of mining coal was improved, making it possible to work even minor veins of inferior coal with profit. But it was not until the increasing cost and scarcity of coal made greater efficiency in its use imperative that effective efforts were made to bring about greater economy in the utilization of coal.

Pulverized Coal as Fuel

Among the most recent of such efforts are the numerous inventions of methods for using pulverized coal as fuel in steam-generating plants. The possibility of accomplishing a considerable saving by the use of pulverized coal in the furnaces of stationary plants and of locomotives has been demonstrated most convincingly by tests which also proved that greater heating efficiency could be obtained by this method.

The economy and reliability of pulverized coal is beyond dispute. This makes it possible to "use up the crumbs" by utilizing the mountains of screenings and tailings accumulated at the mines of the rich coal districts, as well as the low-grade coal, heretofore considered worthless as a fuel, which is found in large deposits scattered over wide areas in different parts of the United States. This coal contains a very low percentage of fixed carbon and volatile heat-producing matter and an extremely high percentage of incombustible mineral matter which forms large clinkers when burned by ordinary methods.

The picture on the opposite page shows the complete installation for burning pulverized coal at the Oneida street power plant of the Milwaukee Electric Railway & Light Company. This plant has been in operation for about two years and has been eminently successful during that period. It was installed without making any change in the settings of the boilers other than a rearrangement of the fire-boxes, and gives them a much higher capacity and efficiency than was obtained by the use of stokers. There are several different systems in use, but that illustrated here shows, in a general way, the principles upon which all are based.

The coal, which may be of very low grade, is unloaded from the car directly into a hopper, from which it is fed to the crusher. The crushed coal is carried by a conveyer to the magnetic separator, which removes the stray iron, such as bolts, pieces of tools, horseshoes, etc., contained in the coal. By gravity or other means the coal is next carried to the dryer, where the water contained in the coal, usually representing from five to fifteen per cent of its weight, is removed by heat.

Another conveyer takes the desiccated coal to the storing bins, from which it is conveyed to the pulverizer. By means of a chute the pulverized coal reaches a bin, from which it is fed by a worm-screw feeder, the speed of which can be accurately regulated, to the combustion chamber of the furnace. If the combustion is well regulated and a sufficient draft of air is admitted, no slag or clinkers will be formed. The incombustible mineral parts of the coal will drop in the form of a coarse brown sand into the ash-bin below, from which they are removed from time to time. The ash contains only about two one hundredths of one per cent of combustible matter as compared with thirty to forty per cent in the ash from other furnaces.

A careful test extending over a

With improved machinery like this, even thin veins of inferior coal which formerly received no consideration may be worked profitably, if the coal is used in pulverized form in the boiler plants of industrial establishments

period of twenty-four hours was conducted by the engineering staff of the Milwaukee Electric Railway & Light Company some time ago, and gave highly satisfactory results. Coal known as Illinois and Indiana screenings was used in the test. It contained about ten and five tenths per cent of water, nearly fifty per cent of fixed carbon, about thirty-six per cent of volatile matter, and when dry averaged 12,000 B. T. U. a pound. Combustion was virtually perfect; there was scarcely any smoke, and no carbon monoxide in the escaping gases. The ash residue represented only from thirteen per cent to fourteen and five tenths per cent.

Efficiency Test

The boiler efficiency was eighty-five and twenty-two hundredths per cent, and the net efficiency, after making deductions for the coal used in the dryer and for driving the machinery of the crusher and the pulverizer, was eighty-one per cent, or from one per cent to two per cent higher than had ever been obtained by the use of unpulverized coal and stokers.

With a consumption of 1,990 pounds of coal an hour an average boiler pressure of one hundred and sixty-seven pounds was maintained during the entire test, equivalent to 546.2 horsepower. During the twenty-four hours a total of 47,775 pounds of fuel was used for changing 393,168 pounds of water to steam. According to these figures one pound of fuel was required to evaporate nine and forty-seven one hundredths pounds of water.

Tests like that referred to, and the experience gathered by careful observations in other plants in which pulverized coal is used, prove conclusively that pulverized coal may be used advantageously in stationary heating plants with a saving of about ten per cent and a possibility of even greater saving in larger plants. The greatest efficiency is obtained with coal so finely pulverized that ninety-five per cent of it will pass through a 100-mesh sieve having 10,000 openings to the square inch.

Showing How the "Crumbs"—Pulverized Coal—May Be Used

This "broken-away" view of the steam-generating plant of the Milwaukee Electric Railway & Light Company illustrates the modern method of using pulverized coal as a fuel. It shows how the coal screenings which are stored in the large bin on the second floor are taken by a conveyer to the tubular drying chamber; thence, having been thoroughly dried, to the magnetic separator on the roof. The separator removes from the coal all particles of iron or steel that may have become mixed with the screenings, to prevent them from breaking the steel cut-

ters of the pulverizer. The cleaned screenings drop into a double bin, from which gravity takes them to the pulverizers, which are driven by powerful motors. Between the rollers and cutters of the pulverizers the coal is reduced to a fine, almost inpalpable powder, which drops on a conveyer and is taken to a storage bin between the second and the ground floors. The fine dust arising is carried by the draft to the dust collectors whence it is returned to the bin. By a worm feed the pulverized coal is conducted to the combustion chamber of the furnace, and ignited

Building a Railway Bridge on a Boat

The tide rises and falls, but the train speeds along at an even level

By P. Schwarzbach

AS you ride through Prairie du Chien, Wisconsin, you hear your fellow passengers talking about the pontoon bridge you soon will cross. At the word "pontoon" you automatically mumble something about Xerxes crossing the Hellespont and try to remember what you once learned about that pontoon bridge he used.

A pontoon bridge is a bridge built on boats, rather shaky and dangerous—a degree better than nothing in an emergency.

It is often made simply of planks laid across rowboats which are placed side by side. While in alarm you imagine your train swaying and finally falling off into the river, you find yourself already crossing the bridge. No jerks, no wabbling, nothing unusual about it! How can that be?

This pontoon is so built that tide variation, and the change of buoyancy due to the weight of the train, are absorbed in its mechanism. Instead of being made up of several boats, it is built in one. The pontoon is a double-decker. On the lower floor is the steam engine, with its drum, that opens and closes the bridge. This is done by means of a heavy steel cable which connects the west end of the pontoon with piling in the river.

Between the lower and the upper floors there are nineteen columns,

each of which contains twenty-eight blocks seven inches thick. The tracks are on the upper floor. As the tide rises some of the blocks in each column are gradually removed by hand. The space between floors is thus shortened

This pontoon bridge, when finished, will have two floors: an upper floor, to hold the railroad tracks, and a lower floor, to carry the mechanism that will keep the pontoon in a constant position

just enough to make up for the tide rise. The blocks are put back when the tide goes down. This operation keeps the level of the tracks above constant. The number of blocks taken out during each tide rise depends, of course, on the kind of tide; the greatest number are taken out at flood tide.

When the blocks are taken out, six

motors on the lower deck, which are run by a turbine-driven dynamo, turn drums that stand beside all the columns of blocking. Wire cables attached to these drums extend up and over pulleys which are mounted on beams on the upper floor. As the drums turn, the cables lift up the tracking and keep it up until the blocks have been removed. Then the cables are unwound and the tracking slips back into place.

The tracks on the pontoon are nine inches higher than the tracks on the pile bridges on either side to which they are hinged. This allows for the sinking of the pontoon when a heavy train crosses it. The tracks approaching the pontoon bridge have a downward incline of nine inches to make transition smooth.

The boat itself is 274 feet long, 56 feet wide, and 6 feet thick. The wood in it was treated with many tons of creosote to keep it from rotting. This pontoon is opened only for very large boats. All small ones must go through spans in the pile bridge built at the east end of the pontoon.

From Prairie du Chien this bridge goes over the Mississipi to Iowa. You may think it a clumsy way to build a bridge; but the Chicago, Milwaukee & St. Paul Railway does not agree with you, for it is building many bridges after this pattern.

This man has pulled blocks out of pontoon bridges for thirty-five years; the blocks are taken out to allow for the tide rise

The tracks are raised by drums and cables when it is time for blocks to be taken out; this takes the weight off the tops of the columns

The Coming Triumph of the Motor-Ship

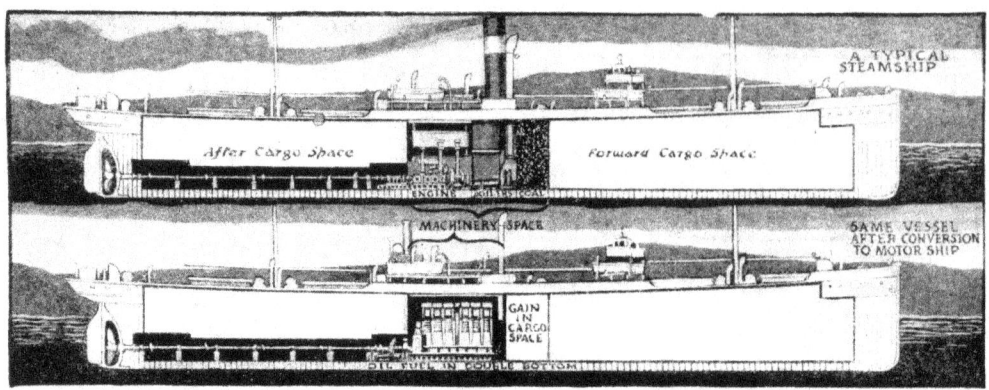

A TYPICAL CONVERSION OF AN OLD STEAMER INTO A MOTOR-SHIP   MANY SUCH CONVERSIONS ARE TAKING PLACE

THE GREAT FUEL BILL BURDEN OF THE STEAMER AS COMPARED WITH THE MOTOR SHIP ON VARIOUS TRADE ROUTES

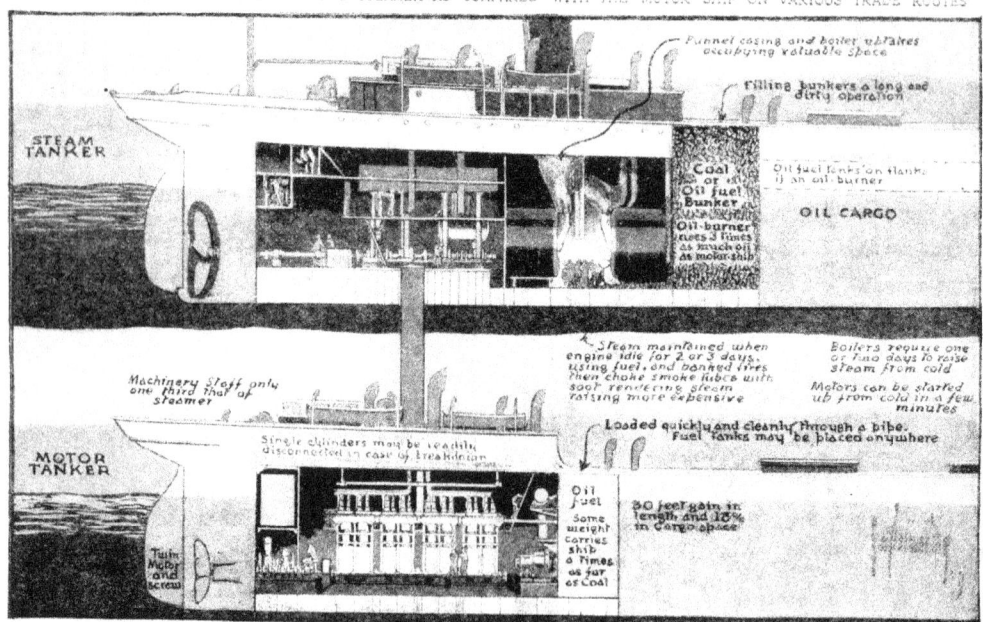

THE big battleships of the world are fired with oil fuel. Now the great passenger steamship companies are following suit. But why use boilers at all? Why not consume the oil directly in an engine, just as gasoline is burned in the cylinder of an automobile? Already the motor-ship is here—but only in small sizes.

For merchantmen the present need is to develop the power of the oil-engine. Today the most powerful units can drive 12,000- or 15,000-ton cargo-ships with twin engines totaling 61,000 horsepower at ten or twelve knots. The oil-turbine appears to be a future possibility. Motor-ships—whose advantages are diagrammed above—are selling at twice the figure given for steamers of like dimensions. Their working, however, costs less than seventy per cent of those of a steamer

A New Lock that Saves Water in the Upper River

The lock shown in this picture is the invention of a German engineer, and seems to solve the problem of overcoming great differences of level in river navigation at a minimum of cost and without draining the river on the higher level of too much of its water. The water-filled tank which holds the load and the entire steel structure supporting it are buoyed by three air-tight cylinders which relieve the burden on the hoisting machinery

Is It Possible for the Railway

By cutting down wind resistance, one hundred

WIND resistance is the big factor in aviation. Not only does it retard airplanes—it also upholds them. It has to be created to make airplanes fly, and it has to be managed with great skill to make them fly well and fast. It is of importance for automobiles at racing speeds. The cars must have low seats, disk wheels without fenders, and a long conical tail in order to reduce the wind resistance. A closed car not at all streamlined, and presenting a front of, say, 48 square feet, would create a wind resistance at 80 miles per hour of $0.003 \times 48 \times 80^2 = 921.6$ pounds, to overcome which, at the 80 m.p.h., would require $\frac{921.6 \times 80}{375} = 196.80$ horsepower, in addition to the power needed for traction. At 40 m.p.h. the power required for overcoming the created wind resistance is only one eighth as large, or about 25 horsepower.

With these familiar facts in mind, many persons are inclined to apply the idea of streamlining to railway trains, and several propositions to this effect have been made. They have fallen on deaf ears. Are the railway authorities blind to their own interests? A brief examination of the conditions will show.

Locomotives Built Since 1905

More locomotives have been built since 1905 than in the entire previous century. In none of them was any effort made to reduce wind resistance, although not a few were intended for speeds of 80 m.p.h. The locomotive is built with a frontage of from 100 to 150 square feet, composed of numerous irregularly projecting minor areas, and there are sharp corners and flat sides on the train it has to pull, all causing wind resistance. An engine at 80 m.p.h. clips a 70-mile head-wind at the rate of 150 m.p.h., and the wind resistance at this rate, if it were realized, would consume from 1,500 to 2,500 horsepower. Since the indicated horsepower of passenger locomotives ranges between the same figures, with few exceptions, nothing would be left for traction, and the actual result is that the train is slowed up at least 10 or 20 m.p.h.

The tractive effort of which locomotives in passenger service are capable has been increased since the year 1900 from 25,000 to 60,000 pounds. Average trainloads have increased from 410 to 700 tons. Maximum locomotive horse-

powers have gone up from 1,400 to 2,800 for passenger trains. The much higher figures for freight and pusher engines need not be mentioned, since these types are not intended for the highest speeds. Taking 2,800 indicated horsepower as the maximum for a locomotive presenting a frontage of 125 square feet, one can figure loosely on the importance of wind resistance in the fastest normal railway operation now.

Big Features Already Streamlined

One square foot of area held squarely to the wind resistance consumes 4.1 horsepower at 80 m.p.h., only 1.73 horsepower at 60 m.p.h., but 27 horsepower at 150 m.p.h. These figures give the keynote in the situation. Even if the frontal area of a railway train were erected squarely, however, at the front of the locomotive, the wind resistance at 80 m.p.h. would be much less than these figures indicate, owing to the elongated shape of the train. In point of fact, the largest cross-section comes at the cab, and the wind is split to a considerable extent by the cow-catcher, the buffer-bar, the high- and low-pressure cylinders, and the smoke-box. When one remembers that perfect streamlining of the whole train would reduce the wind resistance at high

speeds to from one fiftieth down to one hundredth part of what it should be according to the cross-sectional area, and that the train, despite its angular ties, is better streamlined in its big features than any other big thing, excepting only a dirigible, an estimate of the horsepower really consumed cannot be placed higher than two horsepower to a square foot of cross-section. This makes a total of 250 horsepower at 80 m.p.h. But, on the same basis, the wind resistance at 150 m.p.h. consumes a little more than 2,000 horsepower.

The 250 horsepower constitutes less than ten per cent of the power employed, and to save more than half of this waste, reducing it to below five per cent, would require the most radical reconstruction and reshaping of every feature in the locomotive and the train. Railway engineers have much more obvious and plausible means at their disposal for saving five per cent of the fuel bills; and this explains their total indifference to the streamlining of railway trains. They have done a great deal in the past twenty years that virtually reduces the wind resistance factor by reducing its percentage of the total.

Locomotives and trains are much heavier than before, while their cross-section is very little larger. The

How a locomotive will look when railway speeds increase. Contrast its outline with that of the present type as shown in the broken away portion of the illustration

Train to Have Airplane Speed?

and fifty miles an hour is possible

If the railways are to meet airplane competition on a speed basis such modern monsters as this streamlined train will make their appearance. The streamlining is necessary to diminish the wind resistance, which becomes terrific at one hundred and fifty miles an hour

average power is more than doubled. Improved and lengthened fire-flue boilers, compounding, superheating, and increased boiler pressures have been the principal means employed. More than 7,000 of the 65,000 locomotives now in use in this country burn oil, but high oil prices are stopping this development. On the whole, the efficiency is enormously increased. Head-wind storms of from 40 to 50 m.p.h., which would have retarded passenger trains seriously fifteen or twenty years ago, are of no consequence with the new equipment.

The Dirigible Shows the Way

The matter changes completely when running speeds of from 80 to 150 m.p.h. are contemplated. A loss of 2,000 horsepower for wind resistance cannot be accepted. Either 80 m.p.h. must be taken as the natural limit for railway speed, or something radical must be done to cut down wind resistance. It is the mammoth dirigible that shows the way. For, with much larger cross-section than that of a locomotive and with much smaller power, it makes 70 m.p.h. A big seaplane, on the other hand, with its large exposed areas and need of sustentation, consumes 1,600 horsepower for wind resistance alone at a speed only slightly higher. The railway train must do better than the seaplane in this respect. And it can be made to do better, as it is similar in shape to the fuselage alone and can have plenty of power.

There are only two difficulties: To make it stay on the tracks and to make it pay. They are enormous but perhaps not insuperable. When people begin to demand airplane speeds combined with railway carrying capacity, the demand can be met for passenger, mail, parcel,

and perishable freight service operated with special locomotive and car equipment and over tracks that have all changes of grade and all curves located at the stations, where speed is reduced.

Asia, Africa, South America, and Australia will soon be bidding for new fast lines giving an opportunity for new construction of tracks and rolling stock without the immense handicap of the old construction on hand. Then it will be financially possible to reduce a wind resistance calling for 2,000 horsepower at 150 m.p.h. to many times less. The aerodynamical possibilities have already been demonstrated beyond controversy by the dirigible. The United States may jump into the lead with a transcontinental railway of this sort to beat the airplane.

The trains must be units, smooth as an eel from nose to tail. But even eels have gills. Dirigibles have their suspended cars, which break the outlines. The streamlined trains may still have the locomotive drive-wheels exposed just enough to facilitate inspection and oiling. But the fronts must be remodeled. A buffer-bar will not be required. The cow-catcher will reach to the rails and will, in fact, sweep them. It may serve as a snow-plow, and may take care of straying elephants, buffaloes, and automobiles. As little air as possible should be permitted to pass under the train. The crude sand-box will be utterly discarded, being destructive of the supreme perfection of track and wheels. Headlight, smokestack, and domes will be lined up in one covered ridge.

The bell is already a mere traditional encumbrance. An alarm of automobile pattern can take its place. A periscope is better than an outlook to give a view of track and approaches. So the cab

need be large enough only for the comfort of the engineer. The fuel should preferably be oil, and the tender can be made as high as the cab. The long fire-flue boiler could be made slightly conical (we have "conical" boilers now) or it may be tipped a few degrees forwardly.

How Streamlined Trains Will Look

In its outward lines the structure will resemble a big freight locomotive in reduced dimensions more than a passenger locomotive. It will have eight pairs of driving wheels in two sets, to get all the traction possible for its weight in all kinds of weather and to distribute weight and pounding over the precipitous track. It will have superheat but probably only single expansion on account of the speed wanted. The need of springs is due solely to roughness, grades, and curves of the track. When flexible bolts take their place, as used now in the boiler legs, the bodies of tender and cars can be lowered to within six inches of the rails. The wheel trucks can be dispensed with, the axles being mounted in the body with access from within; and the wheels can be removed from the outside for truing up.

The bearings will still be air-cooled. All airbrake apparatus can be enclosed in the false bottom of the cars, and may work on sheaves on the axles. The first car may be for mail and freight, the second for passengers, the third a baggage-car shaped for tailing off the streamline. The train may be made as long as the traffic justifies.

When the railway engineers, with their greater competence in all details, begin to dream and figure and design along these lines of thought, the railway train with airplane speed will not be far away.

Homemade Water Pressure Systems

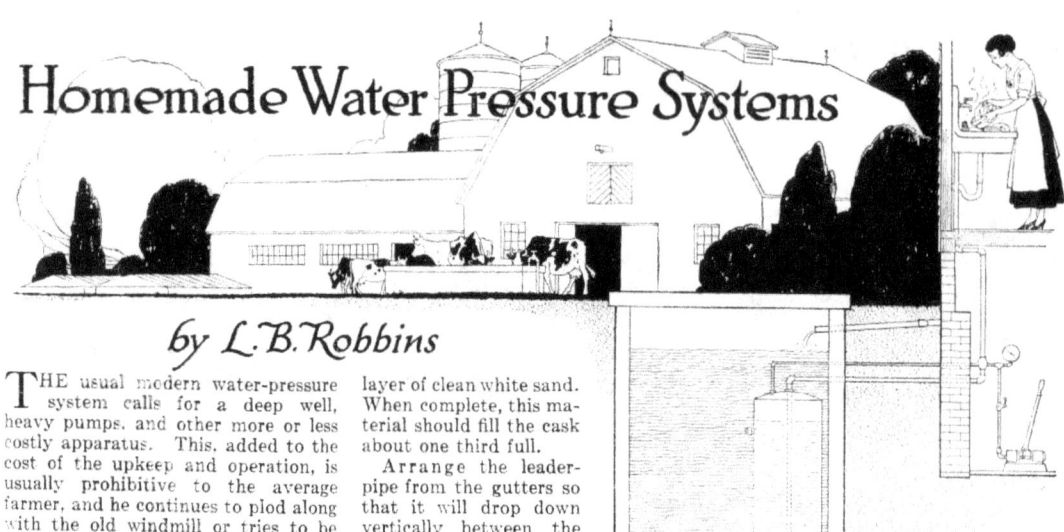

by L. B. Robbins

THE usual modern water-pressure system calls for a deep well, heavy pumps, and other more or less costly apparatus. This, added to the cost of the upkeep and operation, is usually prohibitive to the average farmer, and he continues to plod along with the old windmill or tries to be satisfied with pumping water by hand.

The pressure system here illustrated and described is not the latest thing in water supply, but its cost is low, operation is simple, and, moreover, it is one that can be installed by anyone possessing only a meager knowledge of tools.

A well is not necessary; neither are pumps or a running stream. The only essentials are a simple, old-fashioned cistern, an old but tight water-boiler, and some good piping.

Its Three Parts

The outfit consists of three parts: the filtering system, to insure cleanliness; the pressure-tank; and the apparatus for generating the pressure. The method of filtering and how to procure good, clean water are shown in Fig. 1.

Two solid casks are set upon a platform, about 18 in. apart, as illustrated. Thread a close nipple or a short length of 1-in. pipe, and run it into a hole near the bottom of each cask, and opposite each other. Give each cask a good coat of paint inside and out, using asphaltum. When dry, proceed to fill the bottom of each cask with filtering material. First put in a layer of loose stones about the size of an egg, then a layer of coarse gravel, and last a

layer of clean white sand. When complete, this material should fill the cask about one third full.

Arrange the leader-pipe from the gutters so that it will drop down vertically between the casks and about 1 ft. above them. Pivot the end of the leader to a rocker pipe, as shown (the construction is further shown in the detail Fig. 3). The tee opening should be considerably larger than the inserted end of the leader pipe, so that the rocker can swing through a small arc each way before touching the sides. Pivot with a steel pin secured by washers and other pins. Provide the ends of the rocker pipe with elbows pointing downward to prevent wasting and splashing of the water. Any tinsmith will make up this rocker pipe at a small cost.

Now construct a pair of floats from tin cans, as shown in detail. A vertical iron rod should be attached to each float, and lead through an eye soldered to the side of each end of the water-pipe. A stop on the rod will bear against the eye, pressing that end of the rocker up when this particular cask is filled. The operation will be understood by studying Fig. 1.

Connect the outlets of the casks to a Y-coupling by heavy rubber hose. This is to lead the water from each cask to a main into the cistern.

To Make Connection

The method of installing the pressure-tank and connecting it with the air-pump and house piping is shown in Fig. 2.

The tank itself is composed of a common water-boiler as large as can be conveniently installed in the cistern. If it can be lowered to the bottom of the cistern, so much the better; otherwise stand it vertically, as shown, at one side.

Attach the main water-pipe to the house so that it enters the tank at the

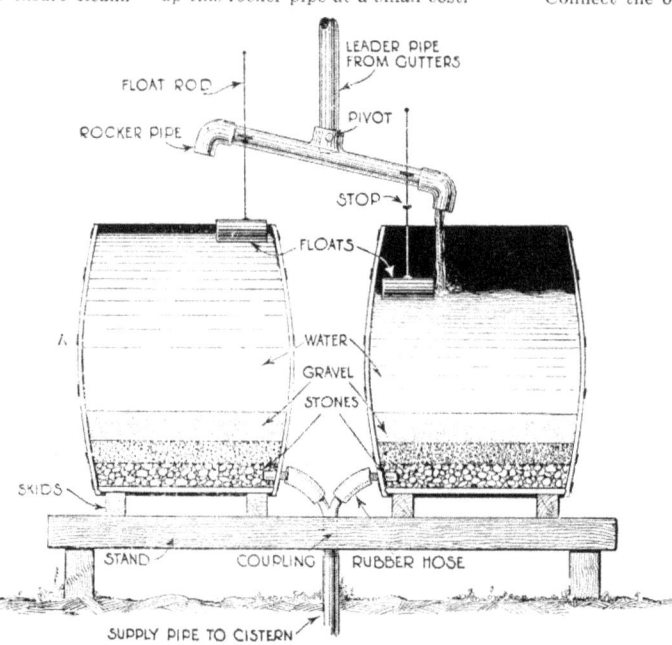

Fig. 1. To filter the water, two barrels are placed on a stand and the house drain-pipe run to the center between them, where a rocking pipe is attached

top, the lower end coming within about ½ in. of the bottom of the tank. Fit this end of the pipe with a strainer. Lead the pipe underground deeply enough to prevent freezing, and place a check valve in the line, to prevent water from running back into the tank. Tap another ¼-in. pipe into the top

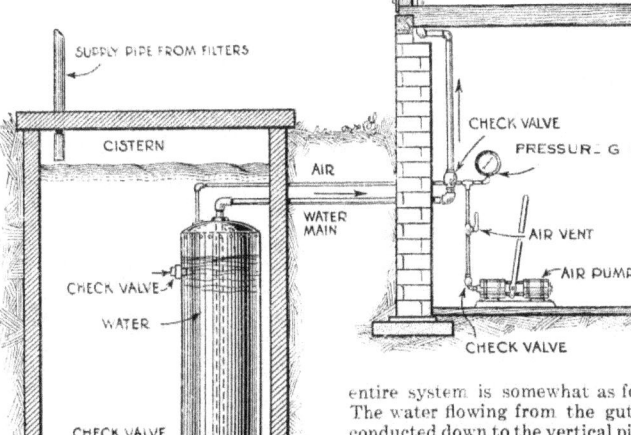

Fig. 2. The pressure tank is set in the cistern or well below the water level

of the tank, and connect it with an air-compressor placed conveniently in the basement of the house. At some point in the air line place an air-pressure gage that will register as high as 50 lb., and between it and the air-compressor fit a relief cock that can be opened to withdraw pressure from the tank. Near the compressor insert an air-tight check-valve. This will retain the pressure when pumping.

How the System Works

The tank should be fitted near the top and bottom with two heavy check-valves with their flaps swinging inward. These will allow the water to enter the tank, but will prevent it from flowing out when pressure is applied. Two valves will facilitate more rapid filling when the cistern is full. Under ordinary circumstances a hand air-compressor may be utilized; but if the farm-house has electric power, an arrangement like that shown in Fig. 3 may be used.

Connect the air-pipe to a small air-compressor operated by a motor of ample power. Extend the pipe out a little way from the gage, and to the end attach a diaphragm pressure-valve such as steam-fitters use. Arrange this so that it will operate at the required pressure. Attach a single knife switch, so that when the required pressure has been reached the diaphragm will force the stem of the valve outward in such a way that it will press against the short end of the switch arm and break the connection at the

other end, thus opening the circuit and stopping the motor. This can be arranged with a double-switch system of control, as diagramed.

The operation of the entire system is somewhat as follows: The water flowing from the gutters is conducted down to the vertical pipe and into the rocker pipe, and flows out of the lowest end to the corresponding cask. When it is full, the float raises that end of the rocker until the water flows into the opposite cask. By the time that cask is filled the first is filtered dry, and the operation is repeated. Thus the cistern is filled with clean water.

The water-pressure in the cistern fills the tank through the check-valves and to within a short distance of the top; this space, being filled with air, prevents the water occupying the entire capacity. Air is now pumped in by hand or power until the maximum pressure is reached to force water

through the pipes in the house. As the water goes down in the tank the pressure lowers accordingly, but can be renewed at will until the tank needs refilling with water. When this point is reached, release the remaining air-pressure by means of the relief-cock in the air line. This will allow the tank to refill with water.

A tank of 50 or 60 gallons' capacity should supply an average family with water for a day. With the hand air-compressor it takes only a few minutes to bring the water under the required pressure, and, with the electric arrangement mentioned, it is necessary simply to throw on the switch and let it pump until the diaphragm-valve automatically shuts off the motor.

A Good Lacquer Finish on Metal Articles

In the production of inexpensive manufactured articles, a lacquer finish is commonly used. People engaged in this work will tell you that there is a different finish obtained on castings and sheet-metal work, although each may be similarly treated. The finish on sheet-metal parts is even all over, but on castings there will be light and dark spots, and many are the conjectures as to the cause.

If the castings are well cleaned and freed from scale and grease, the trouble is due to the structure of the casting. There will be soft and porous spots—spongy—and these, when dipped, will absorb more of the lacquer than the denser portions. Unless these can be washed out the spots will always show up. Only by cutting out the spot, which is not practical, or by washing to remove the excess can an even finish result.

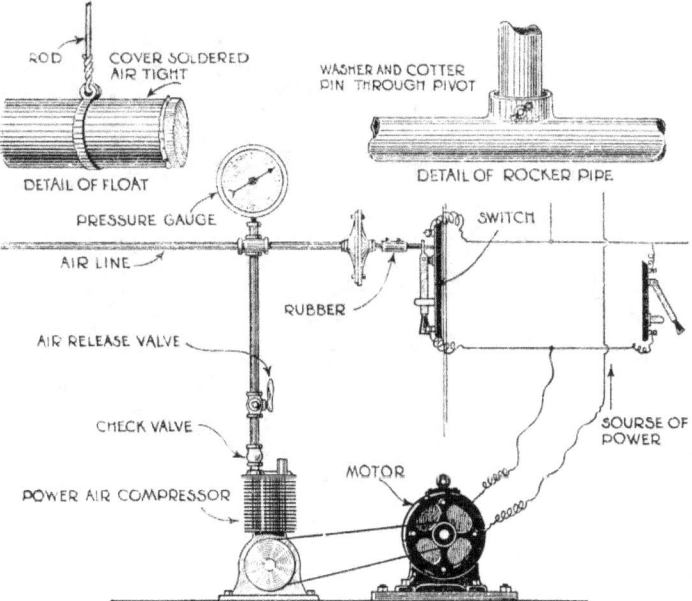

Fig. 3. Where electric power is installed, this simple motor drive may be used for operating the system with air-pressure. It works automatically by the pressure in the tank

Jumping the Dams with Your Ship

A boat-lifting mechanism that would open waterways now closed

A MECHANICAL fish that can jump a dam is the idea of Johann Jargen Richard Haalck, who emigrated from Stelle, Germany, to this country about ten years ago. Many tributary streams would teem with life but for the spots along their course where canals or expensive systems of locks would have to be built.

The new idea would do away with such difficulties. Only a system of dams needs to be constructed, each dam being furnished with the mechanical contrivance for lifting boat, cargo and all, over the obstruction. Standardized forms of river boats would be used.

The boat-lifting mechanism is a structural steel framework composed of two triangular-shaped side members joined at the top by a horizontal transverse pivot-shaft; on this is mounted a bridge-like cradle. The two sides are made rigid by transverse girders, which also provide for mounting the whole upon four four-wheeled trucks running on a track laid on the bottom of the river and extending up over the dam.

The boat is hung from the overhead pivot-shaft by two bridge-like members that are suitably tied together by cross-girders and provided with a longitudinal track on which run two small carriages with pulleys over which the boat-lifting cables extend downward from the drums in the house mounted midway between the bridge members.

In operation the boat is first floated between the side members and under the lifting platform. Then the small carriages on the lifting platform are moved back or forth until the ropes drop perpendicularly, when they are connected with the special cables attached near each end of the boat. This done, the boat with its cargo is lifted by electrical power bodily out of the water. The framework is set in motion, the gear wheels toothed in the track so they cannot slip. Freedom to swing permits the platform to hold the boat in a horizontal position.

The mechanical arrangement for lifting vessels over obstructions in otherwise navigable streams: the boat, cargo and all, is bodily lifted out of the water and transported over the obstruction

Saving $50,000 by Welding

WHEN the huge army transport *Northern Pacific* ran aground in a fog and sustained a fracture in the giant steel stern-post, to avoid delay and an expense exceeding $50,000 in making a new casting, the officials in charge decided to resort to a thermit weld. Now, what is thermit? The greed of aluminum and certain other materials for oxygen is the basis of thermit. When granulated aluminum powder finely ground is mixed with powdered oxide of iron, and when heat is applied in sufficient quantity to the mixture, a reaction is started which rapidly gathers enormous heat and spreads quickly through the whole mass, melting the iron in about thirty seconds. A temperature between 2,600 and 3,000 degrees is raised, second in intensity only to the temperature of an electric arc, and nearly half as hot as the sun itself. A mixture of calcium and silicon and mixtures of other substances also can be used in thermit, the best kind of material being determined by the character of the work to be done. The fracture in the wrecked vessel was 20 feet above the floor of the dry-dock, so a scaffold and platform had to be constructed. The next step in repairing the vessel was to cut a three-inch gap in the steel with an oxyacetylene torch, the gap to admit the entrance of the molten thermit-steel. A wax pattern was built in the gap, 67 pounds of wax being used

The spectacular part of the process came when all was ready and the thermit was ignited. This was preceded by heating the parts to be welded with gasoline torches. The heating required seven hours. Two men, standing on ladders, simultaneously ignited the powder and climbed down to safety. Then the tapping pins of the crucibles were knocked out 45 seconds later, and the molten steel flowed into the fracture, welding it with a fine grade of steel

A mold-box large enough to hold 46 barrels of molding material was constructed, 10 feet in length, 51 inches long on the side parallel with the ship, and 48 inches above the platform. Openings for preheating the fractured parts were arranged, and gates into which the molten steel could flow were provided. The intense heat would come from 1,400 pounds of thermit, divided between two huge crucibles on each side of the mold-box. As a precaution against fire the chains were wrapped in asbestos, and sheets of this material were placed between the crucibles and the hull of the ship, and the platform below was sprinkled with sand. Then the thermit was poured in

Domesticating the Airplane in Crowded Cities

How the problem of providing starting and landing facilities might be solved

By Carl Dienstbach

Just as the commercialization of the automobile depended on good roads, so that of the airplane depends on adequate landing platforms. One suggestion (not to be taken too seriously) is that of a series of ropes, fastened to sandbags, and stretched across a platform. The ropes bring the airplane to a stop

THE airplane has come to stay. It has passed through its probation period in a remarkably short time, and, since the beginning of the Great War, has developed with a rapidity unparalleled in history.

"But," ask the doubters, "will it ever become commercialized, as is the automobile?"

It is quite safe to answer "yes." It is true, the obstacles to the successful commercialization of the airplane are, in some respects, greater than those that had to be overcome in the case of the automobile; but they are by no means insurmountable.

Lack of Landing Facilities

The greatest obstacle to the commercialization of the automobile in the early days of motoring was the lack of good roads. That obstacle was removed in a remarkably short time. The most serious obstacle to the commercialization of the airplane is the lack of adequate starting and landing facilities where they are most needed —that is, in large cities, where wealth and business are centralized.

Now that the aerial mail service between New York, Philadelphia, and Washington has become a permanent and rapidly growing institution, the plan has been proposed to eliminate the delay of transmitting the mail from the post office to the remote starting field or from the landing field to the post office. Why not erect on the roof of the Post Office Building a permanent platform for the starting and landing of the mail-carrying airplanes?

The plan is by no means impractical, in spite of the objections offered by its opponents. A catapult, similar to that which now speeds seaplanes from shipboard, would insure the mailplane a safe start from the roof. A safe landing could be made possible by a device similar to one tried with success some time ago.

On the deck of a cruiser, a series of ropes fastened to sand-bags were stretched across the landing platform. The weighted ropes gently retarded the airplane and it came to a full stop without injury. A railing in front of the propeller would make the device equally applicable to the tractor biplane.

A Revolvable Platform

The device might even be improved and simplified by employing a single "catch-cable" wound around drums supplied with suitable brakes. The landing platform would have to be revolvable, because airplanes must always land against the wind.

The erection of such a platform would tend to promote the commercialization of the airplane. Soon airplane garages would be built, with huge elevators for the machines.

SHIPS AND THE SEA

In the Grip of the Deep

By Waldemar Kaempffert

FULLY eight thousand ships lie at the bottom of the ocean—ships of all countries, which were torpedoed, mined, bombed, scuttled, or sunk by the shell-fire of German submarines. The world needs shipping, despite the wonderful efforts that were put forth in the United States and in Great Britain to keep pace with the sinkings of German submarines.

Let us admit at the outset that most of the ships that were sent to the bottom cannot be recovered by the methods that have thus far been developed. In the open sea it is hardly practicable to recover a ship that lies at a depth greater than one hundred feet. Only in a sheltered bay are greater depths reached.

How are they reached? By actually sending a man down beneath the waves—a man who dons a diver's dress and adapts himself to conditions that nature never intended him to meet.

A Remarkable Invention

In many ways the diver's dress is one of the most remarkable of inventions. Reduced to its essentials, the dress consists of an incompressible metal helmet and a compressible fabric garment for the body and legs.

Air is pumped into the helmet, and thence into the upper part of the dress, through an air-hose connected either with a pump or a tank of compressed air in a surface vessel. It escapes through an automatic valve in the helmet which the diver can regulate.

Blown up with air, the dress containing the diver becomes a veritable bubble. To sink it and keep the man inside in an upright position, it must be weighted. Hence shoes are donned, each of which contains sixteen pounds of lead. Around the body eighty pounds of lead are distributed. Exclusive of the diver, the dress weighs one hundred and seventy - five pounds. In order to prevent the diver from capsizing and coming up feet first, even though he may be thus weighted, the legs of the dress are laced up the back, a procedure recommended by the British Admiralty and adopted by the United States Navy.

Everything depends on the nice regulation of air-pressure. If the pressure is too little the diver feels the water trying to squeeze him; if it is too great his dress becomes too buoyant. So he adjusts the valve in his helmet until the water-pressure without is a little more than equalized by the air-pressure within.

The automatic escape-valve also provides an outlet for the poisonous carbon dioxide exhaled. Bubbles of air constantly stream out of the valve. Since it is automatic, the valve is forced open to let the surface air escape when the pressure within becomes too great. Moreover, it is so constructed that water cannot enter. The diver must adjust the valve to suit his convenience.

How He Goes Down

The diver descends by a "shot-rope"—a rope sunk by a heavy weight or a shot. When he reaches

Study these rear and front views of the Schrader regulation navy diving dress, and you will understand what it means to stay under water at a great depth. The shoes are weighted with sixteen pounds of lead each. About eighty pounds of lead are carried on the body. Exclusive of the diver, the dress weighs 175 pounds. Air is pumped in through the helmet. By means of a valve the outer pressure of the water must be counteracted by the pressure of the air. The legs of the suit are laced up the back, a procedure recommended by the British Admiralty, in order to prevent the diver from capsizing and coming up feet first. Everything depends on the nice regulation of air-pressure. If the pressure is too little the diver feels the water trying to squeeze him; if it is too great his dress becomes too buoyant. The deeper the diver descends, the shorter is the time that he can stay below. The record by the United States Navy is about 300 feet

the bottom he takes with him a "distance-line" attached to this weight. The distance-line helps him to find the shot-rope again; for it is dark far below the surface—darker than the darkest night. He must feel his way. A little electric lamp held in the hand illuminates an object a few inches away. That is all the light that he has.

He reaches the bottom. One pull on the breast-rope tells the attendant above. The diver clears away the distance-line, coiling it in the left hand, and slips the left wrist through a loop at the end, so that he will not drop the line in a moment of forgetfulness. The bottom is often muddy. Therefore he does not flounder about and stir up the silt; he knows that a cloud of mud would prevent him from seeing anything.

For the same reason, the diver keeps to the lee side of his work if there is any current. He spreads himself over the bottom, particularly if it is soft. He does not try to stand. He crouches close to the ground, so as to offer as little surface as possible to the action of the tide.

Safer than It Used to Be

Thanks to the researches conducted by the British Admiralty, diving is a safer profession than it used to be. A man no longer risks his life when he has descended to the great depth of three hundred feet. Time was when a diver was hoisted to the surface after having performed his work below, sometimes at a depth of more than one hundred feet, without any regard to the interval that should be allowed.

When a man is subjected to great air-pressure, his blood and his tissues absorb nitrogen. This nitrogen is not chemically combined with the blood, as in the case of the oxygen that we breathe, but is released again in the form of bubbles when normal pressure is restored. In other words, a man who works in a diver's suit at a depth of one hundred or two hundred feet is like a siphon of seltzer water. So long as he remains at

Why not make a diving armor inherently so strong that it will resist the squeeze of the water at great depths? Here you have the answer in an elephantine contrivance with which some promising experiments have been made. Since the man inside has very little freedom of movement, he must be hoisted overboard by a derrick. The armor has an artificial arm or claw. If a man were to stick out his hand at the great depth to which this suit can descend, it would be crushed into a pulp by the mere force of water-pressure

Getting into the diving armor is a feat that a circus contortionist might be able to perform with celerity. Yet something like the invention pictured below will have to be used if depths greater than 300 feet are to be reached

that depth or is subjected to the equivalent air-pressure, his blood is as quiet as the water in the siphon. But if the pressure be suddenly removed the nitrogen bubbles off from his blood exactly as a gas bubbles up out of the water in the siphon when the spring valve handle is pressed down and a glass of water is drawn. This bubbling up of the nitrogen is extremely painful. What is more, it may cost a man his life. He suffers from what is known as "caisson disease" or the "bends."

Working Under Water

The British Admiralty's investigating committee, consisting of Dr. Haldane, Dr. Boycott, and Lieutenant Damant, discovered that if a diver is brought up gradually from a great depth nitrogen is able to boil from the blood and tissues so slowly that no distress is experienced. A table was compiled which is now consulted whenever it is necessary to send men to the bottom of the ocean. That table tells at once at what rate a man should be hauled up.

The deeper a man goes, the less work is he able to perform. When the United States Navy divers descended to a depth of three hundred feet, for instance, they could do little more than walk about for a few minutes in order to locate the wreck of the F-4. Now, a diver is supposed to work under water. Usually he locates a wreck and patches up the leaks. Since his efficiency is limited by his depth, it is clear that the present form of diver's dress is not all that can be desired.

Why is it not possible to construct a suit of steel—something so strong that it will of itself resist the pressure of water at even six hundred feet? The diver could then breathe air at ordinary pressure. Inventors have contrived new suits —diving armor— that are amazingly ingenious. And yet ship salvors and men who work at the bottom of the ocean will have none of them.

The diver cannot directly use his hands when he is incased in his suit of armor. Indeed, it would be squeezed into a shapeless mass if he could protrude it at a depth of several hundred feet. Hence, ingenious artificial hands have been invented which are worked from the inside of the suit. Sometimes these artificial hands are mere claws, sometimes highly ingenious and complicated mechanisms.

Perhaps some form of diving armor will be invented to meet the salvage engineer's requirements. Certainly the rewards to be earned are rich; for literally billions of dollars' worth of ships and cargos lie at the bottom of the ocean.

"She Starts, She Moves, She Seems to Feel—"
The intricacies of ship-launching explained

By Joseph Brinker

A BROKEN bottle of champagne or water; cheers; the smell of smoking grease; fluttering flags; the excited shouts of invited guests and of the workmen whose toil has completed the vessel as she splashes into the water—these are the outward or visible signs of a successful launching.

To the average spectator at a launching the thing seems very simple. The sponsor gives a signal, it would seem, when she is ready to break the bottle on the vessel's bow, and then something or other happens to start the vessel on her path to the near-by water.

Contrary to the general opinion, the sponsor does not give the signal to the workman when she is ready to break the christening bottle over the ship's bow. Instead, it is the workman who gives the signal to smash the bottle when he himself is prepared to start the vessel on her way to the water.

Rehearsing the Christening

To be sure that no slip-up occurs at these functions, the christening party is usually rehearsed the day before. A dummy bottle is used. But there is a right and wrong way of breaking the bottle. Miss Flora must watch for a signal from the man in charge, and break the bottle on the cord against the bow of the ship as the big

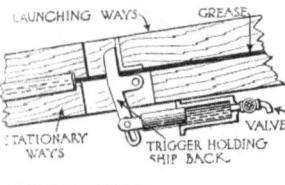

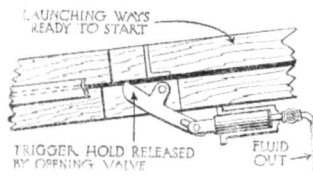

Pull the trigger and launch your ship! This is the latest thing in launching methods. The "trigger," fastened to the stationary ways, holds fast the launching ways until a valve is turned to let out the liquid, thus releasing the trigger and starting the vessel on its first voyage. Each side of the ways has its own trigger

hull begins to move. But she must not wait too long. Unless she swings it in just the right way, the bottle will not break, the hull will get beyond her reach, and the ship will go unchristened.

Seamen believe that a ship is hoodooed at the start if she is not properly

Just taking the water. A typical scene at the launching of a government-built ship. It all looks easy enough to the crowd of spectators looking on

christened. In the launching of a government supply boat, recently, a young woman was saved from this embarrassment by the quick wit of a workman on board the hull. While leaning over the bow, he saw that she had not broken the bottle, and by "snapping the whip" he did the actual christening for her. He received a gift of ten dollars for his quick thinking.

Two Ways of Launching

Two general methods are employed in launching a ship: the one endways, stern first, and the other sideways. Each method has its own particular advantages. Along the Great Lakes more boats are launched sideways than endways.

But, no matter on what plan the

vessel is launched, there are certain fundamental principles that hold good in any case. The first work is the placing of a series of keel-blocks arranged in one line perpendicular to the water in the endways method. The height of the tops of these blocks is so arranged as to form a straight line of support for the vessel throughout its longitudinal length.

Shifting Thousands of Tons

These blocks support the entire weight of the vessel at its center line. The weight at the sides, as shown in the first of the diagrams at the top of page 40, is taken by side cribbing or shores. These keel-blocks and side shores carry the entire weight of the vessel while she is being built. Looking at the keel-blocks from the side of the vessel, those at the front are higher than those at the rear, so that the boat is built on an inclination, in most cases of about three sixteenths of an inch to the foot.

To launch the vessel, its entire weight is transferred from the keel-blocks to the launching set up on either side of the keel-blocks at an inclination of about one quarter inch to the foot. This transference of the vessel's weight may be done by one of two methods: namely, by the wedging up of the launching ways as shown in the figure already referred to, or by the sand-bag method illustrated at the top of page 41.

To understand either of these methods, it is perhaps best first to point out the relations of the launching ways, the stationary ways, and the keel-blocks. The keel-blocks simply support the weight of the vessel while it is being built. The stationary ways run almost the entire length of the ship on both sides of the keel-blocks at points about midway between the keel-block and the sides of the hull, and are constructed of a series of heavy timbers suitably tied together.

These stationary ways are put in place shortly before the vessel is ready to be launched. The tops of the ways are then coated with grease specially prepared to suit the weight of the vessel and the estimated number of days before the launching will take place. The thickness of the layer of grease is also varied to meet

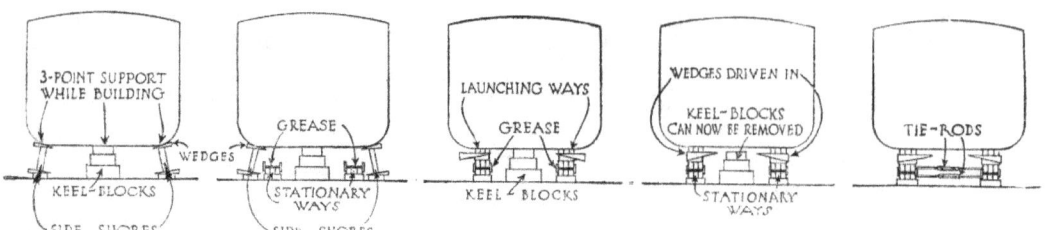

Here are the principal stages in launching a vessel endways. First the weight of the ship is supported by keel-blocks and side shores; then, just before the launching, the stationary ways are put in place and the tops covered with grease; next the launching ways are placed on top of the stationary ways, and wedges and blocks are driven between them and the ship; then the wedges are driven in, lifting the ship from the keel-blocks and transferring its weight to the launching and stationary ways; and—last step except the actual launching—the keel-blocks are knocked out, leaving the vessel ready to slide into the water. Tie-rods keep the launching ways from spreading under the weight

the conditions and the time of the year in which the launching takes place; for it must not melt before the launching date or be forced out when the weight of the vessel has been transferred to the launching ways.

Getting Her Ready to Launch

The launching ways are placed directly above the stationary ways, and are separated from the stationary ways only by the film of grease. The space between the tops of the launching ways and the hull of the ship directly above them is filled in with blocks between which are two wedges, which, when driven up on each other, force the blocks into close contact with the hull. When these wedges on each of the launching ways are driven up far enough, the entire hull is lifted up slightly from an inclination of three six-teenths of an inch to the foot to one quarter of an inch to the foot, and thereby lifted clear of the keel-blocks.

The entire weight of the vessel is at this time supported on the two launching ways, so that the keel-blocks can be removed. The launching ways are prevented from spreading, while the vessel's weight is upon them, by means of steel cables and turn-buckles that join the two ways. The vessel is then ready to be launched.

The Sand-Bag Method

The other method of transferring the weight of the ship from the keel-blocks to the launching ways is by means of sand-bags. The ship is built

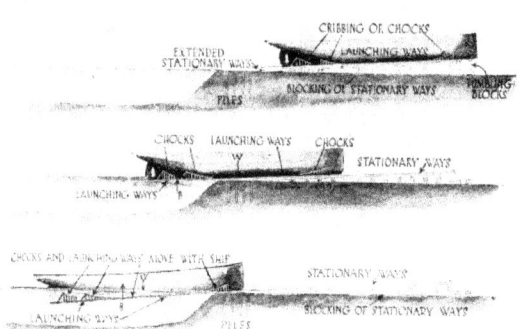

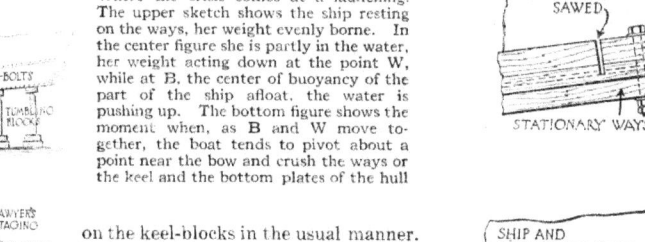

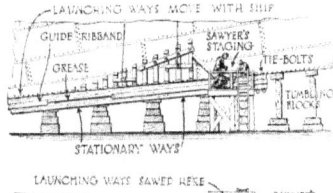

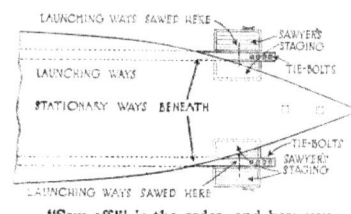

"Saw off!" is the order, and here you can see what is hidden from the ordinary spectator—the details of how the launching ways are cut through to free them from the stationary ways

Where the crisis comes at a launching. The upper sketch shows the ship resting on the ways, her weight evenly borne. In the center figure she is partly in the water, her weight acting down at the point W, while at B, the center of buoyancy of the part of the ship afloat, the water is pushing up. The bottom figure shows the moment when, as B and W move together, the boat tends to pivot about a point near the bow and crush the ways or the keel and the bottom plates of the hull

on the keel-blocks in the usual manner. Before the launching date, the top blocks on alternate keel-blocks are knocked out of place by means of a ram, and bags of sand are placed between the second blocks and the hull of the vessel. Each bag of sand is then wedged up tightly against the hull of the vessel. After bags of sand have been placed on the tops of each of the keel-blocks by this method and the weight is transferred to them, the

stationary and launching ways are set up in the manner just outlined. When the weight is to be transferred to the launching ways, the sand-bags are cut, thus allowing the sand to flow out and lower the hull upon the launching ways. The ship is then ready to be launched. The five steps in this operation are clearly shown on page 41.

The critical point in the operation occurs just before the vessel becomes entirely water-borne, or supported by the water instead of the launching ways.

At this time the boat tends to pivot about a point near the bow, and it is necessary to build up a wood cribbing at this point to distribute the strains over a larger area of the hull, so as to prevent the keel or bottom plates of the hull (generally narrow at that point) from being crushed in. Similar blocking between the launching ways and the rear portions of the hull is employed.

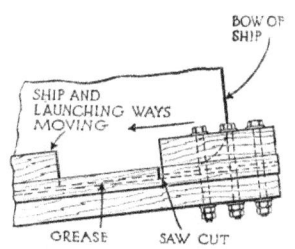

Details of the operation of "sawing through," showing how the launching and stationary ways are bolted together and how the saw cut sets the ship free to slide into the water

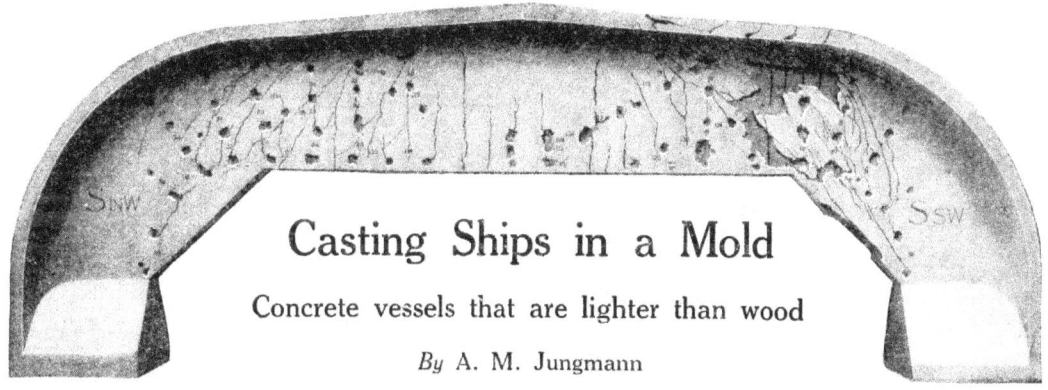

Casting Ships in a Mold

Concrete vessels that are lighter than wood

By A. M. Jungmann

CONCRETE looks like stone, and everybody knows how quickly a stone sinks when it is cast into the water. That probably accounts for the idea that a concrete ship must be very heavy. But according to R. J. Wig, Chief Engineer of the Department of Concrete Ship Construction of the United States Shipping Board, a new concrete material that has recently been developed will enable us to build concrete ships twenty per cent lighter than wood ships.

Why a Ship Floats

As a matter of fact the same inquiry that now arises about concrete ships arose when ships of iron or steel were first built: how can they float, being so much heavier than water? The ability of a ship to float does not depend upon the buoyancy of the material used in its construction, but upon the weight of the ship compared to an equal volume of water. So long as the ship is lighter than the water, it will float, whether it is made of wood, steel, iron, or concrete. A ship floats for the same reason that a china cup or a tin box floats.

While concrete ships are not new, —the first one was built in France by M. Lambot in 1849, —the use of reinforced concrete for the building of a great merchant marine really dates from the success of the Faith, a concrete cargo steamship of about 5,000 tons, which was launched in 1918, and which is now successfully plying the waters of the Pacific coast.

When the Faith took to the water, many persons predicted that she would not last any time at all. But her behavior under all conditions, from calm seas to severe storms, has settled the question as to the worth of concrete ships. She can withstand the heaviest seas, and is as easy to handle as any standard wood or steel ship. In fact, the Department of Concrete Ship Construction of the Shipping Board has made contracts for more than forty ships—cargo-boats of 3,500 and 7,500 tons dead-weight capacity, and tankers of 7,500 tons dead-weight capacity.

Construction Not Like Steel

The construction of concrete ships is something very different from the construction of either a steel or a wood ship. First of all, a wooden mold is made in the shape of the outside of the ship. Then the reinforcement is put in place, and secured so it cannot move. After that the inside wall of the mold is set in place just the right distance from the outside wall to allow for the slab and the ribs. When the molds are completed and the reinforcement is in position, the concrete is poured into the molds, or forms, in such a manner that every bit of the steel reinforcement is thoroughly covered.

As far as possible, the concrete for the bottom, frames, shell, and deck is poured in one continuous operation. When completed, the hull is one solid mass without pieces or patches. It takes about two weeks for the concrete to harden. Then the wooden forms are removed, and a solid ship is the result.

The hull is only one part of the job; for, when that is finished, the machinery and fittings must be installed. This is done in the usual way, except that the fittings are fastened to the concrete by means of bolts that are put in place when the concrete is poured, or through "sleeves" provided for the purpose.

To Protect It from Sea Water

The hull must be covered to prevent the sea water from penetrating to the concrete and the steel framework. A new coating is now being used to give the needed protection against

One of our new concrete ships being built. This shows the arrangement of the mesh-and-steel reinforcement before the concrete is poured in the molds

The largest testing machine in the world, at the Bureau of Standards, Pittsburgh. It is shown testing a reinforced concrete beam 23 feet long and 8 feet 2 inches high, representative of a section of a concrete ship at cross-rib

both the action of sea water and the growth of marine parasites. In the opinion of Mr. Wig, the danger to the hull from these causes has been obviated. He stated recently:

"From our comprehensive tests of concrete structures at sea water, we are convinced that concrete ships will last a minimum of several years without any protection whatever. By the application of protective coatings which are well known to us, we are certain of an extended life of several years additional, and with the further developments of protective means upon which we are now working, I believe the concrete ship can be made as permanent as steel, if not more so."

Concrete ships are not launched end on, as is the usual way. They are launched sidewise, because the stresses caused by the sidewise launching are considered to be much less than those incident to an end-on launching. Another reason for this method of launching is that the ship can be constructed on an even keel with all the vertical lines truly vertical.

The concrete shell is from three to six inches in thickness. In a ship of 3,500 tons dead-weight carrying capacity, the bottom is five inches thick, the side walls and deck four inches thick. The shell is reinforced by a network of bars running both up and down. The bars are so placed that there is a protection of from five eighths of an inch to one inch between the bar and the outside of the concrete.

How the Ships Are Reinforced

The reason for the reinforcement is that a ship is subjected to severe stresses when it is on the ocean. Concrete is very strong in resisting compressive stresses, but very weak in resisting tensile stresses. Therefore the concrete in the ship is reinforced with steel because steel has a high tensile strength. Steel bars are placed in the concrete in such a manner that they will take up the tensile stresses.

Steel bars are used also to help the concrete to withstand the compressive stresses. For example, they are placed in the top and bottom of the concrete ribs and are bound together by steel bands. In this way is built a ship that combines the ability of steel to withstand tensile stresses with the ability of concrete to withstand compressive stresses.

"Hogging" and "Sagging"

When you take into consideration exactly what happens to a ship traveling on the sea, you will perceive the wisdom of this construction. The ship is subjected to severe longitudinal stresses, due to the action of the waves. It is alternately supported by a wave-crest amidships, with the ends of the ship in troughs, known as "hogging," and then supported by wave-crests under the ends of the ship and a trough amidships, called "sagging."

In "hogging," when the support is amidships, the deck of the ship is subjected to tension and the bottom to compression. Under "sagging" the reverse takes place: the deck is under compression and the bottom is under tension. In a steel ship, the plating of the deck, bottom, and sides is relied upon to take up these stresses. Therefore sufficient steel reinforcement is placed in the deck and the bottom of a concrete ship to take up the tensile stresses and to assist in withstanding the compressive stresses.

So far, it has not been possible to make concrete ships as light as steel ships, but every effort is being made to produce a concrete mixture that will be much lighter in weight than either broken stone or gravel. When this has been accomplished, it will be possible to build concrete ships that will be nearer the weight of steel ships.

Not as Light as Steel

Even though concrete ships are not made as light as steel ships, they can compete commercially with steel ships, because it costs less to build them. It is estimated that it costs about one hundred and twenty-five dollars a dead-weight ton to build concrete ships, while steel ships cost two hundred dollars a dead-weight ton.

Another advantage in building concrete ships lies in the fact that, excepting the small amount of steel used in the reinforcement, it is possible to build the hull without calling upon any of the trades or industries that are required in ordinary ship construction.

© International Film

The *Faith*, largest concrete ship afloat. She was launched on March 14, 1918, and has been plying the waters of the Pacific so successfully that all doubt of the seaworthiness of concrete ships has vanished

When at rest the HD-4 is supported by the hull and two outrigger hulls or pontoons connected with the main hull by means of a deck shaped like the wing of an airplane, and therefore made to work its passage. Steering is accomplished by turning the tail set of hydrofoils on a vertical axis

Seventy Miles an Hour on Water

By William Washburn Nutting

A STRANGE new craft, unlike anything thus far developed for air or water navigation, made its appearance recently on the Bras d'Or Lakes, Nova Scotia, where Dr. Graham Bell, inventor of the telephone, has his laboratories. The new glider, which has been startling the natives of Baddeck by tearing about the lake at the rate of seventy miles an hour, was developed from a series of experiments conducted by Mr. F. W. Baldwin and Dr. Bell during the past ten years. She is the fourth full-sized hydrodrome to be built. Hence the name —HD-4.

The new craft is not a seaplane; neither is she a hydroplane in the usual sense of the term. She is the successful application of the principle of dodging the resistance of the water by lifting the hull clear of it on a system of planes not a part of the hull itself. Many have been the attempts to work out a successful application of the idea on paper, as the patent office records will show. Cooper Hewitt and the Italian Forlanini both attained a degree of success with gliding craft using superposed planes several years ago. But the HD-4 is the first really successful embodiment of the idea.

At high speed the hull of the craft is entirely clear of the water, and is supported on small steel planes arranged in groups like the shutters of a Venetian blind. There are three of these sets, one on each side forward and one at the stern, giving a three-point support like that of an iceboat. A fourth set at the bow, called the "preventer" set, is used merely to assist the hull in climbing out when getting under way, and to prevent the bow from diving when traveling in rough water.

The faster the boat goes, the higher she rises from the water, so that she automatically reduces or reefs the submerged hydrofoil surface to just the amount required to carry the load.

This reefing principle is one of the important features of the HD-4. Some day it may be applied to an airplane to reduce the wing area when traveling at full speed and hence increase efficiency.

In spite of the fact that the HD-4 weighs more than ten thousand pounds, she is supported on less than four square feet of submerged hydrofoil surface when at sixty knots, which means that every square foot is lifting more than two thousand pounds. When you consider the fact that an airplane wing supports about ten pounds to the square foot, this seems unbelievable, until you remember that the areas of the supporting surfaces are in inverse proportion to the specific gravities of the mediums in which they act. Salt water is nearly eight hundred times as heavy as air, which explains the fact that the hydrofoil surface need be but $\frac{1}{80}$ of the area of the wings of an airplane to support the same load.

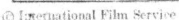

The gentleman who looks like Santa Claus is Alexander Graham Bell, the inventor of the telephone. He has now devised a boat which lifts itself out of the water just as an airplane lifts itself up off the ground

She is not a seaplane, and not a hydroplane. She dodges the resistance of the water by lifting herself clear of it on a system of planes not a part of the hull itself. Her speed is seventy miles an hour. The HD-4 is sixty feet long

Ready to Battle with the Deep

Clad in armor, the diver defies pressure that would crush an unpro tected man to death

DOWN, down, down, the diver goes after sunken treasure. How deep? Can he reach the *Lusitania*, which lies in more than three hundred feet of water? Not with the present diving apparatus. He would be squeezed into a pulp by the relentless grip of the water.

How deep can a diver go? The record is somewhat more than three hundred feet in sheltered waters. Out in the open sea it would be difficult to descend one hundred feet. The billions of tons of shipping and cargo sunk by the Germans lie in depths of more than one hundred feet. They cannot be reached with the aid of the ordinary diver's dress.

After all, that follows from the very nature of the dress. It is but a combination rubber garment that covers the whole body from the neck down, except the hands, which protrude through elastic cuffs, water-tight at the wrists. To this flexible garment is bolted a windowed helmet to which air is pumped from above. Inflated as it is, the dress is as buoyant as a cockle-shell. To sink the diver like a stone and to keep him upright, the soles of his shoes are weighted, each with sixteen pounds of lead. In addition, eighty pounds of lead are distributed around his body. His equipment weighs about one hundred and seventy pounds. Add to that his own weight of one hundred and fifty, and the total comes to three hundred and twenty-five pounds. He is connected with the surface not only by the air-hose, but by a signal- or life-line in which telephone wires are usually embedded.

Air Within Resists Air Without

Don't suppose that air is pumped into the helmet simply for the diver to breathe. Without the air he could not descend as far as he does. It is the air pressure within the suit that resists the water pressure outside. That intense pressure must be nicely regulated, so that he will not be squeezed to death by the relentless water. The deeper he goes the greater is the pressure of the water and the greater must be the opposing pressure of the air. There is a limit to the air pressure that a man can endure. Hence there is a limit to the depth to which he can descend—about three hundred feet under the most favorable conditions.

© Underwood & Underwood

This elephantine diving armor is the invention of Charles H. Jackson. A descent of three hundred and sixty feet has been made in it. It is said to have all the faults as well as all the merits of its predecessors

But this is not the only limitation imposed by the ordinary diver's dress. A man who has gone down as far as three hundred feet may not be hauled up quickly like a fish at the end of a line. The pressure of the air has forced nitrogen into his blood; for nitrogen constitutes about 80 per cent. of the air we breathe. That nitrogen must be released very gradually. The man's blood-vessels are like a corked bottle of soda-water. You know what happens when the cork is suddenly removed; you know that bubbles shoot up.

So it is with the diver. The nitrogen in his blood would bubble off if he were quickly pulled up. The result might be instant death—certainly suffering and a case of the disease that has come to be known as the "bends." The longer he stays below, the more highly charged with nitrogen does his blood become, and the longer must be the time allowed for him to rise. If he has been working for an hour at a depth of two hundred feet it would take four hours to haul him up. Only by such tedious "decompression" is diving made a safe calling.

These being the limitations of the ordinary diver's dress, why not construct a metal suit, a rigid suit of plate mail, so strong that it would in itself resist the crushing force of water beyond three hundred feet? Let the man breathe air in ordinary atmospheric pressure,—the kind that you are breathing now,—no matter what his depth may be. Wouldn't that solve the problem? Wouldn't that enable him to come up to the surface at once

without fear of having his blood boil in the effort to rid himself of its nitrogen? Wouldn't that make it possible to go down five hundred, six hundred, even one thousand feet, and reach hulks that seem now hopelessly inaccessible?

A Negro Mechanic's Invention

Long before Germany began to sink ships by the score, inventors had tried to realize this idea. The latest of these is Charles H. Jackson, a negro mechanic. A descent of three hundred feet has actually been made with his suit. Will it mean a fortune to its inventor, and the recovery of untold wealth that now lies at the bottom of the sea?

Anyone who is familiar with the past history of diving armor will tell you that there is little originality in Jackson's suit. It has all the faults and all the merits of its predecessors. Encased in such a rigid shell, a man cannot climb down a ladder, seize a rope, and lower himself, after the traditional manner of divers. A derrick must actually pick him up and drop him overboard. Jackson's suit, for example, weighs four hundred pounds, and that is light compared with others that have been experimented with. Depths of even six hundred feet have been attained with similar armor. No difficulty in breathing is experienced. But the man within the steel shell has no freedom of movement. He looks like some elephantine creature from another world. He moves about clumsily. His arms are encased in steel, and so are his hands. They must be so protected; for the terrific pressure of the water would crush his fingers into a shapeless mass. Hence, Jackson, like other inventors, employs an artificial hand—a mechanically operated iron claw.

Salvage experts reject all these diving suits—reject them because the joints leak at moderate depths and become tight only when the great water pressure at three hundred feet or more compresses them. Indeed, some inventors actually install automatic pumps in the suit to remove the water that seeps in. The most that can be expected of a man in such a cumbrous apparatus is to locate a wreck and possibly to guide an electric lifting magnet to steel billets or a clam-shell bucket to a pile of loose cargo.

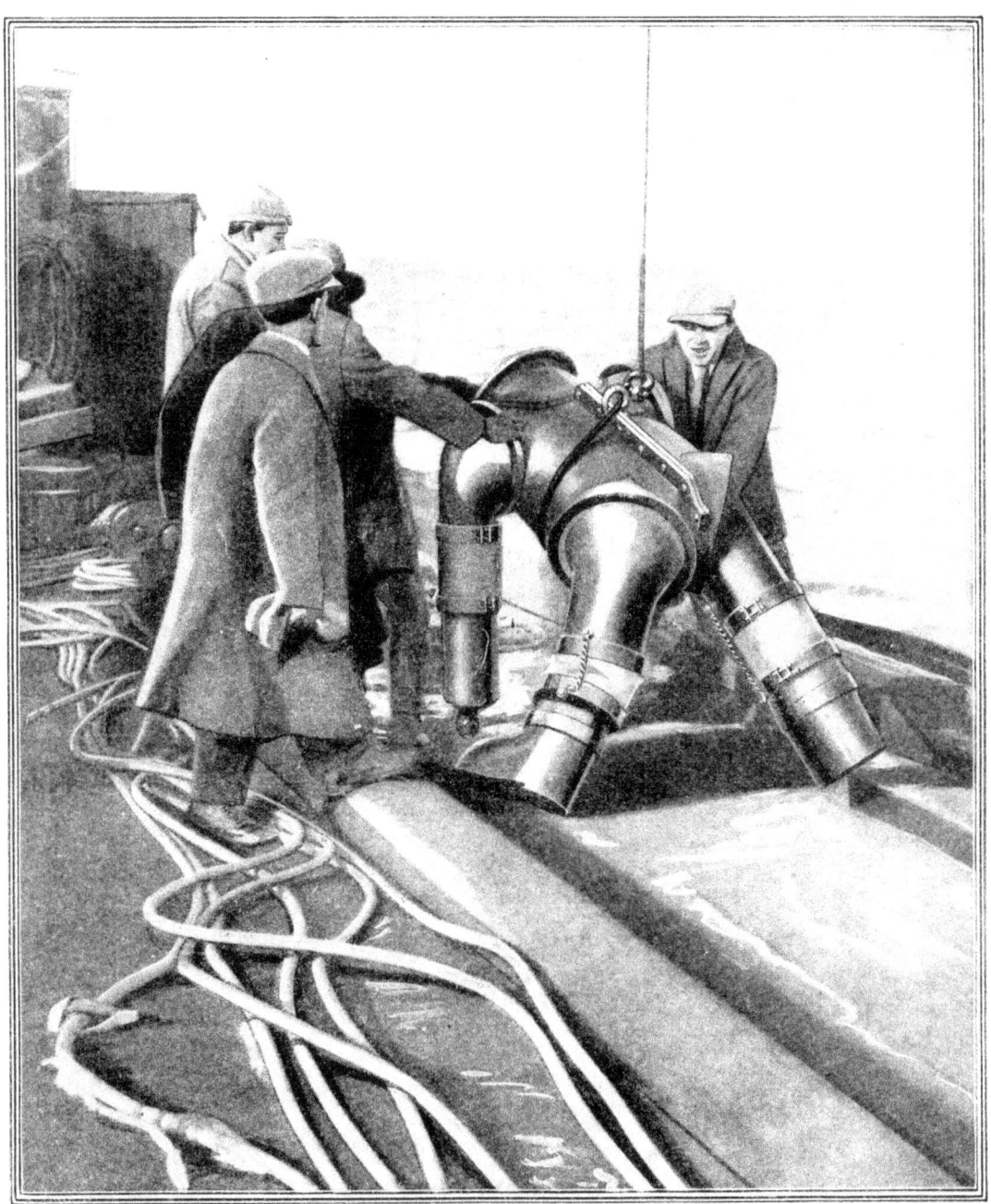

As Helpless as a Fish Out of Water

Untold riches lie in the depths. Can the diver reach them? All depends on the armor. For this modern knight, if he is to secure the treasure, must meet and conquer the dragon of the great depths—crushing, deadly water pressure. That is why he is clad in steel armor so heavy that a derrick is needed to lift him from the deck of the ship and lower him over the side. If he were clad in the ordinary diving dress it would be necessary to fight back the killing pressure of the water by opposing to it the pressure of air forced into the suit by powerful pumps. But there is a limit to the amount of air pressure a man can endure: that limit is approximately the pressure required to fight back the water at a depth of three hundred feet. If men are to explore the greater deeps they must go down clad in armor in itself strong enough to defy the tremendous water pressure, leaving the diver free to breathe air at ordinary surface pressure while he works. Such a suit, the one shown in the reproduction of a photograph above, has been invented by a negro mechanic. A descent of three hundred and sixty feet has actually been made with it.

The diver is ready to adventure in the deeps. The heavy armor will save him from the crushing pressure of the water, but can he do useful salvage work encumbered by this shell of steel?

Lifting Sunken Treasure from the Sea

Ships and cargoes worth billions were sunk by the Germans. Can that treasure ever be recovered?

By Walter Bannard

ACCORDING to an official statement issued by the United States Shipping Board, the Allied and neutral nations lost 21,404,193 dead-weight tons of shipping during the war, which means that Germany maintained an average destruction of about 445,000 dead-weight tons monthly. Most of the ships that were torpedoed undoubtedly lie under water at both low and high tide. Without the aid of a diver, they would remain there until they literally dissolved. And so divers must be sent down—exceptional men, amazing men, who perform under-water feats of physical endurance that seem well-nigh incredible, and who accomplish tasks in carpentry, fitting, and blacksmithing that seem little short of miraculous to the untutored layman.

Man Lives in an Air Ocean

Man is an animal that lives at the bottom of an ocean of air. Like the water of the sea, the ocean of air has weight. At sea-level it presses upon you with a force of fifteen pounds to the square inch. Nature has fashioned you so that you can eat and sleep and walk about when the pressure of the air upon you is somewhat less or somewhat more than fifteen pounds to the square inch.

But there are limits. Some deep-sea fish literally explode when they are brought to the surface: the water-pressure to which they are accustomed has been removed. You will not ex-

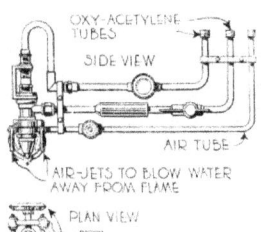

This is the oxy-acetylene torch, which melts steel under water—the invention of Brusch and Beyer. A jet of air blows away the water, so that the flame can reach the steel; it also guides, and steadies the flame of the torch

plode if you are raised by a balloon to a height of five miles, but you are likely to die for lack of air.

All these facts must be borne in mind when one considers what is demanded of a diver. To send him down in order that he may explore a wreck is not so much a matter of preventing him from drowning as it is a matter of air-pressure. He must breathe under conditions that were never considered by nature. A cubic foot of salt water weighs sixty-four pounds.

The deeper a man descends, the more cubic feet of water press upon him. For every foot that he goes down, the pressure increases about half a pound to the square inch. Clearly, a depth is soon reached at which he would literally be squeezed into a pulp, assuming that the water can not enter his body. He would be "imploded"—the opposite of exploded.

The purpose of the diver's dress that will be used by hundreds of men in raising some of the treasure now at the bottom of the ocean is not only to prevent drowning, but to offset the pressure of the ocean by pressure of air.

Why the Diver Is Not Squeezed to Death

A diver's dress consists essentially of two parts—an incompressible helmet and a compressible garment for the body and legs. Down and down the diver goes. The pressure increases. It tries to force the air out of the dress into the helmet. Air must be pumped into the helmet and the upper part of the dress until the pressure within is equal to that without at the

level of the breast. The man breathes; he exhales a poisonous gas—carbon-dioxide. An escape must be provided for the exhaled poisonous gas, as well as to prevent the air-pressure within from increasing beyond a desired degree. So an automatic outlet valve is provided in the helmet.

Blown up with air as he is, a diver is like a bubble: he must be sunk. Hence he carries about eighty pounds of lead on his body and sixteen pounds of lead in each boot. His dress, exclusive of himself, weighs about 175 pounds.

Air is supplied by means of a flexible hose connected with an air-pump in a boat above, or with a compressed-air tank. Besides the air-hose, the diver is usually connected with the

Cutting away wreckage is the chief function to be performed by the oxy-acetylene under-water torch. In the future the torch will be used, wherever it is possible, instead of dynamite

Floating Her with Big Cans of Air

Sometimes pontoons (huge especially constructed steel cylindrical tanks) are sunk beside the vessel and fastened in place by divers. The water in the tanks is then pumped out. Such is the buoyancy of the pontoons that the vessel rises. As a general rule, the diver first tries to clear away wreckage, which is done by means of an under-water oxy-acetylene torch or by dynamite, and then closes all the large holes. How does he stop an opening made by a torpedo? Its general shape and size must be known. With the aid of an adjustable mold of wood, the diver indicates the general size and shape of a patch that will fit the holes. Next the patch is constructed of steel plates or planking, lowered, and drawn into place by cables outside and inside of the hold

A cofferdam is simply a water-tight wall built up from a submerged portion of a ship. When it is pumped out (the leaks in the hull having all been plugged), the ship rises by virtue of her own buoyancy, as well as by that added by the cofferdams

surface by a signal line or life-line in which telephone wires are embedded. He descends by a "shot-rope"—a rope sunk by a heavy weight or a shot.

It is evident that much depends on the air-supply and the air-outlet valve in the helmet. The deeper the man descends, the more air must be supplied, since the pressure must increase with the pressure of the water. If the pumps are supplying one cubic foot of air a minute when the diver is near the surface, they must supply two cubic feet at thirty-three feet, three cubic feet at sixty-six feet, four cubic feet at ninety-nine feet, etc. As the pressure increases, curious disabilities appear. The diver cannot whistle, for example, because he cannot make his lips vibrate. His voice, heard over the telephone, becomes strangely shrill and squeaky.

Moisture condenses on the front glass of the helmet. He cannot see. He sucks in a mouthful of water through a tap (the "spit-cock") and spurts it over the inside of the glass. It is washed down, and all is clear again.

The Record for Diving

How deep can a diver descend? The record is about three hundred feet or thereabouts, made by divers of the American Navy when the F-4 was raised some three years ago in the waters of Honolulu. At that great depth little work can be done; the diver can do little more than grope around.

The air is composed of about eighty per cent nitrogen, a gas that does not readily enter into chemical composition with other elements and which serves the purpose of diluting the oxygen of the atmosphere so that we may breathe it in the proper amount.

The Perils of High Pressure

At high pressure the nitrogen penetrates the blood; it is not chemically converted into something else that is breathed out, as in the case of oxygen. The deeper the diver descends, the more nitrogen is forced into his blood and tissues. If he is brought up from a great depth too rapidly the nitrogen bubbles off; the man suffers agonies; he has all the symptoms of "caisson disease" or the "bends"; he may die.

A table was compiled by the British Admiralty, which is now consulted before sending men to the bottom of the ocean in order to raise a sunken ship. Is the depth 66 feet? The engineer refers to his table. "Twenty-nine and a half pounds per square inch," he reads. That is the pressure of the water. If by any chance the diver descends to 192 feet and remains at that depth over one hour, the process of bringing him up will be tedious. He will be held for five minutes at a depth of eighty feet; for twenty minutes at a depth of seventy feet; for twenty-five minutes at sixty feet; for thirty minutes at a depth of fifty feet; for thirty minutes at

forty feet; for thirty-five minutes at thirty feet; for forty minutes at twenty feet; and for a final forty minutes at ten feet. In all, three hours and forty-five minutes must elapse between the time that the man starts to come up and the time that he is clear of the water, in order that the nitrogen may be slowly and safely released from his system.

Problems for the Salvor

How does the diver ascend? He tightens up his air-valve. His buoyancy is increased. It becomes possible for him more or less to float up the shot-rope. With each foot of ascent the volume of air in the dress expands; the speed is quickened. At first the legs and arms can be moved easily enough, so that the diver can hold the rope and adjust his valve to suit the changing conditions. Soon the dress blows up hard as a board; he is as helpless as if he were shackled; he cannot reach out and adjust the valve. What, then, becomes of the excess air? It escapes through the cuffs.

There is no magic procedure applicable to the salving of every wreck. A salvage engineer must study each case on its own technical merits, and he must invent the peculiar technique that is to be followed in a given instance. Just as a surgeon must consider the age, structure, physical strength, and vitality of a patient, so must the salvage engineer take into account the position of the foundered vessel, the nature of the extent of the damage she has sustained, and the

WATER BLOWN OUT HERE

The air that is pumped in blows the water out through a convenient vent, and sometimes through the very leak through which it entered

character of the salvaging machinery that is at his disposal.

The methods which were employed in salvaging one ship are not likely to be applicable in every respect to the raising of another. The salvor must be able to estimate the weight, trim, and stability of a ship in her damaged condition, and quick to form a plan of operation, taking into account the position of the vessel, the nature and extent of damage which may have been done, the tide and currents of the water in which she lies, and the character of the plant at his disposal.

If a ship cannot be raised, at least her cargo may often be salved. With pneumatic drills or with the oxy-acetylene torch (a blast of air keeps the water away from the flame) a hole can be cut in the plating of a ship as she lies on her side, and through that hole her cargo can be raised. Is the cargo steel? Then the electric lifting

magnet is lowered. It fastens itself with leechlike tenacity to as much as it can carry. Gold, copper, silver,

The pontoons, or barges by which a ship is lifted and conveyed to shallow water are tossed about by the heaving sea, so that the strain on the suspending cables is unevenly distributed. If a stiff structure were used, like this proposed, some of the difficulties might be avoided

must be raised by other means. Orange-peel and similar buckets will raise coal and other bulk cargo.

The Case of the "Lusitania"

The quickest way of entering a hold is to dynamite an opening: but nowadays that method is resorted to but rarely because the tangled mass of wreckage that results imperils the

diver and makes his work more difficult. Perhaps that is the method that may be employed to salve the cargo of the *Lusitania*. The ship herself is probably irrecoverable: for she lies at a depth of 300 feet in the open water.

Ships of under 2,000 tons are usually raised by slipping under them chains or cables, which are fastened securely to barges or hulks at the surface.

As the tide rises the vessels at the surface also rise, lifting the ship below, by means of the chains or cables. The barges are then towed into shallower water, and the chains are tightened. At the next rise of the tide the process is repeated.

Sometimes pontoons (huge specially constructed steel cylindrical tanks) are sunk beside the vessel and fastened in place by divers. The water in the tanks is then pumped out. Such is the buoyancy of the pontoons that the vessel rises. Large ships have thus

Ships of less than 2,000 tons are usually raised by slipping under them chains or cables, which are secured to barges or hulks at the surface. As the tide rises, the vessels at the surface also rise, thus lifting the ship below by means of the cables. The barges are then towed into shallower water and the chains are tightened. At the next rise of the tide the process is repeated. And so, after days and days of patient waiting for the tides and patient tightening of the lifting chains or cables, the foundered ship is at last beached. It is easy then to pump out the ship and to patch her

been raised—for example, the British cruiser *Gladiator*, a ship 320 feet long, displacing 5,500 tons.

As a general rule, the salvor first tries to clear away wreckage, which is done by means of an underwater oxy-acetylene torch, or by dynamite, and then closes all the larger holes.

Weight of the Sea

The sea's weight affects not only the diver but the sunken ship. Pile cubic foot upon cubic foot of water, and it follows that at the deepest part of the ocean the pressure must be measured by tons to the square inch. At sixty feet, for example, the pressure is 26.7 pounds to the square inch; at one hundred and fifty feet 66.7 pounds; at three hundred feet 133.3 pounds. The raising of a sunken ship is complicated not only because its hold is inundated with water, but because of this weight.

How fast can a man in a ordinary dress work under water? A one-inch hole can be drilled in mild steel at the rate of one to one and a quarter inches a minute. An inch hole can be bored into timber with a pneumatic auger at the rate of three to six inches a minute.

How is the opening made by a torpedo closed? Its general shape and size must be known. With the aid of an adjustable mold of wood, the diver indicates the general size and shape of a patch that will close the hole. Next, the patch is constructed of steel plates or planking, lowered, and drawn into place by cables outside and inside of the hold. To secure a patch in place, bolts are employed.

Instead of patching holes and pumping out a ship, the salvor may decide to blow the water out of her. In other words, he opposes air-pressure to water-pressure, exactly as air-pressure is opposed to water-pressure in the diver's dress.

The first step is to brace or strengthen the decks, so that they will not burst as a result of the air-pressure. Port-holes, funnels, all the openings, are tightly closed. The next step consists in providing access to the compartment. What is called an air-lock is applied to the cover of the hatchway. An

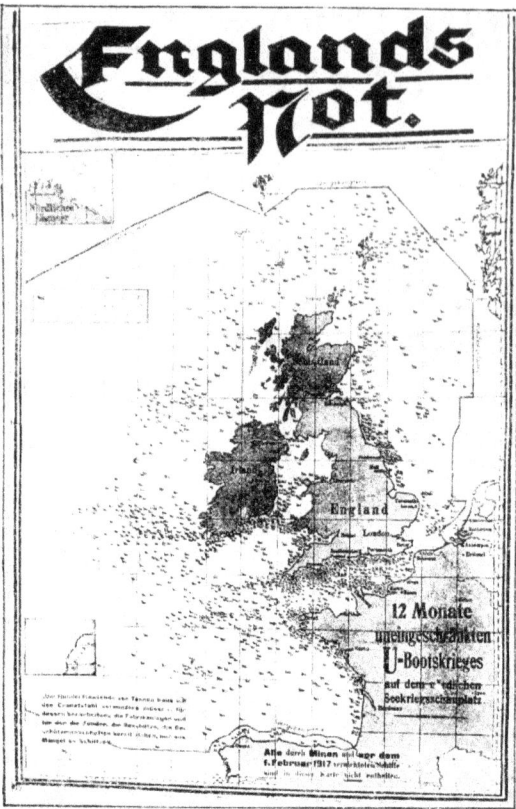

"England's Need"

When German submarines were torpedoing ships at the rate of several hundred thousand tons a month, a German magazine published this boastful map. The little marks indicate the sunken ships. The map serves our purpose to prove how rich is the opportunity of the salvor

air-lock consists simply of a chamber that has a door communicating with the outside and a second door communicating with the hold.

The diver enters and closes the outer door behind him. He then passes through the second door into the hold, and closes that door behind him too; for the hold is full of compressed air which must not be allowed to escape in too great quantities.

These air-pressures must be carefully considered by the salvage engi-

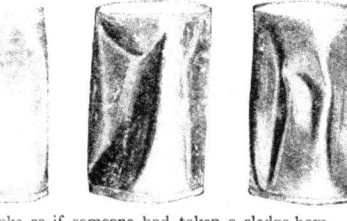

It looks as if someone had taken a sledge-hammer and tried to flatten these cans of vegetables. But it was the enormous pressure of the water and its great depth that "imploded" them. They formed part of the cargo of a torpedoed ship

neer. The decks of a ship are not intended to support thousands of tons of water. Hence it is often necessary to shore a ship so that when she is pumped out they will not collapse. And this shoring must be done in inky blackness.

The openings closed and the air-lock in place, air is pumped in. The pressure of the air must be greater than that of the water. Gradually the water is blown out through a convenient vent, sometimes the very leak through which it entered; but the salvor usually relies upon a scupper-pipe.

Allies' Salvage Plans

That a systematic effort will be made to raise the ships that can be salved may be inferred from the fact that not only have the United States and the Allies studied the problem individually, but that they have pooled their salvage facilities and formed a joint salvage council. Fully one quarter of the world's shipping lies at the bottom of the ocean.

Ships and cargo sunk by mine and torpedo are worth about three billion dollars at present market prices for labor and materials. Up to date the British Salvage Department has salved more than 470 ships. But most of them were recovered before they had actually sunk.

Raising Ships of 2,800 Tons

Before the war it was considered impossible to salve vessels of more than 1,600 tons if they lay in deep water; but since the war vessels of 2,800 tons have been raised.

Most of the ships that went down were sunk in open water. It is considered a feat to raise a ship in a protected harbor, if she lies in a hundred feet of water. But out in the open the depth must be less than a hundred feet, particularly in a region where the winds are high and the sea rough, if salvage operations are to be successfully conducted. Only those ships which have been sunk close to shore in a more or less sheltered position are likely to be raised. The commercial risk in the open is too great.

Ships that Mother Seaplanes

Craft of the "Hush-Hush" fleet may play a part in first trans-Atlantic flight

© International Film

One of the most up-to-date seaplane mothers, *H. M. S. Furious*, of the British navy. Note the flush deck and the tunnel at the stern

THE wind was rising to half a gale, and the pilot of the seaplane felt that it was time to call it a day. "Let's go home to mother," he said to his observer, and set the flying-boat's nose in the direction of the English Channel.

An interval of swift flight, and the observer shouted: "There's mother now!" The seaplane's nose dipped down, and in a long, graceful curve she swung down to the water.

Surrounded by Secrecy

"Mother's" picture appears at the top of this page, and that of another mother to a seaplane brood at the bottom. They were known in navy slang as "Hush-Hush" ships, because of the great secrecy maintained about them while the war was on.

Although credit is given to the Italian and British navies for developing these mother ships, it was our own navy that first succeeded in launching an airplane from shipboard. Long before the war, a slanting runway was built above the forward deck of a warship, and a skilful flyer, starting his plane at the top, soared off.

Another daredevil later succeeded in landing safely on a similar narrow platform; but, even with a broad airplane deck provided, such circus stunts are not en-couraged. To land safely on so small a space needs the extreme of nice judgment plus luck.

When the war saw the seaplane brood increase to large numbers, the British and Italians improvised mother ships from old liners by providing accommodations for hangars and installing special cranes to pick up and lower the planes. Later, craft designed for this particular purpose were built, of which the *Furious*, shown above, is the last word. Her flush upper deck, of unusual width, gives the airplane pilot plenty of chance to taxi off at the start, since the modern seaplane has wheels as well as pontoons. Under the deck are the hangars, reached through the tunnel-like opening at

Although somewhat smaller than the latest "mother" ships, *H. M. S. Argus* also provided comfort for airplanes

the stern. The large traveling crane forward picks up homing planes from the water, and swings them into the tunnel or lifts them out of their nests to the deck, ready for flight.

Mother Ships in Peace

Perhaps the mother ships will play an important part in the first trans-Atlantic airplane flight. Believers in the seaplane as the type of aircraft most likely to succeed in that feat point out that mother ships stationed at intervals along the ocean track would reduce the element of danger to a minimum. Seaplanes suffering from engine trouble, or buffeted beyond endurance by storms while in mid-ocean, could find safe haven and continue their flight; an airplane once committed to the venture would have to go on or go down.

The difficulties of making long non-stop flights is illustrated by the numerous failures attending the attempt to inaugurate an airplane mail service between Chicago and the East.

Not only were the first aviators to make the attempts forced down, but their machines were unable to continue the flight and the service had to be temporarily abandoned.

Trapped in a Sunken Submarine

Tons of water pressed down on the hatch. How could he escape? Recollection of an old physics lesson saved him from death

By Joseph Brinker

THIS is the story of a restless English schoolboy who listened wearily as his teacher talked Boyle's Law and Archimedes' Principle. What was Boyle or the laws that gases and liquids under pressure obey to him?

The time came, ten years later, when they meant life or death to him. It happened during the war, when he was one of the crew of a British submarine.

His boat was lying idly on the surface of the water. All the officers and crew except himself were enjoying a holiday ashore. He went below to work in the engine-room, and left the conning-tower hatch open. Then something happened—the young man never knew just what—and the boat lurched down by the bow.

He thought of the conning-tower hatch, and rushed forward to close it. But the water was already pouring in with such force that he was unable to accomplish his purpose. He returned aft to the engine-room, and locked himself in by means of a water-tight door which separated the boat into two compartments.

In Complete Darkness

He was safe temporarily, but the other half of the boat was filling with water, and the submarine was sinking lower and lower.

The engine-room was in complete darkness, due to the short-circuiting of the electric system. Everything he touched shocked him.

The salt water on the storage batteries generated suffocating chlorine gas. He rushed blindly at the engine-room hatch and tried to force it open. But his strength was no match for the pressure of the water outside; and, even if he had been able to open it, he would have been no better off, for the water would have rushed in and drowned him before he could possibly get out.

He Reviews His Past

Realizing that his last chance was gone, he grew quite calm, and, as is the custom of dying folk, reviewed the things he had done and those he ought not to have done while here on earth.

As he came to that day in school when he unwillingly learned of pres-

sure, he stopped. Water, gas, pressure —what about them?

He Remembered That Lesson

Then he remembered that, in the case of a body submerged in water, the combined force of the gas within and the strength of the body material must equal the force of the water without, or the water would crush the body in. Just so the combined force of the air in the submarine and the steel of the hatch equaled the force of the water pressing down on it. If he could make the force of the air in

the boat equal to the pressure of the water on the boat, then the hatch would be relieved of the strain of holding off the water, and would be easily lifted.

How could he increase the force of the air? Then he thought of Boyle's Law, which he had been forced to memorize on that long-ago school-day —the volume of a gas varies inversely with its pressure. If he decreased the volume of the gas in the submarine, he might increase the pressure enough to make it equal that of the water. This done, the hatch would be in equi-

librium and consequently easily moved.

His next problem was how to decrease the volume of the air. There was only one way to do this — by letting the water in. If he did that he would be taking a last desperate chance; for he did not know how much compressing the air needed before it would exert a force equal to that of the water outside. If this failed him, he would drown without a doubt.

Taking a Desperate Chance

He decided to take the chance. So, groping in the dark, he opened the bilge-valve and let the water in. As it crept up about his waist he philosophically decided that he would much rather drown than suffocate, anyway. But when it reached his waist he grew somewhat nervous and thought he'd better try the hatch.

He pushed against it, and it opened slightly; but it slammed shut again, crushing his fingers, for the pressure of the air was not yet great enough to resist the water without the aid of the hatch.

Nursing his injured fingers, he waited stoically until the water in the engine-room reached the hatch coaming. Then, with just his head above water, his ear-drums strained from the compressed air, he made a final attempt to raise the unwilling hatch.

His patience, bravery, and cool-headedness were rewarded. At last the air was compressed enough to equal the force of the water, and the hatch moved easily.

Physics Sees Him Through

After helping him thus far, physics decided to see him through. The compressed air in the submarine repelled the water which was trying hard to pour in and drown him. The air shot upward with great force, taking the sailor along with it to the surface of the water leaving him there, breathless.

After he had righted himself and shaken the water from his eyes, he looked about and saw, a little distance away, a friendly destroyer, which soon picked him up.

Such an experience must surely force on any man a great respect for school, teachers — and fate.

The air shot upward with great force, drawing the sailor through the sunken submarine's hatch and carrying him to the surface

He Prayed that Air-Pressure Would Help Him to Open the Hatch

Trapped alone in the engine-room of a submarine when it sank, an English sailor was saved by his quick wit and the remembrance of a long-forgotten physics lesson.

It came to him that by letting the water into the submarine the air would be compressed until the pressure inside would equal that of the water outside and he could raise the hatch. He gambled on the chance that the air-pressure would enable him to lift the hatch before the rising water drowned him.

It was a close shave, but he won, and as he raised the hatch-cover the rush of air behind him carried him through and to the surface—alive.

A Visit to Davy Jones's Locker

Going down in a submarine—Walking on the bed of the ocean—How the war's lost ships are to be recovered

By Lloyd E. Darling

THIS article takes you down in Mr. Simon Lake's new salvaging submarine, the "Argosy" and "Argonaut." The character of this vessel, or combination of vessels, is made apparent from the drawing. With vessels of this type the bottom of the ocean is to be explored, cargoes and ships sunk during the war recovered where possible, and salvaging work in general carried on.

*The buccaneers of old could get no more wealth than now lies on the bed of the ocean awaiting our hand. Mr. Simon Lake gave us the opportunity of inspecting his craft and of actually walking on the bottom of the sea. The machine is exceedingly practical.—*EDITOR.

GETTING down into the submarine *Argonaut* is no very difficult process. You clamber up into an open hatchway at the front of the *Argosy*, and find yourself at the mouth of a huge steel pipe, four and a half feet in diameter. Descending the pipe is a good deal like going down an ordinary stairway. Really, it is a kind of half-ladder, half-stairway. Angle-irons riveted to the wall are the steps.

After descending about ten feet you find yourself at an elbow in the pipe. It stops going straight down and heads off toward the *Argonaut*, the particular angle downward depending on the amount she happens to be submerged at the moment. Electric lights illuminate the passageway.

At the far end is a steel wall barring progress. At the right, however, and within the huge joint which holds the access tube to the *Argonaut*, there is a round opening, about two feet in diameter, which leads to the "air-lock," a chamber about seven feet long, three feet wide, and four and a half feet high. At the far end of the air-lock is the door leading to the actual working chamber itself. This door is closed. There is a heavy air-pressure on the other side—needed to prevent the water from entering through the hole in the floor of the working chamber. This is

about four feet by two and a quarter feet in size. Through it the actual operations on the bottom of the ocean or on a submerged ship are carried on.

As is evident, the air-pressure in the little room you have just crawled into must equal that within the actual working chamber itself, or you would never get the door between the two open. This is precisely what this little room, the "air-lock," is for. Like the lock in a canal through which a ship passes to gain a higher level, so this air-lock exists to raise the air-pressure that surrounds you up to the higher level of that within the working chamber. The door between you and the latter is thirty-eight inches long by twenty-three inches wide—say 875 square inches in total area. For every foot that one goes down into the water, the pressure increases by .434 pounds. If one were down 25 feet then, the pressure would be 10.85 pounds per square inch more than

the approximately fifteen pounds per square inch always on the earth's surface. If the door has an area of 875 square inches, as we figured, and there is a push of 10.85 pounds on each square inch, it appears that there is a total of 8,750 pounds pushing on the whole door, roughly—about four and a half tons! Obviously you couldn't get the door open and yourself into the operating chamber without the use of this intervening air-lock. Using the air-lock, one can get up a balancing pressure and overcome the four and a half tons.

Once inside the air-lock, the door is closed behind you. It is locked from the inside. Simon Lake seizes a piece of heavy iron bar lying on the floor, and pounds the door and the locks even tighter. This is to make sure no air escapes. He then turns a valve.

It gives you a creepy sensation—turning that valve. You know how a chicken must feel when he sees the axe descending. But it is only momentary.

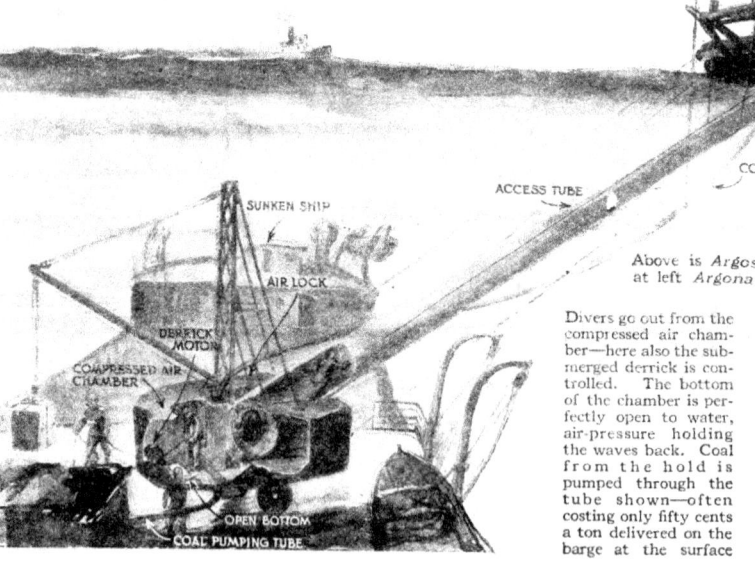

COAL PUMPING TUBE

Derrick shown holding coal tube also elevates cargo to surface.

SUNKEN SHIP

ACCESS TUBE

AIR LOCK

DERRICK MOTOR

COMPRESSED AIR CHAMBER

OPEN BOTTOM

COAL PUMPING TUBE

Above is *Argosy*; at left *Argonaut*

Divers go out from the compressed air chamber—here also the submerged derrick is controlled. The bottom of the chamber is perfectly open to water, air-pressure holding the waves back. Coal from the hold is pumped through the tube shown—often costing only fifty cents a ton delivered on the barge at the surface

In comes the air. *P-tsh-h-h—sh-h-h-h-h-h-h-h.* It is the same kind of a sound, though not quite so great in volume, that a locomotive makes when it draws up to a station and the safety valve starts to blow off because the engineer is no longer using the steam in his cylinders. In that little room the noise is deafening.

Suddenly the air is shut off. In the meantime you

have been swallowing, pinching your nose, and endeavoring to get the Eustachian tubes that lead from your throat to your ears under the same air-pressure as that coming into the room. Otherwise your ear-drums would protest unmistakably against the pressure. Mr. Lake has shut off the air, so you may have a few moments in which to accustom yourself to the increased pressure. Then he reaches for the valve again.

P-t-sh-h-h-h! On it comes again. In a few moments it begins to lose its sharp whistle, eases off into a long-drawn-out sigh, and then ceases altogether. The reason for this is that the air has been coming from the working chamber, and now ceases because the pressures in the two rooms have become equal. Obviously it should now be possible to open the door connecting the two. You find that Mr. Lake is already pounding on the door's latches. A few more pounds, and presto! the door swings open. The air-lock has served its purpose—let you conquer the barrier. You and the others scramble through the opening.

You find yourself in a little steel-riveted, steel-plated room, shaped as to floor and ceiling a good deal like the bottom of a flat-iron, and about seven feet high. In reality it is the prow of the under-water boat. This accounts for its triangular or flat-iron shape.

The Hole in the Floor

But the thing that attracts your attention above all else in the room is a hole in the floor about two and a fourth by four feet. There is water in that hole. There are all kinds of shells lying down there, stones, and many weird objects.

It's the bottom of the ocean you're looking at. Suddenly some kind of sea-denizen that probably calls himself a crab of a sort heaves into view. He has a shell on his back, and four or five legs sticking out on each side, and all told, legs and all, he is about as big as the palm of your hand. Evidently local society down here disdains to go forward and backward as is the custom in regions above, for he or she locomotes sideways, and really can dart through the water at a remarkable pace even though side-stepping it all the while. Interesting customs the natives have down here.

A fish comes plowing into view. He apparently doesn't like the scenery just above him, and veers off and out of sight. It's lucky for him he did, for Mr. Lake is already standing over the opening, waiting, a spear in his hand,

for anything that may come along.

Look! Here comes something crawling over the bed of the sea. It looks like an army tank headed for somewhere. It has a big horseshoe-shaped shell, and you can't see any legs, or wheels, or anything under it. You wonder how it gets along. Down shoots Mr. Lake's spear and Mr. or Mrs. Sea Tank is impaled as easily as if Mr. Lake had been aiming for the middle of a fried egg with a table fork. Up comes spear, sea tank and all. It proves to be a huge horseshoe crab, more than a foot across. What appear to be about two dozen legs wave wildly from each side of the shell as you pull

Simon Lake standing on the bed of the ocean, spearing fish and other sea denizens. Air-pressure holds back water at his feet—otherwise it would fill the room. From such an under-water room sunken vessels may be salvaged, oysters and clams gathered and a variety of other under-water work carried out

it off the spear. You put the crab bottom upward on the floor of the boat where it can kick up its heels to its heart's content, and no harm done, for it is in the same predicament a turtle is when turned over on its back.

It dawns on you that you have forgotten you are under considerable air-pressure. It has all been so interesting, looking down through that opening on to the bed of the ocean, you have forgotten all about anything else. The truth is, the air-pressure within the far corners of your lungs, and at the back of your ear-drums has become the same as that within the room, and, except for a general sensa-

tion of heaviness, it is not noticeable.

Standing in that chamber and looking through the opening in the floor, one can realize very readily how it would be if one happened to pass over a submerged wreck. Nearly the whole vessel would stand out as plainly as a board at the bottom of a brook. It is really remarkable how much light gets to the bottom of the sea at ordinary depths. Most sea water, Mr. Lake says, is quite clear, in spite of the fact that one has a hard time realizing it when trying to look down into it from a boat at the surface. In Southern waters, particularly, one can frequently see two or three hundred feet with ease.

To Salvage Wrecks

Had there been wrecks close at hand in Bridgeport Harbor, it would have been possible to get over one, and perhaps walk on its deck. As it is, you may take off your shoes and stockings and stand in the opening, walking on the bed of the ocean as the submarine moves along. You can get a very good idea of how easy it would be to size up the condition of a sunken ship, get things from its interior, send out divers, and otherwise work with it, almost as easily as if it were but two or three feet under the surface. You have to concede that Simon Lake has turned out an exceedingly practical contrivance; and one of striking originality.

What Mr. Lake is going to do with his *Argosy* and *Argonaut* is perhaps best made evident by the following quotations from his book, "The Submarine in War and Peace":

Somewhere off Bridgeport, Conn., lies the wreck of the old Sound steamer *Lexington*. Tradition has it that she has a fortune in her safe. Many a ship has been sunk in the waters about Hell Gate; search was carried on there for years for the old British frigate *Hussar*, which struck on Pot Rock and sank during the Revolutionary War. They tell that she had four million dollars (£820,000) in gold on board to pay off the British troops, and that she carried this treasure to the bottom with her. There is a cargo of block tin in a barge somewhere off the Battery, New York, and many a ship with valuable cargo lies along the coast from Newfoundland to Key West. The yearly loss in ships and cargoes throughout the world has always run into many millions of dollars, and since the war this has been multiplied a hundred-fold, amounting to billions. The time will come when many of these ships will be found, and such of their cargo as is still valuable will be salvaged.

"Ships like the *Argosy* and *Argonaut* will get this booty," says Mr. Lake.

Compressed Air to the Rescue

How a submarine, sunk in ninety feet of water, was raised and its crew saved

By Robert G. Skerrett

THE British had a rod in pickling for the German Grand Fleet if the Kaiser's sea force had ventured forth again. To be specific, the instrument of chastisement consists of a flotilla of very large submarines of a new type, and the purpose for which these boats were built was the forming of a mobile ambuscade just where the foe would be least likely to look for under-water craft. They were designed to have exceptionally high surface speed and to be capable of accompanying a battle squadron to distant zones of operation offshore, and there to submerge and lie in wait for the enemy's armored squadrons.

The present story has nothing to do with the details of construction of these powerful submarines other than to say that they are not propelled by internal combustion engines, but are driven, instead, by steam-turbines when operating at the surface—steam making it possible to attain greater speed, and possessing other technical and military advantages.

Spectacular Feat of Salvage Corps

Now, one of the conditions that have handicapped the employment of steam in submarines has been the great amount of heat radiated by the boilers, and the well-nigh insufferably high temperature of the boiler and engine spaces; and it is upon this fact that the present story turns. Several months ago, the British authorities gave an outline of some of the spectacular work carried out to successful climaxes by the Salvage Section of the Admiralty. Among these performances was the raising, by means of compressed air, of a vessel which had gone to the bottom in water ninety feet deep. The public report did not describe the type of craft; but it has since become known that the vessel in question was one of the big steam-driven submarines mentioned. The boat

foundered on her first deep-water trial trip: to be exact, she went to the bottom when making her initial dive. She had aboard of her at the time not only a full crew but nearly double that number, and among the extra personnel were certain high officials of the Admiralty. There were fully ninety people inside of her when she was headed seaward.

A Strange Oversight

It is not certain whether the presence of so many inquiring spectators or the newness of the crew led to confusion and oversight at a critical stage of the maneuvers, but it is known that the primary cause of the disaster was the failure to close some of the ventilators over the boiler and engine spaces before the boat submerged!

In all probability, the steam plant heated the craft to an uncomfortable degree, and in order to secure relief the ventilators were kept open aft, and this was forgotten during the execution of certain orders immediately preceding the fateful dive. As a result, the vessel no sooner settled below the surface than water poured in violently through the unsealed ventilators; and, before the descent of the submarine could be halted, hydrostatic pressure increased the speed and force of the invading flood, sending her bottomward like a dropping stone.

Half of the people on board, those

in the space aft, were drowned.

Those forward were trapped and seemingly doomed.

Happily, the submarine was under escort, and the bubbles of air escaping surfaceward steadily from a fixed position occasioned alarm. Divers were promptly brought to the site and sent down. They discovered the foundered boat. Their hammer blows upon the hull were answered by tappings from within the unflooded interior, and signals were exchanged telling the nature of the accident and the predicament of the survivors.

Fresh Air Forced In

The first concern was to save the lives of those in the forward half of the boat, and it was realized that air would have to be supplied them before the confined atmosphere became too foul to sustain life. To this end, hose connections were made through the outer deck by way of existing fittings, and fresh air was forced into the vessel while the tainted air was withdrawn at another point. From time to time, liquid food was sent down through the hose to the imprisoned men. These relief measures were continued for a day or more.

In a somewhat similar manner, compressed air was forced into the flooded after half—entering through the top of the pressure-resisting hull and driving out the contained water through an opening which was cleared in the under side of the submarine's shell. Sufficient buoyancy was finally obtained in this manner to cause the boat to bring her bow above the surface. Then, by means of oxy-acetylene torches, an opening was cut in the exposed deck and a way cleared for the escape of the forty-five survivors within. The submarine was next towed to port, dry-docked, and refitted for service.

This salvage achievement stands without a parallel.

Liquid food was sent down through the hose to the imprisoned men, and finally the submarine was forced to the surface by compressed air

AVIATION

Inventing New Flying-Machines

How the designer, with the aid of mathematics and laboratory instruments, evolves new types to outwit the enemy

ITALY was the first nation that used the airplane at all in war. That was in 1911, during the Libyan campaign in Africa. Aeronautically one-sided as that war was, it was of immense importance in the development of the military flying-machine. Whatever doubts there may have been —and there was enough ink spilled by army officers in military journals to prove that airplanes could never, never take the place of a reconnaissance in force —Europe knew that when the next war came it must reckon with the man in the air. Now, the Libyan campaign, although it taught many a useful lesson, did not by any means reveal all that would be required of a flying-machine in battle. No hostile airmen, no anti-aircraft guns molested the Italian scouts who flew over the hot sands of northern Africa and unerringly noted the positions and strength of the enemy. Rifle-fire alone was indulged in as an offset to aerial reconnoitering.

There was some Italian bomb-dropping—also very useful in the lessons that it taught. But no one could foresee that in 1915 men would be called upon to fight twenty thousand feet above the ground, that special machines would be required for reconnaissance and for bombarding, and that even a special kind of aerial tactics would have to be evolved to make the utmost use of an airplane.

The machines used in the 1911 campaign bear only an outward resemblance to the machines in which the flyers of the Allies are now engaging the enemy. The airplane of 1911 was hardly the scientifically designed machine of today. Its engine was so untrustworthy that flights of two and three hours were extraordinary; its climbing powers were nothing to boast of; and its speed was scarcely sixty miles an hour.

The Scientist Enters

What a change there has been in a brief three years! In our factories we are building machines that it is confidently expected will make one hundred and

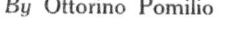

By Ottorino Pomilio

Ottorino Pomilio is one of the foremost aeronautical engineers of Europe, although he is scarcely twenty-eight years of age. A graduate of the École Supérieure d'Aéronautique of Paris, at the outbreak of the 1911 Tripolitan war with Turkey, Italy commissioned him to purchase flying-machines to be used at the front. Later he assisted in reorganizing the Italian aeronautic service. Busy though he was, he found time to write two important books ("Aviation Motors" and "Aeroplane Construction"), and eventually to build the famous Pomilio Aircraft Factory.

eighty miles an hour. Our bombing and reconnaissance planes are now faster than the fastest fighters of 1916. Our engines may be depended upon to drive airplanes through the air for many hours.

This almost breathless progress is due entirely to science. The blacksmith and the carpenter who built the machines of the daring pioneers of 1909–1912 have made way for the engineer and the mathematician. Rule-of-thumb has given place to mathematical formulas; guesswork to absolute knowledge.

A flying-machine is not merely a thing of linen, spruce, and steel. It is in very truth a wonderful artificial bird. Like a bird, it has a body—its fuselage; like a bird, it has a heart and lungs—its engine; and, like a bird, it has brains—its pilot. The flying-machine differs from the bird only in that its various parts perform special functions. The engine of a flying-machine, for example, supplies only propulsive power; the heart of a bird, on the other hand, must also keep the blood in circulation.

How He Approaches His Problem

A flying-machine is conceived by a modern designer primarily as an engineering structure. The speed it is to make, the load it is to carry, the ease with which it is to be maneuvered —all are carefully computed.

To create a new type of flying-machine, a designer may proceed in one of two ways. Whichever way he may decide to adopt, the horsepower at his disposal is his primary consideration. One method may be termed analytic, and the other synthetic.

Suppose that a flying-machine must be built to carry a certain number of passengers, a certain number of guns, a certain amount of ammunition (machine-gun bullets and bombs), and in addition enough fuel, oil and supplies for a definite number of hours.

What is the best form of wing for a given type of flying-machine? Will it be strong enough to resist the air when flying at 120 miles an hour? The airplane designer must answer these questions. He secures his model rigidly in a tunnel; then he blows air against it, to see what resistance it offers

Otto Lilienthal was an engineer who began his gliding experiments in 1891. His apparatus was merely a rigid pair of wings. As the picture shows, he would grasp the handle-bar, run down a hill, and then draw up his legs. Thus he was able ultimately to fly several hundred feet. Eventually he built this biplane glider. Because the wind blows in gusts and never in a steady stream, the "center of pressure," as the scientists say, was constantly changing. Hence, to maintain his balance, he had to throw his weight from side to side. In 1896, while testing a glider in which he intended to mount a motor, he lost his balance and was killed

In such a machine the Wright Brothers taught men how to fly. Modern machines are very different, but the art of flying as the Wrights taught it has not changed

Everybody used to laugh at Samuel P. Langley. For all that, he was the first man to fly a motor-driven flying-machine. That was on May 6, 1896

Away back in 1908, before the Wrights came out into the open, Voisin in France made machines in which a few enthusiastic Frenchmen risked their necks. There was nothing to prevent side-slipping except a few vertical partitions between the wings, partitions that were supposed to offer the necessary resistance but didn't, as this historic photograph abundantly testifies. Voisin is still in the airplane business, but he now builds machines that can fly in any weather

After the Wright Brothers went to Europe and showed the world how to fly, Henry Farman abandoned the crude boxlike Voisin in which he used to fly when there wasn't a breath of air stirring. He frankly copied the Wright machine in its essential features. Here is his 1909 machine. Nowadays the aviator sits comfortably in a beautifully modeled boat-shaped body that offers little resistance to the air

After the Wrights came Curtiss in 1908 with this machine. Wings are not arched in this way nowadays; it's inefficient to do so. The little triangular terminations to the wings are the ailerons, or flaps by which side-to-side balance was maintained

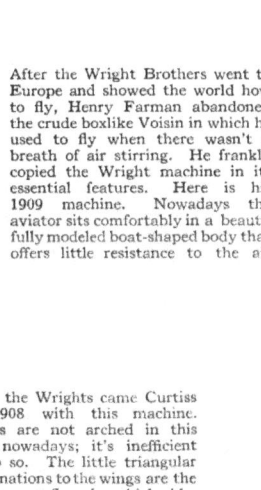

Louis Blériot was a manufacturer of automobile lamps who had tried hard to fly ever since 1900 and smashed machine after machine before the Wright Brothers vaulted into the air in France. He watched them, saw why he was making little progress, and built a monoplane (here it is) in which he flew across the English Channel on July 25, 1909. He made about forty miles an hour; now one hundred and twenty is considered about right for a fighter

When Henry Farman took these three passengers up with him in 1909, the world voiced its amazement. Only recently the giant Handley-Page flew with twenty-one, and no one batted an eye. Few of the present flyers would risk their lives in this flimsy machine. They would not mind seeing the ground between their legs five thousand feet below, but they would mind the unscientific and weak wings. The air resistance was so great that forty miles an hour was the best to be hoped for with such a machine

Suppose, in addition, that the machine must have a definite climbing and horizontal speed.

Given such a machine to create, the designer must determine the minimum amount of power required to get the best results; in other words, he must choose the most suitable type of motor. Hence I call this the analytic method of designing.

Airplane Problems

But suppose—and this is the usual case—the designer is given a certain engine with which he is thoroughly familiar. Around that engine he must drape, as it were, a machine that will carry a certain load of passengers, armament, and ammunition, and fuel for a flight of a certain duration. This I term the synthetic method of designing.

In either case, however, the designer must discover the best climbing and horizontal speed obtainable with the material and machinery at his disposal.

Although there is no hard-and-fast rule to govern the choice of the various elements of which a flying-machine is composed, nevertheless limits are imposed beyond which the designer may not go without fantastic results.

For example, suppose he is asked to create an airplane that will have a speed of three hundred miles an hour, maintain that speed for a flight of from four to five hours, and carry fuel enough for so long a journey, besides three passengers. Such a machine would require an engine of several

When Pégoud first looped-the-loop, the world gasped, but it also wondered what was the good of it. The value of such maneuvers is evident

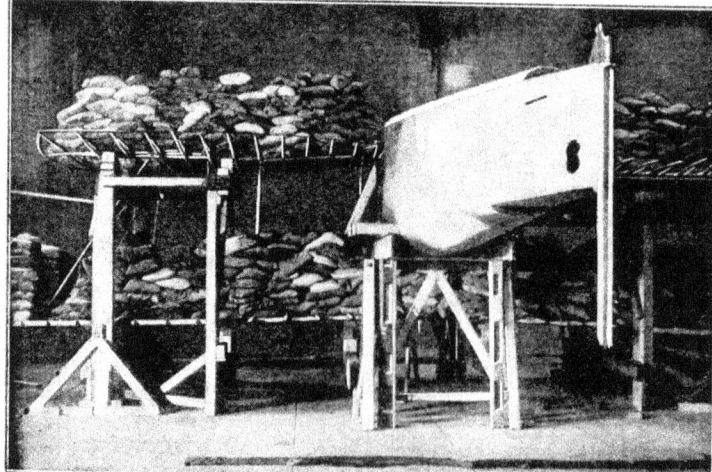

A flying-machine is supported by the upward pressure of the air against its wings. The wings must withstand the blows of the most violent storm. To find out whether they are strong enough, the maker turns the machine upside down and loads the under side of the wings (now turned up) with sandbags of known weight, until a few handfuls of sand cause the wing to sag

thousand horsepower—a wild absurdity at the present time, because the engine required has not yet been constructed, although there is no doubt that it will ultimately be developed.

Reducing Air Resistance

We all know how important is fineness of line in building fast transatlantic steamers or torpedoboats. The hull must be so shaped that it parts the water easily. So, likewise, the flying-machine must part the air with the least possible resistance.

In a sense, it is easier to build a fast steamship than a fast flying-machine. The steamship, after all, is a single huge hull. On the other hand, the flying-machine has not only a hull (its fuselage, in which the pilot sits), but a vast expanse of wing, with its struts and wires.

An airplane is essentially a structure of surfaces—useful surfaces and parasite surfaces. As anyone can guess, the useful surfaces are

in the present war. Every aerial fighter must be an acrobat now. He twists and turns, side-slips, and performs a dozen evolutions unknown to Pégoud

Hence a designer is dependent on a well equipped aero-dynamic laboratory.

In a wind-tunnel provided with the most precise measuring instruments, a model is placed, one hundredth as large as the actual machine that is ultimately to be made. In the wind-tunnel the efficiency of the proposed construction is very accurately ascertained. If the results at first obtained are not satisfactory, the model can be changed, although the mathematical investigation that has preceded its construction is usually so accurate that changes are never very radical.

Importance of the Wing Curves

If a flying-machine is to rise from the ground at all, it must have the proper curve of wing. Hence, the designer concerns himself more with the wing than perhaps with any other one feature of the machine. On the whole, the curve adopted usually resembles that to be found in the wing of a flying bird. The top of the wing is convex, the bottom concave. Like a bird's wing, its greatest thickness lies near the point of attachment to the body; it becomes thinner near the extremity.

The painstaking studies that were made of birds' wings have served their purpose; for most of the successful flying-machines of today have wings which, aerodynamically considered, are derived from those of a bird. There are as many varieties of wing curves for airplanes as there are for birds, and each serves its own function. The properties of each curve have been very carefully determined in the wind-tunnel, so that a good designer can select that curve which gives him the maximum amount of support and the least head resistance for the particular type of machine on which he happens to be working.

Wind-tunnel experiments have shown that exaggerated curves are not conducive to high speed—indeed, that they are best for slow-speed

the wings that support the structure in flight; the ailerons, or controlling flaps, which maintain balance from side to side; the elevator; the rudder; and the fixed tail. The parasite surfaces are the fuselage, or hull, to which the main wings are attached and in which the pilot sits; the struts and wires necessary to hold the wings together; the landing-gear; the radiator; etc. The best flying-machine is obviously that in which the amount of parasite surface is reduced to the minimum.

It might be supposed that the experiments made by Langley, Eiffel, Prandtl, and other physicists answered once and for all the question: What is the most efficient curve of wing? Eiffel and others studied the effect of the wind on small model airplanes in elaborately constructed wind-tunnels; but their findings hold good only for the particular models studied. Supposing that I design a different landing-gear or a different form of wing. I must still conduct tests with models to ascertain whether or not I have made any substantial improvement.

Here are the biggest and smallest machines. The big machine is the bomb-dropping Handley-Page; the small one a 125-mile-an-hour fighter. It might be supposed that the big machine is nothing but an enlargement of the small one. But the wing curves are different in order to secure lifting capacity in the big and speed in the small machine machines. Moreover, they have also shown that trailing edges must be very sharp.

Testing a Model with Sand

Having scientifically determined what form he will give his craft, the designer proceeds with the construction of a full-sized machine. This is all the more important in the building of military machines in which the problem of armament must be considered. Guns, bomb-dropping devices, cameras, wireless apparatus—all offer a certain amount of head resistance. This head resistance must be reduced to a proper disposition of the devices named before the factory can begin to build the machines in quantities.

The designer must find out exactly to what extent its wings can be loaded. What is known as a "static" test is therefore conducted, the purpose of which is to confirm the original mathematical calculations, and to enable the designer to correct unavoidable errors in computation. In what does the static test consist?

The machine is turned upside down, and bags of sand, weighing about a hundred and twenty pounds each, are placed upon them. At last a dramatic moment is reached, when a few handfuls of sand, added to the load already imposed, cause the wings to sag.

And now comes the tense moment when the scientifically conceived and carefully tested machine is to be flown for the first time. The pilot —a man picked not only for his skill as an aviator, but also for his intimate knowledge of the mechanism of a flying-machine, climbs over the cowl

and belts himself into the seat. A mechanician gives the propeller a twist. There is a roar. The machine bounds over the grass like a living thing.

Satisfied at last that he may trust himself to this pulsating thing of pistons and cylinders, the pilot turns his elevator ever so slightly. The machine leaps into the air and soars up and up into the blue. In half an hour

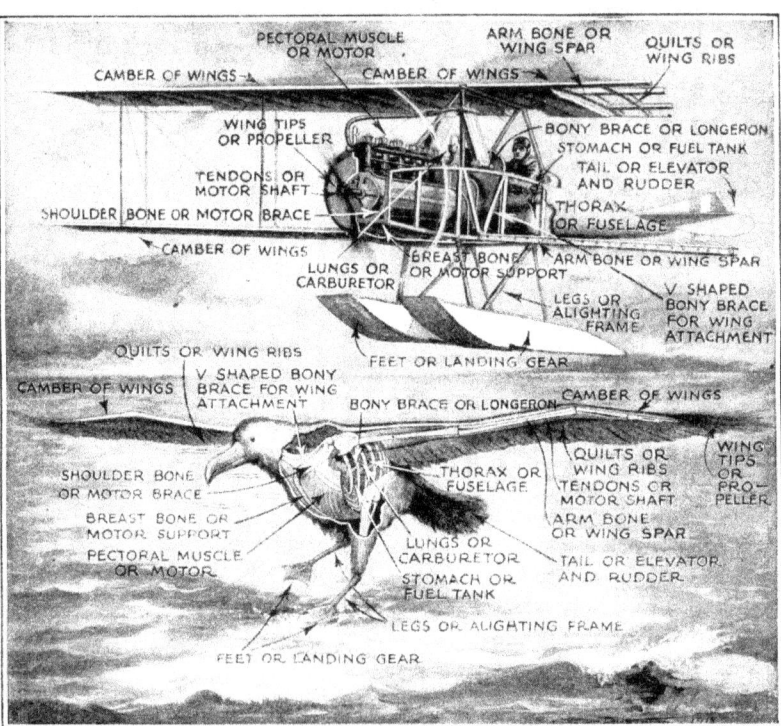

There is more than an external resemblance between the airplane of to-day and a representative bird. The chief parallel is found in the fact that in full flight a large bird does not flap the central part of its wings, but forces itself through the air with more or less fixed wings at a slight "angle of inclination," deriving support exactly as does an airplane, from air pressure. Also, that it flaps the tips of its wings not for support, while doing this, but for propulsion, which it derives by automatically "feathering" them like a sculling oar. The only difference between the wing-tips and the screw-propeller of a flying-machine is that the rotation is reversed continuously, long before a single circle has been completed. Another striking likeness is in the location of the bird's motor (its pectoral muscle) right in front, as in an airplane, and the nourishing of this muscle by the digestive organs and the lungs, much like the feeding of a motor from fuel-tank and carburetor. The electric ignition of a flying-machine, again, closely resembles the incitement of a muscle by the motor nerve. Other parallels between natural and artificial flight become obvious from the accompanying drawing, contrasting an albatross with a seaplane

he knows all the tricks and all the whims of that new machine.

The Airplane Finds Itself

The pilot assures himself that the machine is as practicable as possible, and that the cock-pit is as comfortable as one may expect of a war machine. He drops to earth in a glide or two, and makes his report. The defects to which he calls attention are:

(1) defects of centrage; (2) defects of control; (3) defects of secondary importance.

It is difficult to determine in a flying-machine what is the exact location of the "center of pressure"—the theoretical point in which the entire force of the air, as the machine rushes on, is supposed to be concentrated. The fault is not easily corrected; for it means that the curve of the wing is not what it should be.

On the other hand, the second class of defects (defects of control) are easily corrected by a displacement of the wings relatively to the fuselage, or by shifting the masses in the fuselage (guns, wireless outfit, etc.); or by varying the surface of the fixed horizontal tail.

Those defects of secondary importance which constitute the third class are more easily foreseen by the experienced designer. They are occasioned by the influence of the various parts of the machine upon one another. Thus, the blast of the propeller as it is deflected by the machine itself may react in an undesirable way on the wings or the fuselage.

An astute pilot, who becomes as much an integral part of his machine as a cavalryman of his horse, may be of the utmost assistance to the designer. Preferably he should be a man who has had much experience in the air over the battle-field.

But it must not be supposed that the pilot indulges in but a single flight. He goes up time and time again, each time reveling in some new acrobatic trick of the expert pilot. Unless a machine can loop-the-loop, spin down nose first, and even fly upside down for a time, it cannot be used in an air duel.

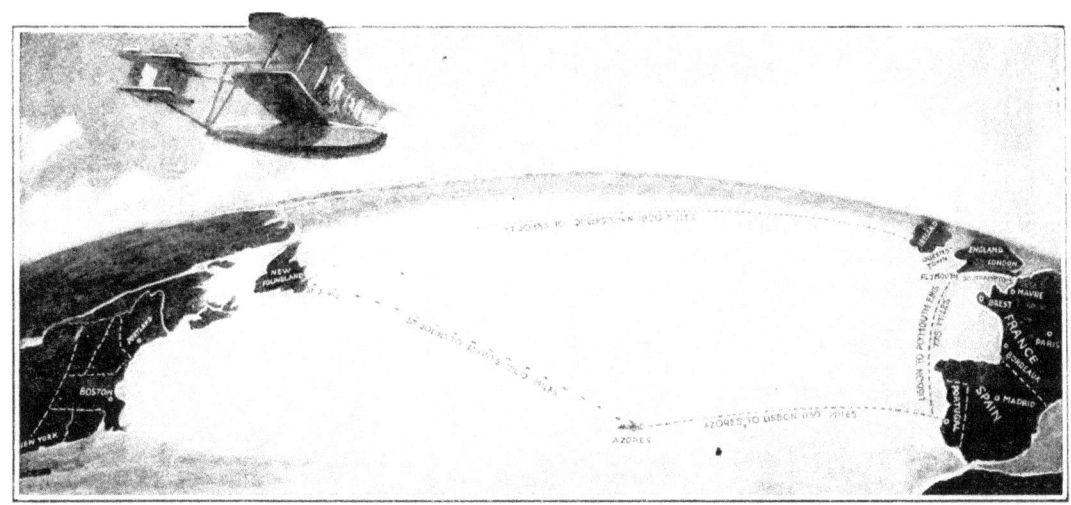

The Azores seem to have been providentially placed where they are in order to assist the transatlantic flyer. Travel to Europe by way of the Azores and it becomes possible to carry a large, paying load of passengers and mail

To Europe in a Flying-Machine

By Waldemar Kaempffert and Carl Dienstbach

IT is not merely by a poetic fiction that the daring men who vaulted into the air from Newfoundland to cross the Atlantic are likened to Columbus. To be sure, they taught us nothing new about the geography of the world. On the other hand, the ocean air through which they traveled is as uncharted as was the ocean of water in Columbus's time. Meteorologically, it was a real voyage of discovery.

The American Navy has reason to be proud of its transatlantic experiment. It was not imbued with the desire to win a rich sporting prize, but to prove or disprove a proposition scientifically. Hence the chain of destroyers that stretched across the path of flight.

The essence of a scientific experiment is control. Every step must be watched. Nothing must be left to chance or accident.

Use Warships as Observatories

For the first time, regular meteorological stations were established in mid-ocean, among them battleships with a radio range equal to that of most land stations. The practice should be continued. What better use can be made of destroyers, of battleships that must be kept in commission regardless of expense, than to employ them as oceanic weather ob-

servatories? Commercial companies, whose enterprise without scientific aid would be as adventurous as that of Hawker, will derive immense benefit from such meteorological work of the Navy; for they will be warned of impending danger.

One great lesson was learned in the great American effort to cross the Atlantic. It is this: The art of flying is in advance of oceanic meteorology. If vessels of the NC type are to be employed for the crossing of the Atlantic—vessels lumbering and slow in comparison with the swift passenger-carrying transatlantic flyer of the fu-

With this drift-indicator—very like a bomb sight—the true course of the seaplane can be established so long as the surface can be seen

ture—we must be able to predict the Atlantic weather twenty-four hours in advance.

The Air—an Uncharted Ocean

Where are the carefully tabulated observations that make that possible? Beyond a few joint experiments in kite-flying made from the decks of the Prince of Monaco's yacht by Teisserenc de Bort and Professor A. Lawrence Rotch, and by government scientists from the decks of the United States Coast and Geodetic steamer *Seneca*, nothing whatever has been done to explore the Atlantic atmosphere.

Before regular transatlantic flying becomes a reality, the United States and Europe must systematically explore the atmosphere over the ocean, just as the atmosphere over the land has been explored from the ground to a height of about twenty miles.

So competent an authority as the late Professor A. Lawrence Rotch predicted, long before our vastly improved flying-machines appeared, that it would be feasible to cross the Atlantic every day of the year in northern latitudes if the permanent western planetary winds that prevail above ten thousand feet could be used. Unfortunately, support is poor at such a height; to carry much cargo there is out of the question. By skilfully selecting his level, Rotch argued, the transatlantic aviator may avoid head-winds; a sufficient number of more or less favorable following winds are always available at medium altitude. This theory of Rotch's, one of the pioneers, be it remembered, must be verified by systematic weather studies and by numerous transatlantic voyages by both airplanes and dirigibles. If the successful trips of many machines vindicate Professor Rotch, commercial transatlantic flying is a matter of but a few years. In the present stage of aviation, it is almost criminal to use land machines over the ocean for passenger service.

We have said that meteorology—transatlantic meteorology—has not kept pace with the demands of aviation. The obstacles offered by fog prove the assertion.

The navigators of the NC flying-boats knew that fog is the aviator's deadliest peril. It may be questioned, however, whether they were quite prepared for the peculiar variety of extensive, obstinate, high-sea fog, combined with a heavy sea, that proved the undoing of two of the planes on the way to the Azores.

It is the old, old story, but is told with a new thrill. The pilot of the NC-4, suddenly deprived of all his subconscious visible indications of horizontal direction, became confused. The semi-circular canals of the middle ear (the spirit level in our heads) tells us whether or not we are on an even keel. Yet, for some curious psychological reason, he could not tell that his vessel was banked—in other words, tilted sidewise so that it tended to run around in a circle.

Wanted—a Substitute for the Human Ear

In a fog a flyer's sense of direction and sense of verticality are so far destroyed that he cannot tell at what angle the machine is banked. He guesses at the angle. The compass is supposed to indicate direction; but when the needle wanders navigation is demoralized. If a machine executes one flying maneuver unbidden and unperceived, it may execute any other equally unbidden, even a nose dive. It was the appearance of the sun through the clouds that saved the NC-4 from running around in narrow circles. Clearly, it is unsafe to cross the ocean in an airplane without the aid of some positive indicator of horizontal direction, some instrument which, unlike the labyrinth of the ear, is not affected by centrifugal force if a mist obscures the sea. Such an instrument is Sperry's gyroscopic level-indicator.

A Great Wireless Station on the Azores

The flight has also taught us the vital importance of maintaining radio communication under all conditions. The commander of the NC-3 had left his ground set ashore. He paid dearly for that. After he had alighted he became dumb, wirelessly speaking. He was able to send messages only in the air with a trailing antenna. Although a half dozen destroyers were near enough to help him, his seaplane was as hard to see on the ocean as a speck of dust on a plate-glass window. Eyes and ears are appallingly limited in range. Only radio communication makes it possible to protect a transatlantic boat flying in distress.

To lay a true course for the small islands in the Azores in mid-ocean is at best a navigation problem of no mean order. But to lay that course through the air in a fog requires all the aid that can be given by modern radio apparatus.

When we consider the mishaps that befell the NC-1, it is evident that without radio apparatus of fair range transatlantic journeys should not be

undertaken. So efficient must the radio apparatus be that it can be used for reckoning latitude and longitude even though the machine may be compelled to float for hours, and even days, on the water.

Since the Azores are likely to become of future aeronautical importance, they will surely become the site of a huge radio station, a veritable beacon of electromagnetic waves to guide transatlantic flyers as surely as a moth flies into a flame. At any moment, when that station is completed, the transatlantic aero-navigator may determine his location and then head for the Azores.

It may as well be admitted that, however romantic it may be to cross the Atlantic without a stop, the direct route from Newfoundland to Ireland is of only academic interest for the time being. Aerial commerce in flying-machines will surely move by way of the Azores, simply because it must be prepared to fight head-winds.

Lieutenant - Commander P. N. L. Bellinger, who captained the NC-1, which was compelled to alight in the ocean near Flores, two hundred miles short of its objective. The crew were saved, but the seaplane sank after being buffeted by the seas for many hours

Lightest and smallest of those who manned the seaplanes, Lieutenant-Commander A. C. Read, of the NC-4, brought the "hoodoo ship" through to first honors after misfortunes had nearly prevented his starting. His was the only one to arrive at the Azores on the wing

Commander J. H. Towers, "admiral" of the seaplane squadron. His NC-3, lost in the fog, was forced to drop into the waves. After riding out a heavy gale she reached Ponta Delgada under her own power, but she was so damaged as to be out of the flight

The Guiding Fingers of Light that Pointed the Way at Night

When Kipling wrote his "Night Mail," little did he realize that his idea of using searchlights to guide the nocturnal flyer would be practically applied in the transatlantic flight of the American seaplanes. Each one of the destroyers stationed in the path of flight to guide the argonauts of the air became a visible beacon at night. For miles and miles the long, rigid fingers of searchlights pointed high in the air to guide the men in the planes.

But the searchlights served not only as mile-posts to indicate the course. They were also used to indicate the true direction of the wind. Every scientific precaution was taken to help the men in the air. Since searchlights cannot penetrate a fog, the commanders of the destroyers received instructions to fire star-shells above the fog. Flares with a candle-power of several hundred power each were also to be used in emergencies

Over the shorter non-stop route the machine best adapted to the task would only waste money. Why is this? Because the airplane has to carry fuel in place of money-earning cargo.

What's Wrong with the Machines?

The American seaplanes were not prepared to fight head-winds. They were slow, heavy, cumbersome. Essentially winged sea-boats, they were not streamlined with that scientific care which is so much in evidence in fast fighting planes. In this respect the Sopwith machine of Hawker and the Martinsyde of Raynham were better.

Although the American experiment is scientifically more valuable than the British, there can be no doubt that the transatlantic flyer of the future will more closely resemble the British than the American machines. Not that this transatlantic flyer will necessarily be a land machine, but that it will place its chief reliance on cleanness of line, on engine power, and hence on speed.

The NC planes are merely large copies of the old Curtiss flying-boat. The engines are mounted outside of the hull—an objectionable procedure because of the air resistance encountered. Aeronautical engineers are well aware of this, but they have been forced to continue the practice because the propellers must clear the waves. But the larger the flying-boat, the less reason for continuing the practice. Any one can see for himself how high the propellers of the NC boats were mounted above the water.

The NC type has proved itself decidedly too slow for transatlantic

commercial flying. Its speed of sixty to seventy-five miles an hour is not what we expect. It is really a flying cruiser intended to detect submarines, and not a transatlantic flyer.

Not one of the machines that earned transatlantic glory in 1919 was specifically designed for the purpose of flying over the ocean. Hawker's Sopwith machine is a fast bomb-dropper. The Vimy - Vickers and Handley-Page biplanes are likewise bomb-droppers. Six hundred miles is the maximum range for which both the British machines and the American NC planes were designed. Much

When nearly in mid-ocean the single-engined biplane in which Harry Hawker and Lieutenant - Commander Grieve dared the Atlantic flight was brought down by a clogged water-cooling system. Their escape from death was miraculous

of the elaborate preparation for the flight can be explained by the provision that had to be made for carrying extraordinary loads of fuel for a voyage far exceeding in length that for which any machine built today is really designed.

Experiments with a Transatlantic Monoplane

Although the Americans collected more practical information on the subject of transatlantic flying than did the British, it must be confessed that transatlantic flying in the future will duplicate more Hawker's splendid

effort so far as speed is concerned. If the business man who must be in London in forty-eight hours is to be deposited in England safely and surely, he must be transported in a machine that will plunge through adverse winds and storms as readily as the *Mauretania* plows through heavy seas.

What will be the aspect of the transatlantic flyer of the future? It must be an evolution of the fast fighting airplane if it is to depend on power and speed. Exactly what form it will take no man may safely predict. This much at least is certain: It will have a completely inclosed body or fuselage: it will have a speed of one hundred and fifty miles, possibly of two hundred miles, an hour.

So many problems will have to be solved before the Ideal Trans-Atlantic Flyer has been devolved. Mr. Carl Dienstbach presents this ingenious solution. You will find nothing revolutionary in his plan; it is based on the experience of the last ten years in flying over land and sea.

Since the machine must cross the ocean, it must also carry much fuel; but the carrying of a heavy load and the attainment of high speed are almost incompatible. An enormous lifting surface must be provided to raise the load, and much of this lifting surface is not only useless in the air, but decidedly detrimental because of the air resistance that it offers.

Since there must be the utmost economy of surface and power, Mr. Dienstbach has decided that the machine ought to be a monoplane of about one hundred feet span. Curiously enough, the late Count von Zeppelin, after devoting the last months of his life to the construction of a mammoth seaplane, also arrived

The NC-1 tuning up at the Rock-away Naval Air Station in preparation for the great flight. All of the NC boats are alike in general appearance

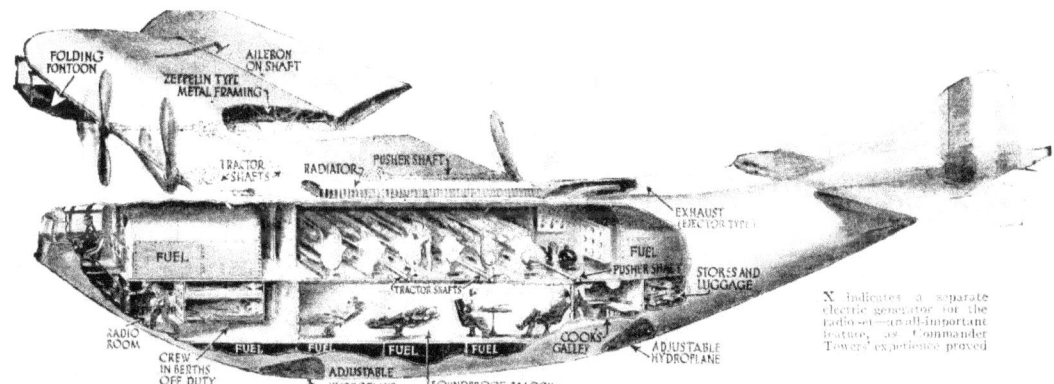

To gain speed the engines and passengers are enclosed in a perfectly streamlined hull. Note that the passengers must sit or recline. There is no room to walk about upright. Weight is precious, so that there cannot be more than three in the crew. Every passenger is weighed before he steps on board. He pays so much per pound for having himself transported

at the conclusion that a monoplane is the correct type—a conclusion of which Mr. Dienstbach was not aware at the time he designed this transatlantic monoplane.

Mr. Dienstbach was inspired by the monoplane of Mr. Grover Cleveland Loening, probably the fastest that has thus far been designed and built. The Loening construction lends itself well for the purpose because it is so stanch, because it does away with fuel-wasting projections, and because it consumes little power for the speed attained.

Now, in any transatlantic flyer, speed is dependent primarily on perfect streamlining, which means that every projecting surface must be so designed that the air is parted with the least possible resistance. It was the French designer Nieuport who was the first to attain racing speed in this scientific way.

The narrow high fuselage of the Loening construction is a great advantage because the engines can be placed within the streamlined hull. Because the machine has but a single surface it encounters less resistance in its onward progress than a biplane. At the start of a flight with full load it is practically impossible to vary speed; but toward the end of the voyage the speed becomes more adjustable as fuel is consumed. Moreover, the flying angle can be reduced from that of the utmost efficiency for lifting to a flatter angle, which means still higher speed.

The mechanical advantage of having the engine inside the hull is obvious enough. But will not the passengers be inconvenienced by the noise, odor, and vibration? Not if they are literally sealed in a mechanically ventilated saloon provided with padded walls. Of course, the quarters will be cramped, but not more so than in the NC flying-boats. They cannot \ upright; but they can lie comfortably on couches. Moreover,

the passengers will be provided with every luxury. Their quarters will not only be ventilated, but supplied with warmed air. If the spectacle of an ocean dotted with ships interests them, they can enjoy it by looking out of their port-holes. There will even be electrically cooked hot food prepared by a small cook, probably a Japanese, who must be a veritable genius. He will act as steward, purser, deckhand, and maid-of-all-work.

It Will Have to Be a Flying-Boat

It is understood that the type of machine must be a flying boat for it is impossible to make a preliminary run at high speed on wheels in so large and heavy a machine. The machine that Mr. Dienstbach has in mind can vault into the air only if it can carry its load at maximum speed.

Suppose the machine were mounted on ice-boat runners, and suppose that the ice area were infinite. In other words, a flying-machine thus mounted on runners would gather speed, travel faster and faster, until finally it would rise into the air. So long as the preliminary run is fast enough, the heavy load can be lifted. A way must be found to run or glide over liquid water as fast as it is possible to slip over ice. Much has been done in this direction by the hydroplane, but nothing comparable with ice-boating.

The fuel-tank of Raynham's Martin—it is large for obvious reasons

This solution is offered: Increase the hydroplane speed by reducing the long, wetted, adhesive hydroplane surface, and give that surface such a shape that the maximum lift is attained with the least drag. This becomes possible if the machine is able to rise clear of the water on one step in front—a step of correct design.

Now, this small hydroplane area will lift the whole load only at very high speed, and very high speed cannot be attained until the whole weight is lifted. We seem to be running around in circles. The only solution is to make this same small surface lift exceedingly even at slow speed, while the hull of the flying-boat is still in the water and offering therefore much resistance to onward progress. This end could be attained by what the airplane designer calls a variable camber surface.

By camber is meant the curvature. It is hard to change this camber in the wing of an airplane; in fact, it has not been successfully done, because the complex construction necessary makes the wing either too heavy or too flimsy. But in a hydroplane variation of camber becomes feasible. Remember that in the water steel can be used, something solid and thick; the construction can be simplified. A hydroplane with a variable surface would act like a kind of crow-bar, a long lever to lift the craft. Hydroplane surfaces can be placed directly below the load and can be controlled from their centers.

You have only to picture to yourself a well in the hull above this adjustable step or hydroplane, much the same kind of a well that receives the centerboard of a cat-boat. In this well is a row of hydraulic cylinders. The pistons emerging from the cylinders are fastened to a flexible steel plate lying normally flush against the bottom of the hull—a plate reinforced at the

Design for the Two-Hundred-Mile-an-Hour Transatlantic Monoplane

If the Atlantic is to be crossed regularly by commercial flying-machines, they must be vessels very different in type from the NC planes. Mr. Carl Dienstbach has designed the two-hundred-mile-an-hour transatlantic monoplane. It is an elaboration of the Loening construction.

fastening points by strong lateral steel ribs.

Pump the cylinders full of water or empty them to a variable degree, and the steel plate can be made to bulge out or to flatten; any degree of concavity can be obtained. The location on the hull is such that if it be given no curvature—only a slight downward slant from front to rear—it constitutes the step so familiar to motor-boat speeders. With the help of subsidiary vertical surfaces adjoining it and moving in harmony with it (vertical surfaces that are slight prolongations of the hull's sides), it performs the familiar function of a step—a step that may be folded in the air against the hull so that it will offer no resistance. If the step be given a steep slant and a suitable curve, that accelerates the impinging water so sweetly downward that no eddies occur (exactly as a high lift wing acts in the air), and a trifle of a reverse "S" curve

at its tail end to dismiss the stream smoothly, it needs must give a strong upward pressure at lower speed as well, while the hull is still in the water. The hollow space behind the lowered plate and the vertical side plates communicates, of course, with the air through the well above.

Getting Off the Water

Let us watch the step in action. We start with a marked curve to the hydroplane to lift the hull out of the water at low speed. When the hull is well out we reduce the curve and increase the speed. Finally all the weight is carried at extreme speed on the small front step. A similar step behind the center of gravity serves as an elevator in the water.

There is also a small vertical rudder which projects into the water and which is low enough to remain there after the tail has been raised by the

elevator. Of course, the aerial controls help these water controls far more than in the ordinary flying-boat because the water speed is so much higher. Once the propellers are running, side balance is maintained entirely by the usual wing ailerons. These ailerons are very long, nearly as long as the wings, and across them blows the strong slip-stream of the propellers.

Since air resistance is so important, we must look for the utmost cleanness of line in flight. We have not an ounce of fuel to waste. The hull of the flying-boat must be given a better streamline than is to be found in the hulls of the famous NC planes. And, above all, it must be possible to fold away the side floats or pontoons that support the outer wings. Of course, this is no original idea. It has been found difficult to fold away the wheels of land machines, because they must be strong to stand severe shocks, and yet light. But floats, which are al-

ways necessary when a single center hull is employed, need not be called upon to withstand heavy shocks. Therefore, as our drawing shows, they can be swung back into recesses in the wings and covered with sliding screens when the machine is in motion. With the slip-stream side control just described, they are required only when the craft is at rest on the water.

The very size of the monoplane solves the propeller problem. The propellers can be mounted out on the wings, so that they can be driven through bevel gears by the athwart-ship engines. Nothing is detracted from efficiency, because the single reduction gear can be embodied in the bevel gear.

So solid and thick is a monoplane wing of the Loening type that there is ample strength and space for the propeller shaft mountings. In order to attain the speed desired—two hundred miles an hour—six Liberty engines and six propellers (four in front and two in the rear of the wings) are specified to attain the best compromise between efficiency and structural strength. The length of tubular shafting required is only a slight disadvantage; on dirigibles such shafting was employed long ago.

It Can Live in Rough Seas

The radiators have been borrowed from the Curtiss models, and likewise the long ailerons. Mr. Curtiss has given us a radiator of minimum resistance that has great cooling efficiency. Mr. Dienstbach has moved it from the sides of the fuselage to the top and shaped it like a long trunk. It will then act somewhat like a fin; but the chief advantage is that it is possible to utilize the exhaust as it is utilized on a locomotive—to send an extra strong draft through the radiator, cooling it in spite of the still further reduced head-resistance.

A craft of the compact type that the monoplane designer has in mind can live in a really rough sea when unable to rise from it. Must it wallow helplessly? Not at all. It proceeds on its own power, runs through a storm faster than it connects with a destroyer. The flyer will always start and alight in moderately sheltered water.

When we fly to Europe we shall not need a string of destroyers across the ocean to help us in an emergency, nor need we wait patiently day after day until the weather conditions are favorable. If transatlantic flying is to be commercial the crossing must be made according to a time schedule. The time-tables may be as elastic as that of a steamship so far as arrival is concerned. If that ideal is to be realized, speed is necessary—extreme speed which makes it possible to cope with contrary winds without lengthening the trip immeasurably beyond fuel endurance.

Next! The Aerial Freight Train

Will this latest project of aeronautics become a practical possibility?

By Adrian Van Muffling, S. A. E.

The aerial freight—at present still a figment of the aeronautical imagination. But the world has grown accustomed to seeing impossibilities changed into facts, and we should probably take it as a matter of course if we were to gaze into the sky some day and see the sight here pictured. The train is linked together by the cable

IMAGINE a string of airplanes, or rather huge motorless gliders laden with freight, traveling over the trackless road-beds of the sky, led by a tremendous "locomotive-plane." It's a picture fairly staggering even to our rather sophisticated modern imagination.

And yet, if reports from overseas are true, the scheme to employ an airplane of enormous power to pull a number of freight-carrying gliders is credited to no less experienced a man than Mr. Fokker, the Dutch aeronautical engineer who developed the first high-speed flying machine during the war and thus gave Germany a temporary supremacy over her foes.

In justice to Mr. Fokker, it should be stated that he first offered his designs to the Allies, and did not enter into communication with Germany until he had been turned down by them.

The most obvious difficulty in trailing airplanes lies in the method of starting and getting the trailers off the ground. Will they be placed

behind one another close to the "loco-motive," and start on their journey as the connecting cable becomes taut? Obviously this method would subject the frame of each unit to stress much greater than it can be built to stand; moreover, the imposition of sudden and increasing loads would result in slowing down the motor unit to below its flying speed. If, on the other hand, the planes were placed at a proper distance apart, with the cable stretched between them, a field several miles long would be required to get up the necessary speed. Two alternatives present themselves. One consists in starting the units closely grouped with taut connection between them that could be paid out gradually so as to increase the distance—a method that is objectionable because of the aerodynamic interference between the units. The other is to accelerate all trailing units simultaneously by means of a moving platform or endless chain, an expedient involving a disproportionate expense.

Another problem presents itself. Conditions in the air would vary materially between points as far apart as the various units necessarily would have to be. For an instant a trailer might travel a little faster than the one immediately preceding, thus relaxing the connection. What would happen when the slack was taken up?

The thought of having each "car" equipped with a reel upon which the cable could wind itself as required presents itself, the tension being kept constant by a compressed-air or spring arrangement after the fashion of the familiar trolley-pole retrievers. The weight of such an equipment would be likely to be equal to that of the average aviation motor.

Difficulties to Be Overcome

Even with the most careful individual control of all units, it would be a very difficult matter to keep them traveling in exactly the same path. The slightest deviation, on the part of the motor unit, from a mathematically straight line would entail a side-slipping effect on the succeeding units. Moreover, the tractive effort exerted by the cable in a forward direction would vary with each change in direction, no matter how slight, and at no time could a constant pull be expected for any extended period of time. That this would seriously affect the maneuvering powers of the train is apparent.

But the greatest difficulty to be overcome is inherent in conditions that govern the relation of power and weight in airplanes. The greater part of the power developed by an aviation motor goes toward sustentation; the rest is absorbed by resistance encountered in driving the machine. If the power be greatly increased the excess will go toward increasing the speed (and it requires roughly four times the power to double the speed).

An airplane has therefore no "tractive effort;" that is, it is inherently unsuited to exert a pull. Now, the motor unit would have to lift itself first of all—no mean achievement if we consider the weight and size of the power plant required, and in order to give any tractive effort the power developed would have to be far superior to that needed for flight, which is equivalent to saying that the machine should be able to keep itself in the air at a speed considerably less than that of which it is capable; a condition which has not so far been realized even in the most modern types of airplanes.

A Remote Possibility

The weight and resistance encountered by a large machine capable of carrying, say, five tons of freight would be much smaller (to a pound lifted) than that offered by five smaller planes of one-ton capacity, each involving a separate set of wings, struts, landing-gear wires, a second pilot, etc. Counting the weight and resistance of the motor unit necessary, but not carrying a "useful load," the power required seems to be at least eight times as much as if the entire cargo were concentrated into one machine capable of carrying it. This is another difficulty added to those mentioned.

These considerations lead one to the conclusion that the aerial freight train is a remote possibility.

Across the Atlantic in a Single Jump

By Waldemar Kaempffert

The Vimy-Vickers machine in which Alcock and Brown crossed the Atlantic is much larger than either the Sopwith or the Martinsyde airplanes. It carries two Rolls-Royce engines of 350-horsepower each. The fuselage, or hull, is forty-three feet in length; the wing span from tip to tip is sixty-eight feet. The gasoline was contained in nine tanks, which were emptied one after another. The largest of these, a seventy-gallon tank, was fitted with life-lines, so that it could be used as a float if the machine fell into the sea, being released by removing two clips. It could have sustained the weight of the two flyers. All told, eight hundred and seventy gallons of fuel was carried in the tanks. Distributed as it was, the gasoline could not wash from side to side in a large mass. Fully loaded the machine weighed fourteen thousand pounds

"CAN it be done?"

Never again will that question be asked when the subject of flying across the Atlantic in a single leap comes up.

There were good reasons to doubt the feasibility of the flight. Had it not been proved by theorists, with their ever-ready pencils, that the maximum range of the airplane is twelve hundred miles? The problem of flying across in one stage resolved itself into one of providing fuel-carrying capacity.

A Battle with the Elements

The speed at which the crossing was made (sixteen hours and twelve minutes) was due largely to the favorable following winds that prevailed. Much of the journey was made at a height of more than a mile, and some of it at even more than two miles. Above ten thousand feet, the meteorologists tell us, the wind sweeps continuously from west to east. Evidently the Vimy-Vickers entered the great permanent planetary westerly swirl.

As meteorologists expected, the cold at that height was intense. Think of a machine covered with sleet! Think of Brown unstrapping himself to chip ice with his knife from the gages and instruments! Think of radiator shutters and thermometers encrusted with ice for four and five hours! Think of an airspeed indicator so clogged with ice that it refused to work! Is there no romance in flying across the ocean? Clearly, the clippers of the air will wing their way to England, not in a dull, mechanical way, but in battling at great heights with the elements.

The experiences of the Vimy-Vickers are curiously like the experiences of the American NC planes. Towers and Read, as well as Captain Alcock, found it utterly impossible to get their bearings in the fog. Says Alcock: "We scarcely saw the sun or the moon or the stars; for hours we saw none of them."

With the radio direction-finder out of order, with only a compass to guide them (and the compass tells nothing of the machine's drift in the wind), what agonies of doubt they must have suffered! It was the providential appearance of the sun that saved Read on the trip to the Azores, and it was an equally providential glimpse of the stars that enabled Brown to determine his position and to bring the craft back to its course.

The flight of Alcock and Brown was meteorologically unthinkable before the war. In 1912 a pilot would toss a handful of torn tissue paper into the air, and if the wind was too strong he would not go up. Alcock and Brown set off in the teeth of a gale.

Fog—the Aviator's Dread

All this speaks volumes for the engines. Alcock did not use the full power of his engines. "I never opened the throttle once," he assured us—meaning, of course, that he never opened it to the full. Nevertheless, the Vimy had an absolute speed of ninety miles an hour. Aided by a following wind, she averaged one hundred and twenty for the voyage. That record is not likely to be lowered for some time.

Fog has always been the aviator's dread. It destroys a pilot's sense of verticality. Alcock believes that for a time he was flying upside down. Other pilots have had that experience. To be sure, it cannot last long. Stand on your head for five minutes and you will be aware of your position simply because of the rush of blood to your brain.

"We looped the loop, I do believe," Alcock tells us, "and did a very steep spiral. We did some very comic stunts, for I had no sense of the horizon." I

Ardbear Bay, near the Clifden wireless station, where Captain Alcock landed. If Lisbon is the San Salvador of flying, Ardbear Bay must be the Plymouth Rock

may be doubted if he looped the loop in a machine so heavily burdened.

As it was, Brown was able to take only four readings: one from the sun, one from the moon, one from the pole-star, and one from the star Vega. That Captain Alcock actually did fly upside-down for a brief moment or two is not a mere supposition on his part. It happened near the water, after a drop from four thousand feet. As soon as he saw the waves he realized his predicament and shot up again. "That period lasted only a few seconds, but it seemed ages," to use his own words. "It came to an end when we were within fifty feet of the water with the machine practically on its back.

Not for the Land Machine

The flight of Alcock repeats the lesson taught by the NC planes: Clouds combined with thick weather and rain render it impracticable to follow the land practice of flying at an altitude of from four to five thousand feet. The pilot must either climb very high in order to locate his position by the stars or the sun, observations which are at best only a rough guide, because of the aircraft's tremendous speed; or he must fly near enough to the surface of the water to see very clearly the direction he is pursuing with reference to the waves.

There is no vital objection to flying low. The sea air at low elevations is steady, and alighting in a flying-boat becomes possible at any time. Indeed, there is a certain advantage in flying low, because the air near the surface has greater supporting power. The one reason for flying very high (ten thousand feet or more) would be to take advantage of the westerly winds that

always blow aloft. But this counts for little with machines that are faster than either the Sopwith that Hawker used or the Vimy-Vickers that Alcock piloted so successfully.

Captain Alcock reinforces the conclusion reached in this article that transatlantic flying is not for the land machine. "The flight has shown

The men who made the big jump. Captain John Alcock, wearing a soft hat and a smile, was "Jack" to everybody at Brooklands, London's flying center. He has been an aviator since 1912. Lieutenant Arthur W. Brown (in uniform), although in the British service, is an American. Mechanics has been his passion since boyhood

that the Atlantic flight is practicable. It should be done, not with an airplane or a seaplane, but with a flying-boat."

The lessons taught by the American transatlantic voyage, and also by the non-stop flight, are alike in emphasizing the need for more efficient radio apparatus to assist the transatlantic flyer. The truth is that radio-communication has proved a lamentable failure in every transatlantic flight thus far made. Alcock tells us that the little wind-driven propeller, by which his generator was driven, blew off. Surely there is need

of an independent motor driven radio set—something that can be relied upon when the craft is in the air as well as when it rests on the water. It may be urged that the additional weight required militates against such an apparatus. Let it be remembered that Alcock landed in Ireland with only two thirds of his fuel consumed. It is evident that the carrying capacity of even a bomb-dropping plane of the Vickers type would not be strained beyond the maximum if an independently driven radio set were provided.

The crippling of the radio apparatus left the Vimy-Vickers machine in a far more serious plight than may be imagined. Not only was it impossible to send any signals of distress had the occasion to do so arisen, but it was probably impossible to use the radio compass—that marvelously simple instrument which has been invented for the safety of air explorers, and by means of which the navigator in the densest fog or the blackest night can determine his position by reference to some powerful radio stations on land or on sea.

Advantages Over the Cable

There is more than may be supposed in Lord Northcliffe's prediction that some day the newspapers of London will be selling on the streets of New York on the same day of their publication. In a sense, the day of the international newspaper has dawned. What anxious moments would have been spared us during the peace conferences in Paris if full accounts of the proceedings had appeared simultaneously in the newspapers of the cities of Paris, London, and New York.

Comparing the "one-jump" route taken by Captain Alcock with that by way of the Azores followed by the NC flying-boats, we see at once that the Azores route

will be the choice of the transatlantic air-lines of the future. Fewer miles to fly means less weight to carry, and less weight of fuel means more paying freight

Flying Straight Up with Whirling Wings

The helicopter's most recent advocate is the scientist and inventor, Peter Cooper Hewitt

By Marius C. Krarup

A NOTED scientist, Peter Cooper Hewitt, inventor of the mercury-vapor electric lamp, recommends flying on the plan that Leonardo da Vinci said he would adopt if he had the power. In other words, Cooper Hewitt advocates a helicopter, which means that his machine is to be lifted by a screw propeller which rotates horizontally to obtain lift and obliquely to obtain both lift and propulsion. By twirling a toy propeller in the fingers with its shaft at forty-five degrees, one observes that when the blade on one side is in a position that gives lift, the other blade is in a position to give propulsion. By varying the angle of the shaft, lift and propulsion can be obtained in different proportions. The machine goes straight up when the shaft is vertical, and it flies horizontally when the shaft is inclined so that the lift produced is just sufficient to keep the machine suspended.

If only one helicopter screw were used, the engine driving it would spin itself around in the opposite direction. Half the power would be wasted and the pilot much discomfited. The plan therefore calls for two screws mounted on telescoping shafts and driven in opposite directions. For inclining the double shaft at the will of the pilot, one or more auxiliary small screw propellers may be used. When the engine and body of the machine are hung considerably below the helicopters, as in the testing apparatus shown herewith, the push required at the level of the engine for slanting the main shaft can be very moderate. Once the machine flies horizontally by this means, the wind resistance against the body helps to maintain the slant. If it is reversed by the auxiliary propellers, the machine flies in the opposite direction. All steering, up and down and sideways, can thus be accomplished by managing the slant, but Mr. Hewitt does not as yet disclose the mechanism that he proposes to employ for this purpose.

When a surplus of power is suddenly applied, the machine jumps in the direction in which it is aimed, as it has only its weight but no great areas of body and wings to retard it. If the power gives out, the machine falls; but the helicopters prevent spinning, having opposite pitch, and break the velocity of the descent somewhat.

So far, the helicopter plan has had no followers except in theory, mostly because the lift thrust obtainable from an ordinary propeller screw is very small. It is difficult to get more than a 200-pound thrust from an eight-foot propeller at one thousand revolutions a minute. This small thrust will operate at great forward speed when given the chance, as is done in the propulsion of an airplane, but that does not help to lift the much greater weight of a machine straight up, ever

© International Film Service Co.

P. C. Hewitt's helicopter test machine. When the electric motors are replaced by engine and auxiliary propellers for starting the helicopter mast, and when room is made for a pilot, this apparatus becomes a flying-machine

so slowly. For that purpose the thrust must exceed the weight. And to use many small helicopters in one machine is practically impossible. It is particularly difficult to slant them all, for propulsion. Besides, small helicopters revolving at high rotary speed are at best very inefficient. The centrifugal forces are troublesome and wasteful.

The extensive experiments made by Mr. Hewitt show the only way in which helicopters can be used. They must be very large. Those shown in the illustration are fifty-one feet in diameter. The four blades are twelve feet long and two feet six inches wide. They are really airplane wings mounted to be whirled around. The maximum rotary speed used is seventy revolutions a minute, but this gives a linear speed of one hundred miles an hour at the middle of each blade.

To get a lift of 2,550 pounds with this machine it is necessary to rotate the helicopter at 70 revolutions a minute and this requires 126.5 horsepower. At 46.5 revolutions a minute the power is more efficient, since at this speed 44.4 horsepower lifts 1,300 pounds, but this lift is not of any use if the machine weighs more. What the lift must be when part of the power is used for propulsion does not seem to be determined as yet. If 126 horsepower is required for supporting a machine weighing 2,550 pounds and more for flying it, the helicopter plan does not yet promise an efficiency equal to that of an ordinary airplane.

The following table gives results obtained with the Peter Cooper Hewitt helicopter testing apparatus:

Revolutions per minute	Lift in pounds	Horse-power	Pounds of lift per horse-power	Velocity in feet per second at 25-foot radius
46.5	1300	44.4	29.3	97.5
56.5	1800	75.4	23.9	118.3
63	2150	98.4	21.8	132
70	2550	126.5	20.2	146.5

© Press Illustrating Service

During a recent test of his helicopter flying-machine, Mr. Hewitt was accompanied by Thomas Edison, who displayed a keen interest in the invention

Voyaging to Europe in an Airship

The skyscraper hitching-tower of the future

By Carl Dienstbach

BECAUSE the giant Zeppelin made no brilliant showing in its raids on England, the average man is apt to dismiss it with a wave of the hand, saying, "It's a failure." When he thinks of air transportation he thinks of airplanes.

It would profit him to read the report of Admiral Jellicoe on the great naval battle off Jutland, in which the Zeppelin showed how effective it can be as a naval scout, and to study the testimony recently given by Admiral Sims of our own navy before a Congressional Committee, pleading for the construction of titanic Zeppelins in this country. It would also profit that average man to ponder the fact that Great Britain now has a fleet of Zeppelins, and that one of them, in all likelihood, will be sent across the ocean to prove the practicability of transatlantic trips through the air.

The truth is that the great rigid dirigibles, to the development of which Count von Zeppelin devoted the last twenty years of his life, are the *Mauretanias* and *Leviathans* of the air. When it comes to voyaging from New York to London with something like the comfort and luxury of a fast ocean steamship, the business man of the future is more likely to buy his ticket for a transatlantic Zeppelin than for a transatlantic seaplane.

Zeppelin Difficult to Handle

Let us admit at once that it is more difficult to handle a Zeppelin, both on the ground and in the air, than a giant flying-machine. Half a dozen men at most are required to swing a huge Caproni or a Handley-Page airplane on the ground so that it heads in the proper direction. But when a Zeppelin comes down to be berthed in its 700-foot shed, a highly trained ground crew of from one hundred to two hundred men, each man knowing exactly what he has to do, must take the huge envelope in charge.

More Zeppelins have been wrecked on the ground than in the air. A fabric composed of a wonderful lacelike aluminum framing cannot withstand terrific collisions with the ground. And so, Zeppelin after Zeppelin was wrecked until a special ground-handling technique was evolved.

Now, a real ship of the air should no more require a shed than does a ship of the sea. It should be weather-proof. Indeed, this was Count von Zeppelin's main idea. In a sense, it is just as ridiculous to run a dirigible into a shed as it would be to house the *Mauretania* whenever she touches port.

Problem Solved by British

But what of the problem of docking this huge dirigible of the near future? There will be no problem—at least, not in the present sense. A shed for airships should be required only for repairs.

Before Europe was embroiled in the bloodiest of all conflicts the British showed us a way out. They built towers of steel and tethered their small non-rigid dirigibles to them by the nose. Thus moored at a height from the ground so great that they could not be dashed against the ground, the vessels swung like flags in the wind.

The idea may be carried out on a more grandiose scale when the transatlantic, weather-proof Zeppelin of the future arrives.

Imagine every important American city fringed with titanic steel towers that resemble the famous Eiffel Tower of Paris—hitching-posts to which Zeppelins are tied. They form integral parts of the steel framing of the skyscrapers that constitute their lower portions; for in the populous city of the future the utmost use must be made of every square inch of ground.

You go up, of course, in an elevator. You step out on a broad platform that surrounds the tower. A gang-plank runs from the platform to the moored Zeppelin, stretching away for eight hundred or even a thousand feet.

At the top of the tower is a curious semi-globular, vertical cup, gilded so that it flashes in the sun. Into this cup the nose of the Zeppelin is received and lashed to the tower.

Platform Swings with the Airship

The cup on top of the tower and the platform from which the gang-plank runs to the ship constitute a structure independent of the tower. The cup and the platform revolve together.

You can see at once what this means. As the Zeppelin, with its nose poked into the cup, is swung by the wind, just as an anchored ship swings with the current in a river, the platform swings with it. Even though the craft may be in the act of swinging in response to a passing gust, you step on board by way of the gang-plank just as you would climb the stairs lowered down the sides of a ship anchored in a stream.

An aeronautical engineer, familiar with the vagaries of gas, realizes that it is not enough to hitch a Zeppelin to a high tower and leave the craft to itself. Gas and ballast must be controlled at anchor as well as in the air. But that is no difficult matter. Water ballast can be pumped through pipes wherever it may be needed, and gas can be conveyed in a similar manner. Hence, we must expect to find pumps on the towers so that an anchored vessel may be trimmed. Indeed, the pumps can be automatic. As for pipes, a part of the aluminum framing of the envelope may be used without adding an ounce of weight to the vessel.

When It Reaches New York

But what of currents of wind? They are almost negligible at the top of a lofty tower. But if they should occasionally make themselves manifest, and that so rapidly that it is impossible to shift ballast or gas quickly enough, then the vessel's own elevating rudders can be used. Oh, the crew must always be on duty, you argue? Not at all. The elevating rudders can be operated automatically. A mercury level will switch electric servo-motors on or off.

What a sight it will be when a transatlantic Zeppelin arrives at New York and heads for its tower, a long, beautifully modeled, slim cylinder. Down she slides. See, she comes up slowly into the lee of the tower and against the wind. Thus she uses the wind as a kind of brake. She must fight it with her own engines, and fight it she does. By nicely regulating his own speed, the commander literally crawls up to the tower. Ah, a line has been thrown out and caught by the Zeppelin. The rest is easy.

Curiously seamanlike, too, will be the airship's departure for Europe. She casts off. She drifts away with the wind, and sinks a little with the stern down (a kind of "tail slide" in airplane parlance), until the propellers reverse the motion past the tower. Then she goes up, up, up into the air, and away to distant London, where her passengers will breakfast day after tomorrow.

Here she is in her shed—a giant British Zeppelin. Will the rigid dirigible of the future be thus housed whenever she finishes a voyage? We doubt it. It is very likely that in the future sheds will be used only for purposes of overhauling and repairing. This ship is the R-33, the largest of British Zeppelins, which has fuel capacity and engine power enough to cross the Atlantic

Just like the bridge of an ocean steamer, you say. And you are right. All the navigating instruments and appliances of the *Mauretania* are found on board a Zeppelin, with a few more such as height indicators that the captain of the *Mauretania* never needs

The engines are inclosed in separate cars, as this picture shows. When helium instead of highly explosive hydrogen is used to inflate the bags in the envelope above, it will be possible to put the engines right in the hull and thus give the craft cleaner lines

All Aboard the Air-Liner

Think of battling storms a mile up! Think of cooking with the exhaust of engines! Think of landing in America with a parachute!

By Waldemar Kaempffert *and* Carl Dienstbach

DO you remember that old story by H. G. Wells, "The War in the Air," in which he describes minutely an airship trip from Frankonia to New York? It will strike those who read it in 1908 as one of the truest prophecies ever penned—only, the trip took place at the *end* instead of the beginning of the "great war."

At Mineola the R-34 looked for all the world like a big (yet not over-large) sea-vessel in a harbor. The paraphernalia scattered about lent color to the view—apparatus analogous to buoys, boats, coal lighters, etc. There were lengths of heavy hose for water ballast, lighter hose for injecting hydrogen into the depleted gas-chambers, huge mounds of steel hydrogen bottles, auto-trucks with fuel, and wheeled ladders for inspecting the hull and rudders, not to forget a ring of search-lights with their heavy electric cables trailing over the ground.

In the Wheel-House

Step into the forward gondola of the R-34. How like the wheel-house of a ship it is! See, here are the usual speaking-tubes, electric bells, and signals. What are these wheels on the side? They control the elevators when the ship is to climb or to glide down. This little cubby-hole is the navigator's office. As you see, it has a table on which the charts can be spread.

Over the chart table Scott and Cooke spent much of their time with protractors, dividers, and many navigational text-books, measuring angles of drift and calculating course made good. Aerial navigation is more complicated than navigation on the surface of the sea. Note that in a separate sound-proof compartment back of the forward engine-room is the radio station.

Where does the crew sleep? In the big envelope above. Would you like to see for yourself? Here, slip on these felt shoes. Why? Well, you will have to climb a ladder, and the nails in your leather shoes might strike a spark. Remember that this great craft is supported in the air by thousands of cubic feet of highly inflammable hydrogen. A spark—!

Step this way, please. See, this is the way up into the envelope—this pipelike vertical passage, in which an aluminum ladder is placed. You are in the huge envelope now. Better light your hand flash-lamp. It's dark, as you see. This foot walk along the keel on which you stand is six hundred feet long, illuminated by only two electric lamps. It is very, very narrow—only fifteen inches wide. It would be hard for two men to pass each other.

In this dark, narrow passageway the men sleep sleep in ten light Italian hemp

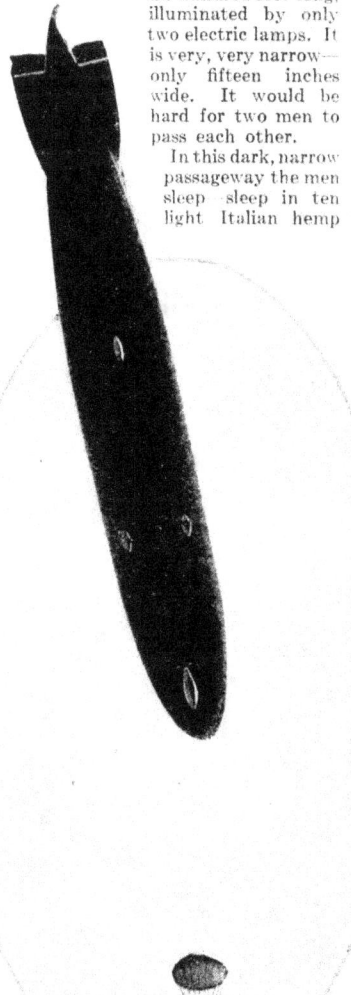

© Keystone View Company

Major Pritchard, on board the R-34, didn't wait for the ship to land. He jumped overboard with a parachute

Major G. H. Scott is thirty-six years old. Five years ago he was flying airplanes at Hendon, England. He personally superintended the construction of the R-34 at Inchinnan

© Press Illustrating Service

hammocks suspended from the framework. Each man shares a hammock with another. One man sleeps while his mate is on duty. Only a man with strong nerves can sleep in this fashion a mile in the air.

One of the passengers, General Maitland, writes in his diary:

Getting into one's hammock is rather an acrobatic feat, especially if it is slung high, but this becomes easy with practice. Preventing oneself from falling out is a thing one must be careful about in a service airship like the R-34. There is only a thin outer cover of fabric on the under side of the keel on each side of the walking way, and the luckless individual who tips out of his hammock will, in all probability, break right through this and soon find himself in the Atlantic.

Running along the 600-foot cat's-walk, inside the envelope, are the parachutes and pneumatic life-belts, kept in canvas containers. Both parachutes and life-belts are always available. If the men should be forced to leave the ship while it is over land, the parachutes will save them. Over the sea, parachutes and life-belts would be used. Every man dons a life-belt almost as soon as he steps on board.

Where the Gas and Fuel Are Stored

Up above you are eighteen hydrogen-gas compartments. Nine of them have automatic valves, which open when the pressure within them becomes too great. The other nine are equipped with hand valves. When the great craft rises, the hydrogen expands. To prevent the compartments from bursting—they are made of gold-beater's skin—the valves must be opened. But why both automatic valves and hand valves? Merely a precaution. Should the valves refuse

In 1492 Columbus sailed across the Atlantic in seventy days. The flagship was the *Santa Maria*, a decked ship of one hundred tons, with a crew of fifty-two men, commanded by Columbus in person

The R-34, first airship to cross the Atlantic, is 640 feet long. From the top of her lowest car to the top of her gas-bag is about 92 feet. Her five engines develop a total horsepower of one thousand. The useful load is distributed as follows: Fuel—4,900 gallons (weight

The British ensign flew proudly at a regular halyard from the stern, as the picture below shows

If the R-34 could stand on end beside the Woolworth Building, she would be topped by the building's flagstaff one hundred feet. But England has announced her intention of building an airship eleven hundred feet long

to Cross the Atlantic Ocean

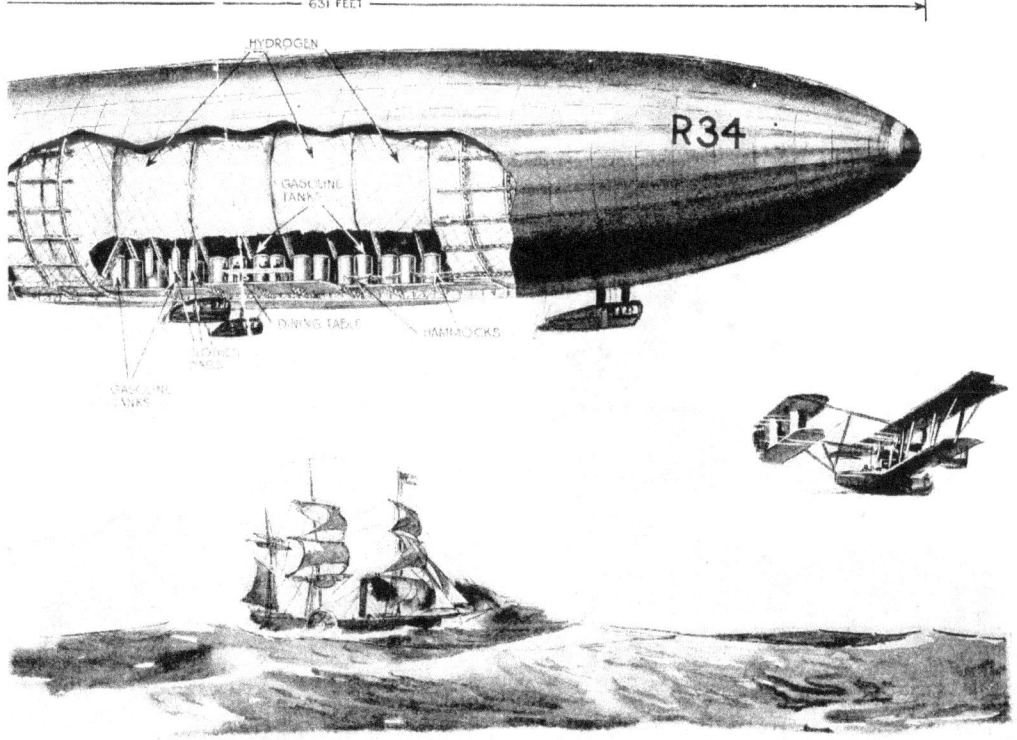

The first steamship to cross the Atlantic, which was one hundred years ago last spring, was the *Savannah*. She was 130 feet long; her horsepower was only 90; her speed but 6 knots. It took her thirty-seven days to cross

The NC-4 has a wing span of 126 feet, almost as great as the *Savannah's* length. It took her about twenty-five hours to cross the Atlantic Ocean

35,300 pounds or 15.8 tons; oil—2,070 pounds (.9 tons); water—3 tons; crew and baggage—4 tons; spares—550 pounds; drinking water—800 pounds. The total is 24.32 tons. The R-34 had thirty-one souls aboard. The crew included two wireless operators, who had several sets of instruments at their disposal. The broken away section of the picture shows how comfortable the passage must have been. Good hammocks, hot food cooked on simple devices attached to the exhaust pipes of the engines—what more could be reasonably expected? The R-34 is a Zeppelin with a vengeance. The arrangement of cars, propellers, rudders, etc., is exactly like those of the large German Zeppelin L-49, which was captured in France intact

To reach the envelope from the forward car, you climb up an aluminum ladder through a pipelike passage. Felt shoes must be worn, lest you strike sparks from the rungs. The huge envelope contains eighteen compartments of explosive hydrogen. The vessel was as safe as a powder factory

The two cars in the middle, of which this is one, were not in danger of bumping while the ship was tethered near the ground, because they are higher than the cars at the end. These cars are all engine. Cylinders, flywheels, valve-gearing, and all else crowd the interior

to work, a fire might start and the crew might have to leap for life. Even flying over a warm city causes expansion. Over the cold sea the gas contracts. Thus, while flying over the ice-floes around Newfoundland, on the voyage to America, there was a contraction of forty per cent. Hence the final inflation takes place at the minimum temperature—in other words, after sunset.

The fuel (5,000 gallons) is contained in eighty-one tanks in the envelope. Each tank has a capacity of seventy gallons. Sixteen of these tanks are fixed to the framework. The others are called "slip tanks," and may be filled with water as ballast. As a rule the water ballast is carried in canvas containers. The addition of alcohol prevents it from freezing at high altitudes. In addition there are eight emergency containers, four forward and four aft. They carry, all told, about three thousand pounds of water, which is pumped in as the atmospheric conditions may demand.

The drinking water is drawn from three tanks. Besides these there is a reserve tank containing distilled water, which may be used in case there is a shortage of drinking water. On the way from England the men were limited to a pint of water at the end of each day.

Where do the men eat? Here in the dining-room, right in the center of the envelope, the only place where the cat's-walk is wide. But where are the tables? They are fastened on hinges to the framework. All the old naval class distinctions are swept aside; for the officers and men use the same dining-room—if room it can be called.

Hot Food for Dinner

The dining saloon is large enough to permit half of the crew at a time to eat. The food is cooked or heated, as it is wanted, on stoves attached to the exhaust pipes of the engines. Hence the engine-room serves as a kitchen. Chairs? There are none. The diners stand at a cupboard.

The meals are good. One day there was an excellent stew—the meat cooked just right, with potatoes and carrots,

A cat was smuggled on board, for luck. Nearly every member of the crew had a mascot, from the engineer officer, who wore one of his wife's silk stockings around his neck, to Major Scott with his gold charm "Thumbs up"

served on an aluminum plate with knives and forks, and, last but not least, one cup of fresh water.

Lieutenant Shotter says in his diary:

The mess cooks are now preparing the midday meal—cold roast beef, cold potatoes, bread, cheese, and chocolate. We were to have had beefsteak, but the two wing engines which are fitted with cookers are stopped for general adjustments, so we are taking cold lunch instead.

Later he writes:

Just had tea at four o'clock with General Maitland, Majors Scott and Cooke, and Lieutenant Harris.

The men in the gondolas are not cold. The cars are heated by the radiators of the motors.

The two side cars, just large enough for two men to do their work in, contain each one engine. Hence the engines are distributed as follows: one in the forward car, two in each of the central side cars, and two

At night searchlights in the neighborhood of New York swept the sky, to guide the R-34 as well as to find out whether she had arrived after sunset

in the stern car. There are, however, only four propellers, the two engines at the stern having only one propeller. Two men work at the engines, one man operating and the other oiling.

General Maitland entered one day in his diary:

Slight trouble with starboard amidship engine—cracked cylinder water-jacket. Shotter, always equal to the occasion, made a quick and safe repair with a piece of copper sheeting, and the entire supply of the ship's chewing gum which had to be chewed by himself and the two engineers before being applied.

It was difficult for the crew to realize that they were on a transatlantic voyage. The trip was actually tiresome. Lieutenant Shotter, who is responsible for

the performance of the engines, notes in his diary:

I am certain that passengers will be fearfully bored unless amusements are organized. One can see at the best only a vast expanse of sea, and at the worst fog, or an awfully downcast sky, which the Atlantic alone can bring. Our gramophone has been a God-send to us during meal-times.

The men had exercise enough. Along the 600-foot "cat's-walk" they would run from end to end, a mile in the air. There are also steps in a vertical ladder to the top of the ship for those who are energetic or have duty up there.

How It Feels to Travel through Clouds

They saw little of the ocean through mist and clouds, having often a sea of clouds below and a sky of clouds above. What is it like to a true denizen of the air? Only the ocean traveler knows the cumulative effect of waking up, eating, going to sleep, and waking again day after day on the water. What must it feel like to become similarly a part of the atmosphere, to drive on and on with only clouds, stars, moon, sun, the blue sky, mists, and an occasional glimpse of the distant ocean below?

Officers and crew have sketched in words the charm and peril of their journey—sunset tints like paintings, magic mirages in the fog, phenomena that have never before been so thrillingly witnessed.

Groping in the Fog

The R-34 voyaged to these shores in a period of comparatively favorable weather; yet there were fogs over the north Atlantic, originating from icebergs or from chilly water flowing south from the Bering Straits. As the result of a series of curious meteorological circumstances, the R-34 had to stem temporarily strong contrary winds and several thunder storms. Her navigator complained of trying up and down currents, "bumps," near the end of the voyage. No wonder they should occur at the very edge of a memorable heat-wave. Such is the power of the engines that these difficulties were easily overcome.

Hundreds of steel bottles containing compressed hydrogen gas were shipped to Mineola in order to inflate the gas-bags of the R-34

The lack of meteorologic knowledge was disconcerting. Lieutenant Harris, the weather observer of the R-34, remarked:

We learned that our old boat was as fickle as a bug, and insisted upon going over the high-pressure areas and plunging and diving through atmospheric changes. It took a good sailor to come over, and it will take sea-legs to get back. The task of navigating an airship is hard, because the air-currents are invariably invisible to the eye and must be sensed by delicate instruments and calculations.

They saw the Atlantic only a few times after they left the Irish coast. It was necessary for Cooke, the navigator, to use the clouds as a horizon to make his observations.

The cloud horizon is not clearly defined; the observer must guess a little as to the true line. Hence an error of fifty miles may occur because the observer has not guessed very well.

Shotter Sees His Double

By means of the shadow cast on the water, and also by special instruments, the navigator was able to measure the drift angle. Sometimes, when the R-34 plunged through the clouds, it was dark, cold, and wet. The windows had to be shut. Periodically the fog cleared away, so that the men on board could see the sea for a minute or two. Eyes smarted because of the bright light reflected from the fog.

The fog played some curious tricks. Read this from Lieutenant Shotter's diary:

Just had a shock. Opened the engine-room window. It is a large affair. Two men can lean out. I looked out and saw another engine-room opposite and a man's figure looking out. Ye gods! My heart beat frantically. Only the reflection from a bank of very thick white fog. We get extraordinary sights up here in this beastly fog—uncanny sights, I think.

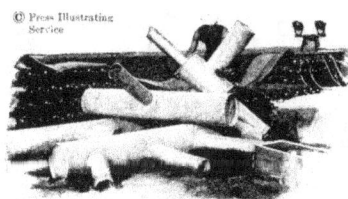

© Press Illustrating Service

The field at Mineola was cluttered with pipes, joints, and paraphernalia of all kinds to refuel the R-34 and replenish her gas compartments

To escape the clouds above and below, Major Scott had to climb at times to an altitude of four thousand feet from two thousand feet. After the airship had passed into Fundy Bay, Scott turned east off his course to dodge a storm, putting on all engines. In this he was very successful. The R-34 passed through the outer edge of it.

It is no easy task to keep the mammoth dirigible on her course. Lieutenant-Commander Lansdowne, the American observer on board, did a two-hour trick at the wheel, and found it hard work

We had a very bad time indeed [says General Maitland] and it is quite the worst experience from a weather point of view that any of us have ever experienced in the air. During the storm some wonderful specimens of cumulo mamatus were seen and photographed.

General Maitland wrote constantly of fogs in his diary:

Suddenly we catch a glimpse of the sea through a hole in the clouds, and it is now easy to see we have a slight drift to the south, which is what was estimated by both Scott, the captain, and Cooke, the navigating officer. A few minutes later we find ourselves above the clouds, our height still being fifteen hundred feet, and beneath the cloudy sky are clouds at about eight thousand feet. We are, therefore, in between two layers of clouds, a condition which Alcock and Brown found themselves on more than one occasion on their recent flight from west to east.

There were remarkable rainbow effects on the clouds. One complete rainbow encircled the airship itself, and the other—a smaller one—encircled the shadow, both very vivid in their coloring. General Maitland writes:

To the west the clouds have lifted, and we see an extraordinarily interesting sky—black angry clouds giving place to clouds a gray mouse color, then a bright salmon-pink clear sky, changing lower down the horizon to darker clouds, with a rich golden lining as the sun sinks below the surface. The sea is not visible.

Still Larger Airships Are Coming

As the trip's slowness resulted partly from general eastward trend of North Atlantic weather, it follows that to cross from Europe to America an airship ought to navigate in the trade-wind regions or that an airship much larger and more powerful than the R-34 be employed. A dirigible's speed and reliability grow with size—a view which has often been expressed by Carl Dienstbach and which is now reinforced by General Maitland's announcement of an airship, now in course of construction, double the size of the R-34.

The commander of the R-34 kept in constant communication with our Weather Bureau during his brief stay. At the slightest indication of danger he would have slipped his mooring and floated away. As it was, the R-34 left at the urgent suggestion of the Weather Bureau. A storm was coming east from the Great Lakes, and the R-34 was in positive danger.

The voyage of the R-34 has strikingly demonstrated that Zeppelins must be navigated like sailing vessels, strictly in accordance with the winds, permanent or temporary. Even much faster Zeppelins—the British plan a hundred-mile-an-hour rigid dirigible—will pay commercially only if they follow the easiest route. They should cross from Europe on the Columbus course, aided by the trade-winds. The Antilles and possibly the Cape Verde Islands will serve as useful fuel stations.

We in America needed this demonstration of a Zeppelin's capabilities. We have been too skeptical, although our vast country is better suited for atmospheric cruises in giant airships than any European country west of Russia. That long period of gentle, quiet weather from June to October is ideal for air navigation.

Roosts for City Airplanes

Solving the difficult problem of landing on a roof

By Carl Dienstbach

LIKE any soaring bird of prey, an airplane must be in motion before it can fly. It must run along the ground before it can vault into the sky. But that is not all: it must make its running start in the face of the wind. This explains why it is possible to confine a condor or a vulture in a narrow cage open at the top.

Street No Place for an Airplane

You can see from this that the problem of flying to and from your office in New York or Chicago reduces itself to the providing of suitable platforms on which you can alight (also in the teeth of the wind) and from which you can start. Clearly, city streets, flanked by high cliffs of architecture, lend themselves about as well for airplane landing and starting as they do for ice-boating. Flying must adapt itself to the wind as much as ice-boating.

Even if a flying-machine could vault into the air from one of the canyons in the Wall street district of lower New York, it would be dashed to pieces against the tall buildings on either side. In order to avoid a catastrophe in starting from a street, the wind must blow exactly in the right direction not only during the preliminary run but also for another minute or so. Within a city, airplanes must land and start above the roofs. Skyscrapers must be avoided as if they were mountains.

There is clearly nothing for it but to use the roofs—a suggestion that has been made time and time again. But this means either a building with a very large roof or a number of buildings of uniform height connected by a common roof. Suppose a number of buildings could be so connected by a roof. Obviously, the common connecting roof must not cut off light and air from the streets below. Some form of iron grating, light and strong, suggests itself.

It is rarely that a group of buildings of uniform height is found, particularly in the skyscraper districts of large cities. To be sure, a solution may be found in the invention of an airplane that can rise almost vertically from the ground. But the present article is concerned with the roofs and the machines of today.

One of the most practical suggestions is by Mr. H. T. Hanson who proposes what seems a method of solving the problem which is worth serious consideration. He would build the platform in the form of a circular, high-banked track —a track that would be constructed of light but strong iron gratings, so that sun and air would still find their way to the streets below.

Landing Against the Wind

Imagine yourself winging your way in an airplane from your country home eighty miles to your city office. An hour after you have started from your own grounds, the lower part of Manhattan looms in sight. There are the three great suspension bridges that span the East River.

Somewhat to the east of the financial district downtown you see the banked track on top of the building on which you are to alight. You head for it. Spiraling down, you touch

The idle machine stands on the circular elevator platform, which fits into the track. Thus the aviators elevate and lower their machines from and to the garage below

the bottom dead against the wind—a feat always possible with the circular track, no matter from which quarter the wind may blow. You run on the landing wheels around and around until at last you come to a very easy stop. By manipulating your elevator you prevent yourself from soaring off again.

In the illustration above, it can be readily seen that the part of the roof on which the track is built becomes a lift. It is the circular part that fits into the bottom of the track, thus you run directly from the track upon the lift or vice versa, as the case may be, which simplifies to a great degree the landing and starting problem.

As Easy to Start as to Land

It is just as easy to start from such a circular track as it is to alight upon it. You have but to tilt the elevator in the rear at the precise moment that the wind is dead ahead and to fly off at a tangent.

To carry out this maneuver successfully, a signal is required which automatically shifts on the track to mark very distinctly that portion which happens to be parallel to the wind, but only on that side where the wind is opposite to that of the circling plane.

This, in turn, means that the planes must always circulate clockwise.

The circular landing track could be carried on stilts above several buildings, without cutting off light or air

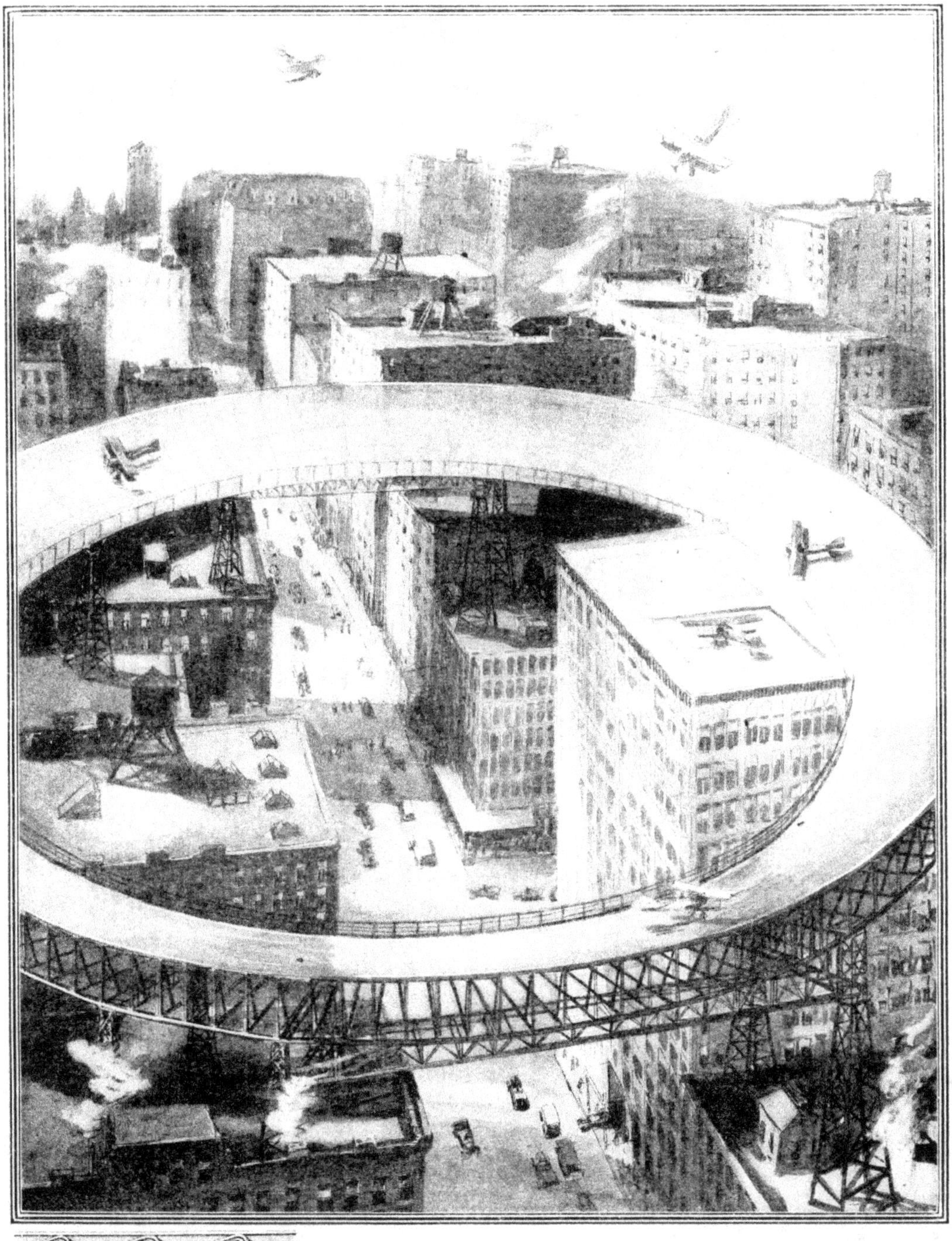

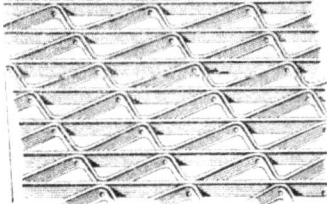

Would This Circular Track Solve the Landing Problem?

The problem of providing a suitable landing platform for flying-machines in our large cities has always puzzled engineers. This is Mr. H. T. Hanson's interesting solution. A banked track of open iron gratings (its construction is shown by the detail drawing at the left, is carried on latticed towers over a group of buildings. On such a track the pilot can start and alight dead against the wind, as he always must. At one point of the track he runs off upon an elevator platform flush with the roof of a convenient building. By means of the elevator he descends to the garage below

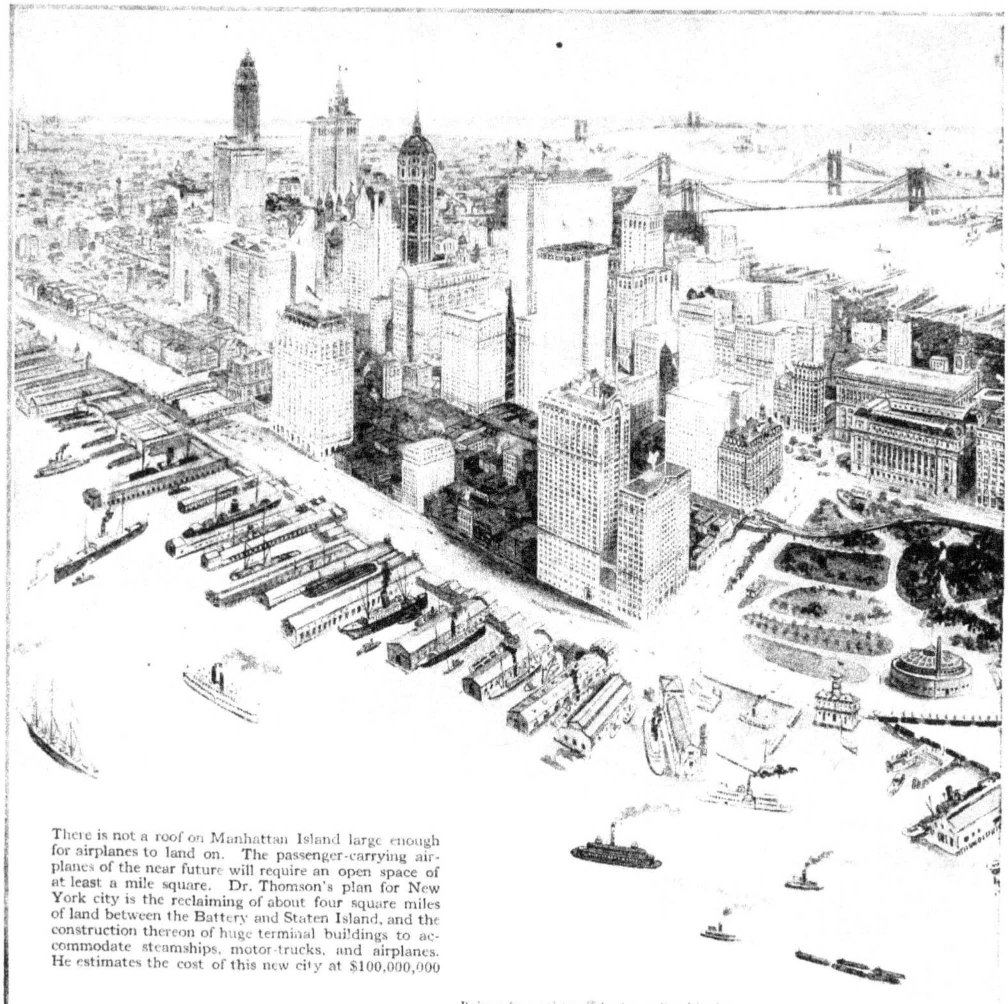

There is not a roof on Manhattan Island large enough for airplanes to land on. The passenger-carrying airplanes of the near future will require an open space of at least a mile square. Dr. Thomson's plan for New York city is the reclaiming of about four square miles of land between the Battery and Staten Island, and the construction thereon of huge terminal buildings to accommodate steamships, motor-trucks, and airplanes. He estimates the cost of this new city at $100,000,000

Redrawn from a picture © by Success Postal Card Co.

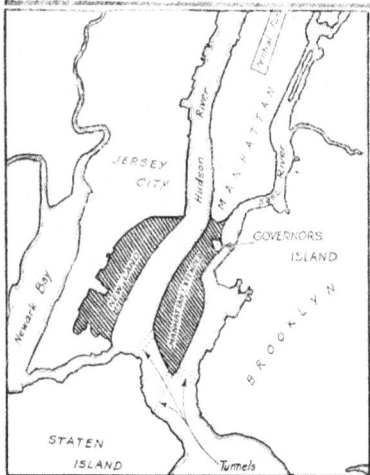

A Plan to Extend New York

By Dr. T. Kennard Thomson,

[*Dr. T. Kennard Thomson, whose daring plan for making New York a great port for airplanes is published here, is an acknowledged authority on pneumatic caissons. He has been retained as consulting engineer in the construction of many New York skyscrapers, and has built pneumatic caissons for important bridges over many of the larger rivers. He was one of the board of fire engineers in charge of the construction of the New York Barge Canal.—*THE EDITOR.]

THE problem of the automobile is roads; that of the airplane, landing-places. And, just as good roads have followed the invention of the automobile and the automobile has multiplied with good roads in a never-ending succession, so airplanes in commercial use will multiply with properly laid out chains of landing-places and the landing-places will multiply with the demand of the airplane.

At present the location of landing-places is controlled by circumstances. Mineola, set in the plains of Long Island, becomes the airplane port of New York. But it is as if ships coming to New York had to dock at Sandy

to Make a Great Airplane Port

Consulting Construction Engineer

Hook and leave their passengers to make their way to the city as best they could. It frequently takes the passenger or the mail package coming by airplane from Washington to New York longer to reach Manhattan after landing on Long Island than it did to fly from the capital to the landing-place.

Wanted, then—a good landing-place in Manhattan. That means a smooth surface about a mile square. For at present the most favored types of airplanes have to run for about a thousand feet along the ground before taking the air, and, having to land at one half their maximum speed, they need space in which to settle down to earth again.

There is, of course, no such space to be had at present on Manhattan Island. Central Park is sacred as the city's one big breathing-space, and anyway it is not suited to the purpose. The answer, then, is—build such a place. My plan is this:

Build a great station for airplanes, steamships, barges, and motor-trucks by extending Manhattan Island at the Battery. It can be done for $100,-000,000, and the land thus reclaimed would be worth many times that amount, for it would create a new city of some four square miles.

My idea for a Manhattan extension is to construct a sea-wall from the Battery toward Staten Island in such

a way that the new Battery will be about the same distance from Staten Island as the present Battery is from Jersey City, and to reclaim the space between these sea-walls—more than four square miles.

Wherever rock is near the surface the space between the sea-walls, or coffer-dams, will be pumped out; but where the rock is more than sixty or seventy feet from the surface, the space will be filled with sand pumped in through a suction-dredge having a capacity of 50,000 cubic yards per day.

Eventually adequate landing-places will be built on top of the buildings of the new city. The buildings will cover entire blocks, and will be narrow enough to make air-wells unnecessary. These blocks can be any length—1,000 feet, one mile, four miles. There will be arcades at the street level, say every hundred feet.

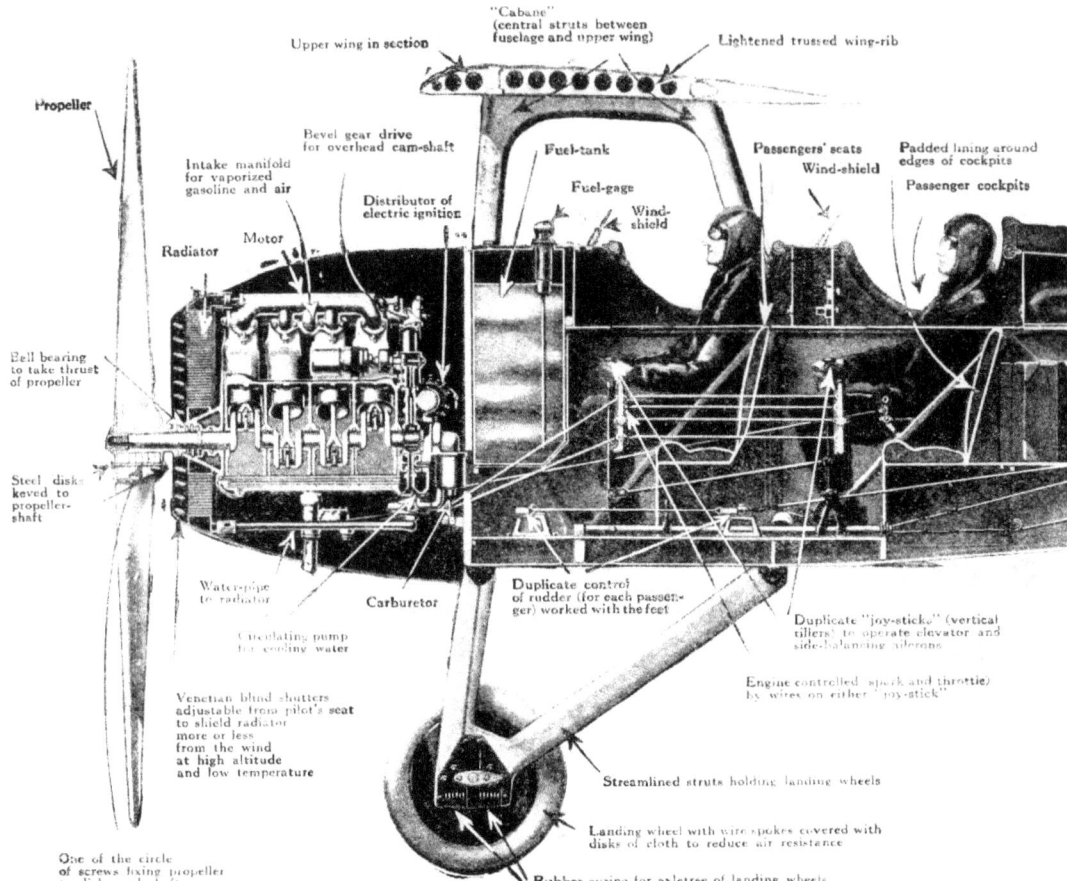

Upper wing in section

"Cabane"
(central struts between
fuselage and upper wing)

Lightened trussed wing-rib

Propeller

Intake manifold
for vaporized
gasoline and air

Bevel gear drive
for overhead cam-shaft

Fuel-tank

Fuel-gage

Passengers' seats
Wind-shield

Padded lining around
edges of cockpits

Passenger cockpits

Distributor of
electric ignition

Wind-
shield

Radiator   Motor

Bell bearing
to take thrust
of propeller

Steel disks
keyed to
propeller-
shaft

Water-pipe
to radiator

Carburetor

Circulating pump
for cooling water

Duplicate control
of rudder (for each passenger) worked with the feet

Duplicate "joy-sticks" (vertical
tillers) to operate elevator and
side-balancing ailerons

Engine controlled (spark and throttle)
by wires on either "joy-stick"

Venetian blind shutters
adjustable from pilot's seat
to shield radiator
more or less
from the wind
at high altitude
and low temperature

Streamlined struts holding landing wheels

Landing wheel with wire spokes covered with
disks of cloth to reduce air resistance

One of the circle
of screws fixing propeller
to disks and shaft

Rubber spring for axletree of landing wheels

How the Airplane Is

The new clippers of the sky, like their ocean-born

REMEMBER the pictures over which you used to pore in the dictionary—wonderful cross-section views of line-of-battle ships, every gundeck showing, and all the secrets of life on the deep revealed? Here is the same scheme applied to the airplane. Even the expert may like to see the things he knows thus visualized.

There appears a striking likeness between the airship of today and a steamship. The same trim, shapely hull, with rudder and propeller, is there, as is the engine-room with its familiar rows of cylinders. But the comparison shows that real mechanical efficiency has been more essential in the air than on the water. The shortness of the propeller-shaft and the location of the engine-room in the extreme front end are its visible evidence.

Its Enormous Propeller

On the other hand, the comparatively enormous size of the airplane's propeller makes it quite as noticeable that the hull (fuselage) of the airplane meets a relatively higher resistance in the air than the hull of the steamer meets in the water.

True, the airplane propeller has also

to drive wings through the air; but the resistance, the so-called "drift" of the wings, although it furnishes the whole support in the air, is considerably less than that of the rest of the machine.

How Air Is Like Water

To understand why it takes so much power to drive the airplane hull, we need only to imagine it driven through water as a submarine is driven when submerged. At once we are struck with the awkwardness of the lower appendage—the frame sticking out for carrying the landing wheels.

We are so accustomed to make little of the effort it takes to drive small things through the air that this appendage did not worry us much, especially considering the attempt made to reduce its resistance by sharpening the struts and "disking" the wheels. But when we think of this appendage as on the bottom of a

motor-boat we feel a strong desire to be rid of it.

Who will be first to find a way of folding the landing gear into the hull during flight, as a gull on the wing tucks up its legs?

The instinctive carelessness about the resistance of things in the air led to those wasteful early designs in which engines, radiators, tanks, men, and what not were scattered over the lower surface and left exposed. Yet the Wright brothers and the Farman brothers were bicycle makers and riders, and might have been expected to remember the effect of a head-wind on pedaling, and the gain in speed in the well known experiment of building a streamline shell around the rider.

An All-Metal Machine

In addition to the landing-gear big open cockpits on top of the loom up as formidable speed the moment we're

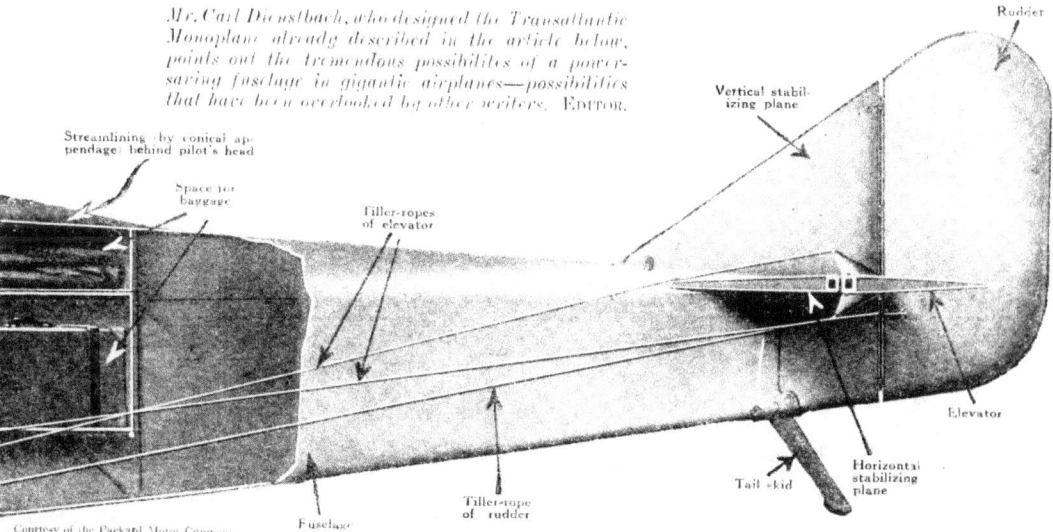

Mr. Carl Dienstbach, who designed the Transatlantic Monoplane already described in the article below, points out the tremendous possibilities of a power-saving fuselage in gigantic airplanes—possibilities that have been overlooked by other writers. EDITOR.

Rudder

Vertical stabilizing plane

Streamlining by conical appendage behind pilot's head

Space for baggage

Tiller-ropes of elevator

Elevator

Horizontal stabilizing plane

Tail skid

Tiller-rope of rudder

Fuselage

Courtesy of the Packard Motor Company

IT is a far cry from the early airplane designs turned out by the Wrights and the Farmans—not so much in point of time as in point of progress—to the trim-bodied flyers of today. More and more the airplane takes on the proportions and appearance of a true ship of the air.

With the multiplication of parts and gear there has grown up a language of the air as unfamiliar to the landsman as is the salty talk of the deep-sea sailor. Yet "fuselage," "cabane," "joy-stick," "ailerons," and words of their brotherhood have taken out citizenship papers with the dictionary-makers. Soon we'll all be saying them.

It is to aid our readers to use them correctly and to read with understanding, that we present this sectional view of a modern airplane with all of its secrets exposed and its parts named and explained.

Increasing Its Speed

sisters, owe their swiftness and power to hull design

submerged and moving through water. From the same point of view, the many struts and wires between biplane wings are seen at their proper negative value, and any design is welcomed which—like, for instance, some of the plans for all-metal machines recently described—does away with these, without lessening the strength and practicability of the airplane.

Merits of Large Airplanes

We get a still more direct grasp of the problem by remembering that if one raises his hand above the windshield of an airplane traveling ninety miles an hour (not at all a fast air speed) it feels exactly as if it had been dipped into the water from a fast motor-boat, so palpable is the resistance. Pondering this fact, we 'll be prepared to correct an over-common to many who have ' the merits or demerits of 'rplanes.

It has been endlessly repeated that, since the weight increases with the cube and the lift only with the square of the wing dimensions, large sizes offer nothing but difficulties. The critics forgot that, within limits, the faster you go the less wing-space you need, and that, while the power-saving of the fuselage increases with the cube, its cross-section or air resistance increases only with the square of its dimensions. In other words, if you make the airplane twice as large you can tuck away everything, except the wings and rudders and propellers, in a fuselage that requires comparatively a trifle more power to drive through the air than that of a machine half as big.

Motors in Separate Fuselages

The designs of the larger machines—the Handley-Page, the big flying-boats, the Caproni, etc.—do not yet take proper advantage of the fact that, to a considerable extent, increase in size is as beneficial to an airplane as to a dirigible or a ship.

This advantage is shown indirectly by the fact that it has been possible to put the motors into separate fuselages on the large machines without as much decrement to efficiency as there would be in a small machine. The separate fuselages have, in their turn, also less resistance, relatively, when built on a large scale. It should be recalled, in this connection, that the discoverer of the vital part played by the fuselage was Nieuport, who deserves to rank with Lilienthal, Maxim, Langley, and the Wrights.

No picture can show all the progress that has been made in airplane designing, because the progress has been along the line of correct proportioning of all surfaces acting on the air, and correct distribution of weights and stresses. What is known about the correct proportioning of an airplane we owe to the modern air-tunnel, in which experiments are made with exact reproductions on a small scale of every existing or conceivable machine. It is exactly this proportioning which makes today's machines stable and airworthy in bad flying weather.

Modern Knights Errant Create
They blazon their devices on airplanes as

An interesting feature of the Great War was the revival by the air fighters of traditions and practices of knight errantry. Just as the knights of old wore distinctive devices, so the chivalry of the air placed distinctive markings on their airplanes

We have to show this interesting beast upside down because the Hun plane he decorates fell that way. The ancient griffin, or gryphon, guarded treasure

One French flyer sought inspiration from the ancients, choosing as his device the symbol of speed and wisdom

Fritz likes boastful heraldry—here we have the Bavarian lion chasing the French cock. How that rooster came back!

The coats-of-arms of which ancient families are proud had their origin in the same spirit that led these fighting airmen to blazon the rose upon their plane

A Lafayette Escadrille flyer, who, while short on heraldry, evidently believed implicitly the magic combination of the mystic lady's name, and the United St

a New Heraldry of the Air
warriors once did on shield and armor

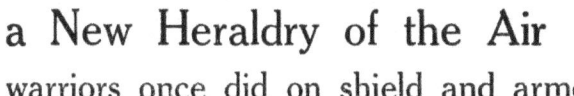

The "sporting" element of individual combat, quite missing from modern war on the ground, still lives for the air fighters, as this German plane indicates

A Belgian version of the Indian head. When knighthood was in flower, this device would have been handed down from father to son, and would have been emblazoned on the flag flying from the family castle. It's hard to be picturesque and democratic at the same time

Like the "one-shield" knights of romance, the Stork Escadrille flyers carried one distinguishing emblem on their plane

Does the dove seem out of place on a fighting plane? Well, the stork is supposed to be in other business, too

Not only appropriate in its suggestion of a "shooting star," but also, in its character as a comet, a legitimate heraldic device

The "Tête Noire" (black head) device and fighting name of the old Norman is recalled by this Indian head on our modern hero Lufbery's machine

The Flying "Circus" and How It Fights

The lesson that the wild geese taught us and how it is applied

By Carlyle F. Straub, Aviator Pilot. *Late of the British Air Force*

IN the early days of the war a combat in the air was much like the jousting of medieval knights—a struggle between two champions. In a sense, the General Staff placed its entire reliance on a few extraordinary flyers. This kind of fighting was peculiarly suited to the British temperament. It brought out the very qualities that have made Englishmen great sportsmen.

While the German airmen cannot be accused of being cowards, they are rarely as good fighting flyers as Englishmen, simply because they have not cultivated sports in the British way. Hence we find that the Germans did more, at least in the early days of the war, to improve the fighting machine than the fighting man. When the German Fokker appeared, it seemed for a time as if the machine were more important than the man; but when the Fokker was outclassed by faster machines developed by the Allies, the individual superiority of the sporting British flyer once more became apparent.

How the Flying Circus Originated

When it became evident that the Allies could build machines in larger quantities than the Germans and that their flyers were individually the more adroit and daring, a new tactical policy had to be discovered. That policy was eventually the adoption of formation flying by the Germans. It proved so brilliantly successful in enabling flight commanders to make the most of a limited number of airplanes, that it has been adopted by all the armies in the field.

It was Bölcke, one of the best flyers that Germany ever had, who seems to

The author of this article is a young American who went to Canada before the United States entered the war, in order to join the Royal Flying Corps. During his course of training he received a good deal of practice in bombing squadrons. Since bombing machines usually fly in formation, he writes on his subject from first-hand knowledge.—The Editor.

They dive straight into the beam of the searchlight. It seems ridiculously mothlike, but there is a good reason for the maneuver. The projectile from an anti-aircraft gun flies in a curved path: the beam of the searchlight is straight. By following the straight line, the men in the machines are safe

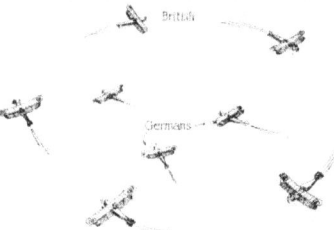

Surrounded! A hail of lead pours in on the Germans. The only possible escape for them is to dive and throw their machines into a spin

Ten thousand feet in the air, above a concealing cloud, flies the British circus, as shown at the left. Beneath the cloud a solitary Englishman soars. He is so much bait. He looks temptingly alone. Two Germans, regarding him as good prey, begin to nibble, as we show at the right. If the bait can take care of himself the circus above the cloud keeps serenely on its course. But if more Germans should appear, down it plunges, and at once there is a "dog fight"—an indiscriminate combat in which all appearance of formation is lost, and in which each flyer takes care of himself

have organized the first fighting squadron to offset the individually stronger English fighters. Formation flying was a brilliant success. What were the chances of one or two Englishmen, however brave and skilful, against a whole squadron? For a few weeks Bölcke swept everything before him. Then the Allies adopted formation flying. And the old days when a single man, possibly two men, soared up looking for a German to fight, were over.

Reason for the V Formation

One of Bölcke's best men was Baron von Richthofen, a former cavalry officer. After Bölcke was killed in a collision with one of his own men, von Richthofen stepped into the limelight. He organized a squadron of his own. It must have been allowed extraordinary privileges. Von Richthofen himself, for example, always flew in a machine painted a vivid red. The other machines of his squadron were painted in riotous colors. It was this fancy-dress aspect of the machines, coupled with the fact that they were ordered from one sector of the German front to another as they were needed, that led some imaginative Englishman to call the squadron "von Richthofen's traveling tango circus." Now "circus" is the accepted term for such an organization.

I do not know by what method air commanders arrived at the conclusion that it was best to fly in approximately "V" formation. At all events, the distinguished English aeronautic critic, F. W. Lanchester, pointed out some years ago that fast migrating geese and ducks adopt the V. As Lanchester puts it, "the air immediately in

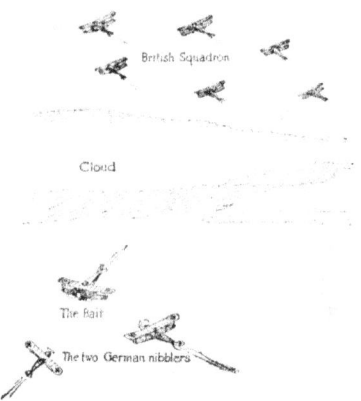

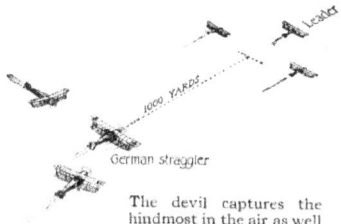

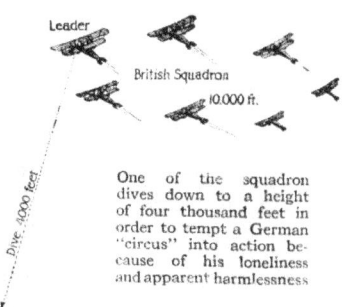

Formation flying requires the closest team-work. It is not so easy as it looks to keep your station in a flying squadron: it requires constant practice. But, when it is once learned, the pilot has a better chance of escaping death than if he went up alone.

Leader

British Squadron

10.000 ft.

Dive 4000 feet

6000 feet

One of the squadron dives down to a height of four thousand feet in order to tempt a German "circus" into action because of his loneliness and apparent harmlessness

The devil captures the hindmost in the air as well as elsewhere. Two British devils detach themselves from the squadron. Then woe to the straggler!

the wake of a bird in flight has residuary downward motion, and so is *bad* air, from the point of view of the bird following. On the other hand, the air to the right and left of the leader has residuary upward motion, . . . and so is *good* air; consequently the V formation arises from each bird seeking the air which gives the best support."

A "Dog Fight" in the Air

It must not be supposed that formation flying begins and ends with keeping station to form a V. The enemy must be outwitted. Three machines climb to a height of 12,000 or 15,000 feet in V formation. Four or five others follow and maintain an altitude of 8,000 or 10,000 feet. At 6,000 feet a single machine—a Nieuport — flies deceptively alone. That lone machine serves the same purpose as the worm on the end of a fishing-line: it is so much bait. Its pilot is a cool, daring, experienced fighter. Suppose that a single German machine swoops

down on him. The rest of his squadron, probably concealed for the most part behind a cloud, leaves him to his own devices for a time, confident that he will beat his assailant. But suppose that he manifestly is outmaneuvered? A machine from the squadron detaches itself, plunges down, straightens out, and sideslips into such a position that the enemy machine is caught between two fires.

The enemy seeks refuge in a nose spin, which makes it almost impossible to hit him. His companions see his predicament, and rush to his assistance—exactly what the Allied squadron far up in the air desires most. The leader of the Allies dives first, followed by his right and left guard. Three machines are left above to watch for more enemy machines. Now a general combat ensues, in which each man must fight independently. All semblance of formation is lost; the mêlée is called a "dog fight."

How the Bombers Fly

On bombing raids, which on the English side are usually undertaken at night by giant Handley-Page machines, formation flying is all-important. The individual pilots must co-operate if the raid is to be successful, which means that they must follow the leading machine captained by the commander of the squadron.

But it must not be supposed that the commander hugs secret plans to his bosom. Every detail of the raid is discussed thoroughly with the navigators of the squadron. Maps of the territory to be traversed and attacked are minutely studied. With the assistance of his subordinates, the commander formulates a plan of action. With the aid of red and green wing lights, the machines keep their station in the V formation. If a stabbing enemy searchlight ferrets out the expedition, the leader plunges straight into the beam, followed by the other machines.

When the squadron is a few miles from its goal the commander gives a signal, and the V formation gives place to a single file, so that each machine may follow the leader over the target.

When the Germans have been lured into the trap so cunningly prepared for them by the British, there is a kind of Kilkenny riot which, in the airman's vernacular, is called a "dog fight"—every man for himself

An airplane propeller is built up of layers of carefully selected wood arranged fanwise and glued together. From five to eight slabs of wood are thus glued together. The density of the wood, the species of tree from which it was cut, and its moisture content must all be taken into consideration by the propeller-maker

The brains of the scientist, and the ingenuity of the in producing the propeller

After the layers of wood have been glued together they are placed in a "glue-press." The glue is applied hot. The propeller, or rather the layers of glued wood out of which it is to be fashioned, is held in the glue-press for a period of twenty-four hours

After the glued layers of wood come from the press they are "roughed out." The corners are usually removed by a bandsaw. Then comes the shaping of the blades. This is still a hand operation in many factories, but during the war machines were introduced, like the one that is shown in the lower picture on the right, to speed up the work

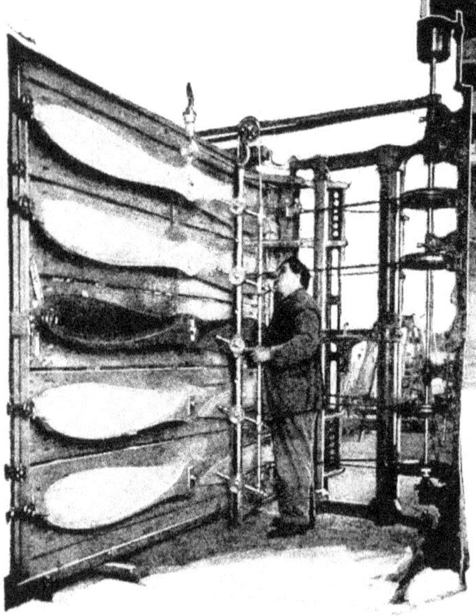

Nearly every propeller factory has a machine like this—a form of pantograph. Such machines have long been used to make duplicates of statues. In the center of the machine here shown is the propeller to be copied. A tracing tool guided by the operator travels over the surface of the propeller. Cutting tools remove from the reproductions excess wood to just the depth indicated by the tracer on the master propeller. The tracer feels, as it were, the surface of the master propeller and communicates the results of its feeling to the other tools

Upon the Skill of Airplane Propeller

the art of the wood-carver, mechanic—all are combined that drives the airplane

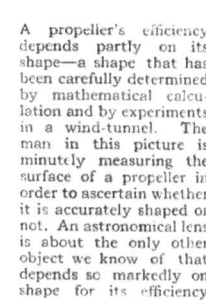

A propeller's efficiency depends partly on its shape—a shape that has been carefully determined by mathematical calculation and by experiments in a wind-tunnel. The man in this picture is minutely measuring the surface of a propeller in order to ascertain whether it is accurately shaped or not. An astronomical lens is about the only other object we know of that depends so markedly on shape for its efficiency

A propeller must be so smooth that a fly would almost slip and break its neck if it crawled over the surface. It was Sir Hiram Maxim (from Maine, if you please) who first realized that an air propeller must be as smooth as glass in order to reduce skin friction. No piano has a finer gloss than that which is given to an air propeller by the most expensive spar varnishes and by very careful hand polishing

And now comes the most delicate operation of all—balancing the propeller. A propeller whirls around at the rate of about fourteen hundred revolutions a minute in the air. If it is unbalanced the strains to which it is subjected will tear it apart. The finished propeller is mounted on an axle that turns on nearly frictionless bearings. The blades must balance in any horizontal and vertical position, which means that the center of gravity must be in the exact center. If the propeller is not balanced in all positions a little wood is rubbed off here and there. Sometimes the touch of a varnish brush is enough to correct the faintest perceptible error

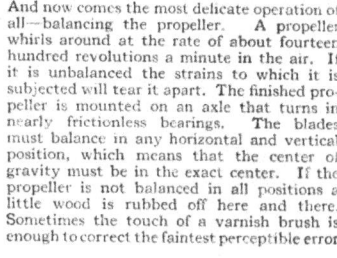

This picture summarizes the making of an air propeller. At the bottom, 1 shows the five layers with which the propeller-maker starts; 2 shows the five layers glued together as they come from the glue-press; in 3 the excess glue has been scraped off; in 4 roughing out has been started, as indicated by the appearance of the hub; in 5 roughing out is finished; 6 shows the propeller after carving. The process here shown is that developed by the Forest Products Laboratory

When an Airplane Perches on the Roof

It's a sign that the day when all of us will fly is not far off

The machine leaped over a huge sign and landed on the top of the building

Jules Vedrines hunts eagles and canvasses election districts in his airplane

PARIS is familiar enough with the roar of airplane engines. But the roar that it heard soon after midday on January 19 last was so much louder than anything that it had ever heard before, so obviously in the midst of the city, that it rushed from its cafés and its houses and looked up at the sky.

Then it understood why this particular roar was different from anything to which it had been accustomed by German air raids.

A biplane was skimming the chimneys. Never had a machine flown so low over the city. Surely, thought Paris, the pilot must be in peril. No sane man would dare to take such chances.

Then a wonderful thing happened. The machine literally leaped over a huge advertising sign, just as a horse would jump over a fence, and landed on the roof of the building behind one of the largest department-stores in Paris.

In a minute the streets around the store were thronged. Out of a side door emerged Vedrines, one of the most daring airmen France has produced, a man given to spectacular performances in the air, a man who has shot eagles from his flying-machine in the Pyrenees, a man who actually canvassed an election district in an airplane when he tried to become a member of the French Chamber of Deputies.

And what was the meaning of all this? Simply to advertise in a characteristically sensational way a proposed trip by air around the world.

Vedrines' achievement, very properly, attracted international attention. For the first time in the history of flying, a man succeeded in landing on a city roof — a space thirty yards by twelve, in this particular case.

To be sure, an American pilot, years ago, landed on a smaller platform built over the deck of one of our warships. But the feat of Vedrines was even more hazardous.

The atmosphere, if you could but see it, is a turbulent, heaving, billowing fluid. It is full of whirlpools, eddies, currents, and counter-currents. It dashes against a row of houses, a forest, a mountain, as the surf dashes up against a seaside cliff. Over Paris the air must be tossed hither and thither like the water in a whirlpool.

These, then, were the aerial conditions that Vedrines had to combat. It required not only courage but extraordinary skill to brave such atmospheric perils. To land as Vedrines did, with practically no injury, speaks volumes for his adroitness—all the more so when it is considered that his was an antiquated Caudron biplane, a pre-war model, driven by an 80-horsepower rotary Rhone engine.

Vedrines, we have said, wanted to attract attention to himself as a possible aerial circumnavigator of the globe. He did more than that, wittingly or unwittingly. He symbolized the pilot of the future.

The day is not far distant when every large city will have its landing platform on the roofs of specially designed buildings. From the suburbs of New York, Paris, and London business men will fly to and from their offices quite as regularly as they are now whisked back and forth by train.

For the first time in the history of aeronautics, a man had succeeded in landing on a city roof

Here's a Bare-Back Rider of the Air

Lieutenant Ballough fears the bucking airplane as little as a broncho-buster does a "bad" horse. But then, airplanes don't really buck any more. Modern flying-machines always right themselves if they are high enough. Only on an inherently stable machine is Lieutenant Ballough's feat possible. Of course, the Lieutenant took good care that he did not throw the center of gravity from its correct position

Although the Lieutenant takes many liberties with his airplane, he has to keep close to the machine's center of gravity. If he were to venture too far from that center, all balance would be destroyed and something would happen quickly

If he steps off it's two thousand feet down; but he has learned to trust the stability of his machine. He can step out on a wing (not very far, to be sure) and walk forward and backward—something that was entirely unthinkable ten years ago

The Flying Mail's Big Debt to War

America again takes front rank in the air with machines of mighty power

By Carl Dienstbach

Built for war, this new American airplane will be used in the mail service. It represents the wonderful progress made by the science of flying in the past four years

THE accompanying pictures illustrate the representative result of what four years of war-stimulated progress have done for the flying-machine. It is the up-to-date representative of that medium-sized biplane with which the Wright Brothers inaugurated flight because it was inherently the most efficient type. It still has room for only two flyers; but that it has lift enough for twice that number becomes apparent from its formidable battery of eight machine-guns.

The essential mechanical advance is seen at a glance: the torpedo shape of the whole, and a display of power that suggests a locomotive. Fortunately, this machine, which is the equal if not the superior of the best in the world, is again as purely an American product as was the original Wright flyer. But the Wrights flew with 30 horse-power, while the biplane of to-day flies with 435!

The Liberty Motor

Yet this is one reason why Americans are again in the front rank of flying. It is in the big things that we excel. The watchlike finish required in smaller perfected motors of the highest grade was long unattainable under American conditions; and finally, at great expense, the French Hispano-Suiza motor was transplanted to this country.

On the other hand, the big Liberty motor aimed at once at many hundreds of horsepower, and American ingenuity in designing won over labor conditions. The war's necessities for extreme speed and extreme lift, as required in the poorly supporting thin air of extreme altitudes and for very rapid climbing in emergencies, were responsible for the astounding increase of power from 30 to 435.

At first this would seem the height of extravagance in a machine devoted to any but military ends. But humanity is to be congratulated in that war development was, after all, identical with peace development. Only great reserve power (as required for war climbing) makes flying safe under adverse conditions; and only by extreme speed can flying compete with other modes of locomotion. Only in the air is such speed useful, because only there does it find a clear road.

Now that the war is over, the airplane shown above will transport mail in place of its heavy battery, and if it had been designed for that purpose it could not be better fitted to its present purpose. In mail-carrying speed is

The original Wright machine flew with 30 horsepower, while this Liberty motor roars out with 435. The engine, shown here at close range, was built for war, but finds use now in the air mail service

everything. The only concession to peace requirements should be the addition of a carbureting device to permit low-priced fuel of high heat value, such as crude oil, to be consumed. It seems by no means impossible, especially as the engine might be started and warmed up with gasoline.

The extreme torpedo shape of the airplane is the natural outcome of extreme power and speed.

Such extreme speed could formerly have been attained only if (by reducing wing surface and camber) the landing speed were likewise raised to a dangerous degree. But now it is different. Together with the greatest power, machines are also given plenty of wing surface and an ample camber of the wings, making the wings very lifting. As this would make them climb rapidly at top speed (while insuring sufficient support at a low, safe landing speed), they are simply driven "downhill"—thus transforming climbing into horizontal motion—to fly horizontally at top speed. Driving "downhill" would seem like pulling down with the propeller. But the modern tractor propeller in front sends its slip-stream against the planes, and that slip-stream acquires a lifting component at the same rate that the propeller pulls downward.

A Seeming Contradiction

In the machine shown the radiator in the slip-stream directly under the upper wing seems a surprising contradiction to the "torpedo shape" of the whole. It especially seems to ruin the lift or that portion of the lower surface of the upper plane which is directly behind it, and of the corresponding upper surface of the lower plane. But the lift of the lower surface is at least less than one-fourth of that of the upper one. And under the radiator there *is* no upper surface of a bottom plane but a fuselage. Also a radiator could not be placed in a better position for maximum cooling than into the strongest part of the slip-stream. Hence it is ridiculously small for 435 horsepower, and saves further resistance by being unusually deep. The tremendous fan effect makes its long channels cooling enough.

After going through the test in the "altitude chamber" the airplane engine will be at home in any altitude. Its power is measured by an electromagnetic dynamometer outside the chamber

High Altitudes Brought to Sea-Level

How airplane engines are tested

IF you wish to ascertain whether a substance will melt at a temperature of 400° Fahrenheit, what do you do? Simple enough: you make a practical test by heating the substance to a temperature of 400 degrees.

It is not always easy to make practical tests under the correct conditions. Take engines, for instance. It is a simple enough matter to try out a stationary engine. Its environment remains practically unchanged. There may be some variations in temperature, air-pressure, humidity, and other atmospheric conditions; but these changes are, as a rule, neither abrupt nor formidable enough to be of material influence upon the efficiency of the engine.

Tests for Airplane Engines Different

It is different, however, with the engine of an airplane. Under ordinary working conditions, the engine of an airplane rising to high altitudes may pass from one extreme of air-pressure and temperature to the other in a few minutes. If we wish to ascertain what effect on the efficiency of the engine such varying conditions may have, we must try out the engine under conditions closely resembling those under which it would be expected to do

its work. It is obviously desirable to make those practical tests before the engine is permanently installed in an airplane.

For the purpose of testing airplane engines under conditions closely resembling those of actual flight, the Bureau of Standards at Washington has established a testing plant commonly known as the "Altitude Laboratory."

In the Altitude Chamber

The testing chamber, in which the engine is mounted in a manner reproducing its mounting in an airplane, is a room fifteen feet long, six feet wide, and six and one half feet high, with concrete walls a foot thick. These walls are insulated on the inside with cork.

The dynamometer for measuring the power developed by the engine is outside of the chamber, and so are the scales for weighing and continuously registering the amount of fuel used, the water-meter for the cooling water, the gas-meter for the air intake, the gage for showing the exact degree of air-rarefication within the chamber, and the dials of the electrical thermometers registering the temperatures at different points on the inside of the chamber.

Pipes and wires connect the mechanism in the chamber with the outside world

The chamber is chilled by an ordinary ammonia refrigerating plant, and, to produce a wind effect, the icy air inside is violently agitated by electric fans. Two electric motors of forty and fifty horsepower drive the refrigerating and rarefying plants. To assist in the maintenance of a low air-pressure in the chamber, cold water is injected into the exhaust.

The dynamometer, which measures the power of the engine with wonderful exactness, consists of a generator driven by the crank-shaft of the engine. The pull exerted by the revolving armature of the generator against its movable magnetic field is transmitted by levers to a scale which indicates the force of the pull in pounds and fractions on a dial. The current from the electric generator is absorbed in resistances or fed to the regular line wires from a switchboard, and may be used to drive other machinery.

Speech Is Difficult in an Airplane

"LET go that joy-stick!" says the instructor in the observer's seat to the student pilot in front. The student promptly lets go, thereby saving himself, the instructor, and the airplane from a crashing fall. Had the instructor shouted through the air his student would not have heard him, even though they were only a few feet apart. The roar of the engine, the rush of the wind, both would have drowned his shouts.

He used some sort of telephone, you say. No, a telephone wouldn't do. What he did use was an airphone, so called because it is an air-phonetic device—not an electrical one. It was invented by Benjamin F. Miessner, of New York. A speaking-tube is used instead of a telephone

The listening airman wears a cap that boasts of pneumatic ear cushions, these being connected with the speaking-tube

In spite of ear-splitting airplane noises, the pilot hears plainly what the observer says through this speaking-tube

wire, and this eliminates the electrical transmission of external noises from the mouthpiece to the ears of the listener. This tube was invented by Mr. Miessner after he had made a careful study of airplane noise conditions—wave lengths, resonance, vibrations, and such things.

For instance, the motor exhaust sounds that occur with fairly uniform frequency are the noisiest ones of the airplane. Mr. Miessner figured out their wave lengths, and varied the length of the tube accordingly, so that there would be no amplification of sound in the tube because of resonance. The material for the tube was picked for its lightness, durability, flexibility, and its ability for preventing the escape of voice sounds and the entrance of airplane sounds. Its dimensions allow for maximum clearness.

The speaker talks into a simple mouthpiece which is not hampered by diaphragms and forced vibrations, as is the telephone transmitter. The sound travels smoothly along to the ear-pieces of the listener. These ear-pieces are really pneumatic cushions in the aviator's cap which surround the ear openings and are in air-tight contact with the head. This bars outside noises.

Of course absolute silence does not reign in the tube when the speaker isn't speaking, but at least the airplane noises are shut out sufficiently to make the voice perfectly clear.

Real Ships of the Air

THE first airplanes were "all wings." Pilot, passengers, engines, tanks, radiators, and what-not were simply dumped over the lower plane. The writer recalls vividly a conversation among the members of the Aerial Experiment Association at Hammondsport in August, 1908, about the designs for the *Silver Dart*, the forerunner of the first practical Curtiss machine. A body to house the engine, etc., was suggested, but dismissed as useless and weight-wasting.

Today the body—the fuselage—is the most essential part of the whole design, since the genius of Nieuport has shown that the time-honored belief that an airplane's wing surface increases only as the square and its weight as the cube of the linear dimensions holds true only for the antiquated "all-wing" type. Since the fuselage came into its own, it has become recognized that the fuselage obtains the same advantages from an increase in size as does the hull of a ship. The larger it is, the more space relatively for housing all sorts of things, and at the same time the less the head resistance.

Now we know, at last, that the lift of a fast airplane depends far less on the relative size and weight of the wing surface than on the total head resistance, and that its head resistance, just as in a ship, becomes relatively smaller as the craft grows larger.

Designed to carry bombs to drop on Berlin, this airplane is easily transformed into a commercial airplane. The multiple motors of giant airplanes (*above*) are placed in little bodies of their own

Cranking the Airplane

It is an awkward moment that science has heretofore neglected

An automobile motor is connected by a chain drive with a shaft that terminates in the propeller hub when the motor is started the propeller turns over

"HOW like a bird!" you murmur soulfully as you watch an airplane glide swiftly through the air. But if you saw the ugly, mechanical way in which its motor was started back there on the field, you'd change your tune to "How like a flivver!" For the airplane's motor is cranked; and this cranking is no easy flivverish job either, since the dangerous sharp-edged propeller is right on the spot all the time.

How is it cranked? One clumsy method is shown in the picture at the top of the page. An automobile motor is released from driving the wheels and is connected by a chain drive with a shaft that terminates in the propeller hub. When the motor of the automobile turns over so does the propeller.

Even this clumsy method is a distinct improvement over the earlier one in which the propeller was turned by hand. That was so "mechanically indecent"!

The pilot and passenger were, in a way, as passive as children in a baby carriage; it was the "gang" that "shoved them into space"—the crowd of mechanics that first took hold of the tail and both wings, and especially that daring matador who started the motor by tackling the deadly propeller. He caught hold of the sharp blade, gave it a twist, and jumped back in time to save himself from being chopped up.

A propeller is shaped the very reverse of a crank, having a sharp edge exactly where there should be a handle. The moment the motor starts, it becomes necessarily as dangerous as a striking serpent.

"Why not use a self-starter?" you ask. Its weight while the airplane is in flight greatly hampers the airplane's power. However, a detachable starter invented by M. Odier, a Frenchman, is now being used on many French airplanes. It is mounted on a bipod, and can be worked by one man.

A tube filled with liquefied carbonic acid is attached by a metal pipe to a long steel cylinder containing a piston. These are located on the longer leg of the bipod. A pulley is fastened to the piston, and over it stretches a cable that comes from the cylinder. The cable travels on, and is wound four times around a grooved drum, being fastened then to an elastic cord. The drum is mounted on a shaft together with a bell-shaped projection that is made to fit into the hub of the airplane propeller. Four bolts on this bell-shaped projection slide into grooves in the hub.

To start the engine, a man presses a lever that releases the carbonic acid. The piston then shoots out, and with it goes the pulley. The cable gives the drum a twist sufficient to turn the engine over.

Now that the engine is started, how is the starter detached? You remember the bolts on the bell-shaped projection and the grooves into which they fit. These grooves are cut at such an angle that the bolts are forced out when the propeller whirls. Then all the attendant has to do is to carry the starter away.

An automobile is used to start the motors of this monster airplane

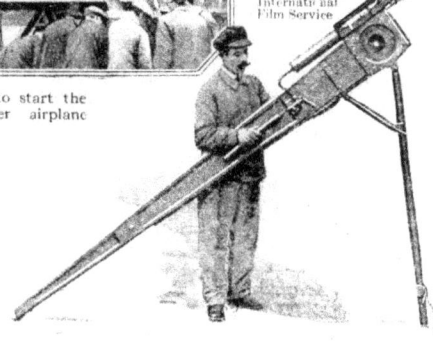

A Frenchman invented the detachable airplane starter at the right. It is operated by the release of carbonic acid against a piston, which whirls a drum connected with it by a cable. The drum is mounted on a shaft, together with a bell-shaped projection that fits into the hub of the propeller, as shown on the left. When the drum whirls around, so does the propeller

A shadow phenomenon observed by aviators and invested by them with various superstitious meanings

Specters that Haunt the Clouds

THE accompanying picture represents an optical phenomenon that has been familiar from time immemorial to the birds of the air, but that has, until recent times, lain somewhat beyond the purview of humanity. Even at the present day, though it is well known to aeronauts, few of them are able to explain it or to give it a name. The picture shows the shadow of an airplane cast upon the clouds, and surrounding it is a ring of rainbow-colored light. The appearance of this light has been likened to that of the emblem painted on the wings of the Allied aircraft, and superstitious aviators regard it as an omen favorable to their cause.

Like the "Ring Around the Moon"

This phenomenon bears, however, a German name, for it is known as the "specter of the Brocken." True, the traditional *Brockengespenst*, which figures in German legend and in the narratives of mountaineers, is the shadow of the observer seen on a *vertical* bank of cloud or fog, and hence visible only when the sun is low; but it is identical in character with the shadows that balloonists and aviators see below them on a horizontal sheet of cloud, with the sun overhead.

In both of these cases the shadow may be surrounded with one or more rings of prismatic color,—the "Brocken-bow,"—though this is not always visible. This bow is due to light that is reflected back to the observer after penetrating the cloud a little way, and is broken up into its component colors by the water drops or ice particles. The process involved is called diffraction, and it is the same that produces the reddish corona, or circle of light, seen around the sun or moon when passing through fleecy clouds.

The weird shadows of themselves seen by mountaineers are often described as of gigantic size; but this is an illusion caused by the impression that the shadow is distant, when it is really only a few yards away. It would be interesting to know whether aviators cherish the same illusion concerning the shadows of their aircraft.

Our Picture Is Faulty

Another mistake about Brocken specters is perpetuated by the various drawings that have been made of them, including the one here reproduced. The mountaineer can see his own shadow, but he cannot see that cast by another person at a distance of more than a few feet from him. Similarly, the artist who made the picture above has drawn the shadow and ring as they appeared to the occupant of the airplane shown in the picture; but, from the assumed angle of vision of the artist himself, the phenomenon would have been invisible.

AUTOMOBILES
$633,000,000 for Highways

THE United States government is planning this year to spend $633,000,000 in building permanent highways. Owing to the vastly increasing use of the motor-truck and automobile, the road problem has become of great importance. Heavy loads at rest on the road surface exert but little effect, except where the surface is too soft to bear the load. A wheel load of 8,500 pounds at rest on an eight-inch concrete slab on a rather wet clay subgrade exerts a fiber stress in tension of only 34 pounds to the square inch directly under the load, as compared with the modulus of rupture for the average concrete road surface, which ranges from 400 to 600 pounds a square inch. Hence the only danger of serious injury to the concrete from loads at rest occurs at the corners of the slabs.

Impact is a frequent cause of road failure, and, with a view to ascertaining everything possible concerning the impacts exerted by various kinds and sizes of motor-trucks operated under different conditions, the United States Bureau of Public Roads is conducting some very interesting experiments.

The scheme in use is to deliver the impact of a moving truck to a small copper cylinder, the blow deforming the cylinder to a definite amount, depending on its intensity. A concrete pit is constructed in the road surface in the track of the motor wheels, and a hydraulic jack is placed in the pit. The plunger of the jack is enlarged at the top with a platform of suitable size for receiving the blow of one wheel of the truck. The copper cylinder that measures the blow is set directly under the plunger of the jack, the impact being transmitted through the plunger to the copper cylinder. The copper cylinders are each turned from half-inch copper rod, and are half an inch in length, being heated and standardized so as to be absolutely uniform in

This concrete pit, with a hydraulic jack and copper cylinder, measures the impact of the blow from the truck wheel passing over it

The copper cylinder that measures the blow is set directly under the plunger of the jack, the impact being transmitted through the plunger to the copper cylinder

physical characteristics before being used in impact-testing work. The impact deforming the copper cylinder is then compared with the impact from a resting load necessary to deform it to the same amount.

Impact is measured in different ways, the truck in motion being caused to fall through different heights before striking the plunger of the jack. The truck is also made to strike obstructions of different heights

placed directly on the plunger of the jack. Different sizes and makes of trucks, operated at different speeds and using solid rubber or pneumatic tires, have been tested out to ascertain the impacts they deliver.

A standard three-to five-ton army truck loaded with five tons of sand, having a total weight of 7,750 pounds on one rear wheel and an unsprung weight (that is, not absorbed or broken by springs) of 1,837 pounds on one rear wheel, operated at a speed of fifteen miles an hour and falling through a distance of three inches, exerted a maximum impact pressure of 42,000 pounds, which was 5.4 times as great as the pressure exerted by a rear wheel when the truck was at rest. When the truck, running at a similar speed, dropped from a height of only half an inch, the impact pressure produced was 28,000 pounds, or 3.6 times the static load pressure.

This indicates that the higher the fall, the greater is the amount of impact pressure, and also shows that trucks with great unsprung weight produce high impact pressures. Similar tests, in which the rear wheel of the truck was forced to strike an obstruction, showed that the impact pressure increased with the velocity and with the height of the obstruction.

Wedge-shaped blocks were also placed on top of the plunger of the jack and the truck driven over them, with the consequence that the impact pressure always increased with the angle of inclination of the block and the speed of the truck. These conditions are duplicated in actual road travel when the truck rolls into a depression of the road.

All of these results will be considered in the future in designing and constructing federal highways, which, as far as possible, will be built to withstand the impact pressures of the heavy traffic that passes over them.

How to Get More Out of Your Car

Automobile Stove Heated by Exhaust

COOKING on the road is made possible by an automobile stove placed on the running-board and heated by the exhaust from the engine. The stove is sheet steel, with the heating compartment all around it. The back pressure from the muffler serves to force the exhaust through the stove. A valve in the pipe between the exhaust pipe and the stove turns the heat off and on at will. Dinner may be cooked while the car is bowling along at thirty miles an hour.

Turning Your Ford into a Wagon

ONE of the most ingenious dual-purpose bodies yet designed to convert a Ford roadster into a delivery vehicle, consists of a metal telescoping load-pan inserted under the usual Ford rear apron. The load-pan remains permanently in place, and is unrecognizable when closed. It can carry 1800 pounds.

Doing Away with the Fly-Wheel

THAT an automobile engine with four or more cylinders will run smoothly without a fly-wheel was demonstrated recently by Dr. Robert T. Williams of Detroit. The inventor, instead of using a fly-wheel, distributes its weight equally among the connecting rods of the cylinders. The balancing weight is placed in the connecting-rod collar fitted on the crank-shaft.

How Would You Like the Life of a Gypsy?

TO most of us the freedom of gypsy life appeals more or less. To be free from all care and conventionality, sleep when you wish, eat when you like, and spend your time as the spirit moves you—what could be more desirable? And if you are rich enough to afford an automobile and one of those new-fangled trailers that unfold into a large and comfortable tent with all the comforts of home, you can eliminate all the drawbacks of the ordinary gypsy life.

The trailer runs on two wheels, is eight feet long, four feet wide, and weighs about 750 pounds. It carries a big waterproof tent, two folding beds with mattresses and blankets, an ice-chest, a folding table, and a folding camp-stove with all necessary cooking utensils. The tent is permanently attached to the trailer, and unfolds like the top of an automobile. One person can set it up in about ten minutes.

Run It on Coal-Gas

IF you want to cut down your gasoline bills, give up gasoline and use coal-gas for fuel. The cost of petrol has risen so high in England that some substitute had to be made if people were to continue to run motor vehicles. That substitute seems to be coal-gas. It costs about one third as much as petrol.

The greatest difficulty in using coal-gas for motorcycles is the question of storage.

Here is a four-ton truck hauling a fifteen-ton locomotive

Why This Steam Locomotive is Riding on a Motor-Truck

THE illustration above is a relic of the famous Priority Order No. 2 of the War Industries Board, which prohibited the use of freight-cars for hauling road-building materials such as stone, and cement, and machinery.

When the Road Commissioners of Wayne County, Michigan, found that they could not get materials to several important roads that were under construction, they turned to the motor-truck for assistance. With the help of a special trailer, a four-ton truck carried the fifteen-ton steam locomotive from job to job, even beating the time by which it had previously been shipped by rail.

The engine was delivered direct to the job with but two handlings, loading and unloading. By rail it required four. First it had to be loaded on a wagon, hauled to the railroad station, and then unloaded. At the end of the run it had to be unloaded on to another wagon from the car and handled the fourth time from the wagon to the ground at the site of the job.

How to Make a Radiator Cool Itself

IN a temperature controlling device for automobile radiators, by Harris S. Coy, of Indiana, the front of the radiator housing is provided with a series of shutters having at one end forked radial arms and springs engaging in a vertically movable bar.

A casing in the upper tank contains a fluid that is quickly expanded by heat. When this happens, the pressure acts upon a piston which causes the rod to open the shutters and admit air to the radiator. When the water becomes cooled, the contraction of the fluid draws the piston upward, closing the shutters.

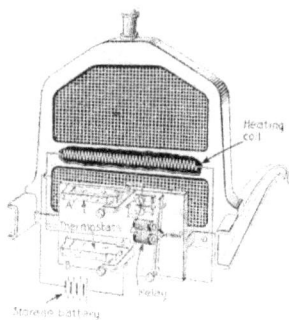

When the radiator gets too hot, the shutters in front of the housing open automatically and cool it off

Tractor motors are tested at different angles to prepare them for a strenuous future existence

Testing Motors for Tractors

OWING to the varying positions in which a motor for tractor use is operated, the motors shown here are tested on a tilting stand, and must start, control, and operate just as satisfactorily on a longitudinal inclination of 45° as they do on a level plane. During the test on such a stand, this motor showed a fuel economy of .54 pounds per horsepower hour.

It is a four-cylinder engine, bore 6½ inches by stroke 7¾ inches, and is practically an eight-cylinder patrol model motor cut in half, with the oiling system and various other details adapted to tractor work. The motor develops from 120 horsepower at 900 revolutions a minute to 160 horsepower at 1200 revolutions a minute. The total weight is 1,580 pounds.

To Moisten Air for the Engine

FOR some reason, moist air gives slightly greater power to automobile engines. A device for moistening the air consists of a small tank mounted under the hood. The water is kept at a level just below the top forming a vapor space through which the moist air passes into the engine.

The air enters through two holes in the top, and follows the course shown in the diagram below.

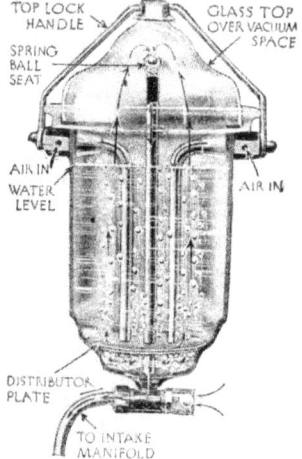

Does your automobile run better in damp weather? Feed the engine moist air with this device

How to Keep Your Truck Radiator from Freezing

ANTI-FREEZING compounds of various combinations have been tried with partial success, but it remained for W. Bonson, of Altrincham, England, to patent the simple device shown below. It is made by fitting a tube of about 1-inch or 1½-inch internal diameter crosswise inside of the bottom tank of the radiator.

The tube is open to the air at both ends, and is made water-tight by brazing into the ends of the tank. One end of the pipe is threaded and covered with a cap having small holes to admit air to the inside of the tube from that end. At the center of the tube at the cap end is screwed a second smaller pipe which extends along the inside of the first tube for almost its entire length. The inside tube is attached to the acetylene tank used for lighting or to the fuel tank.

Whenever the truck has to be parked when the temperature is below the freezing-point, the fuel or acetylene valve is simply opened and the inside tube filled with gasoline or acetylene gas, which is lighted with a match. The air required for combustion comes from the ends of the open tube, and the heat given off keeps the water hot. While the device was originally planned for war trucks, it can be applied with equal success on commercial motors, where a frozen radiator is apt to mean cracked cylinders.

Detachable side windows have turned this old touring car into a comfortable limousine

A tube in the radiator is filled with gasoline and lighted to keep the water from freezing

Detachable Windows for Old Cars— A Cold-Weather Comfort

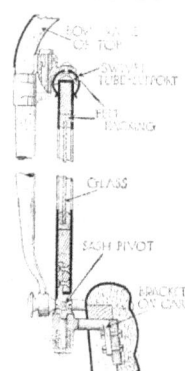

OWING to the curtailment of automobile manufacture, your car may be worth as much as you paid for it and it may pay you to provide it with modern conveniences. E. A. Armstrong, of Highland Park, Ill., has patented detachable side windows, obviously preferable to the ordinary side curtains.

Each window is inclosed in a metal sash. The sashes in the way of the doors are fastened to the doors, while the others are fixed in place, although free to weave or move as the top moves because of the car's vibration.

First Loaded Truck Trailer Over the Alleghenies

THE first loaded motor-truck trailer ever hauled over the Allegheny Mountains early in December completed a round trip between Akron, Ohio, and Boston, Mass., a distance one way of 740 miles.

Motor-truck haulage over the roads between Akron and Boston began early in 1917, and has continued ever since. Several attempts had been made to employ trailers in the work, so that the tonnage carried per trip could be increased to five or eight tons to reduce the cost per ton hauled.

However, these had to be given up, because the trailer could not be held in place when descending the long steep grades.

The trip was finally made possible by equipping both the tractor and the trailer with a new type of pneumatic air-brake which permitted the outfit at times to coast down some of the steep grades at a speed of over thirty-five miles an hour, with the trailer in complete control.

The brake equipment is the invention of A. L. Parker, of Cleveland, Ohio. It consists of a small four-cylinder air-pump driven from the engine; a reservoir tank on the tractor; another reservoir tank on the trailer; a distributor valve; an unloader valve corresponding to a safety valve on a boiler; and the necessary piping and flexible connections between the tractor and the trailer.

This flexible connection is carried along the draw-bar, so that it does not interfere with the operation of the outfit when rounding corners.

Pneumatic air-brakes on both tractor and trailer insure safety on steep grades, making it possible to haul freight over mountain roads

The achieving of a homemade tractor is possible if one happens to own a stationary engine and if there is a junk-yard in the neighborhood

You Can Make a Tractor from a Stationary Farm Engine

THE owner of an eighty-acre farm in Lehigh county, Pa., recently constructed a farm tractor, starting with a fifteen-year-old one-cylinder stationary gasoline engine.

The truck and other parts of the tractor were taken from the scrap-pile. Adding part to part, this mechanical horse soon began to assume definite proportions, and the finished product has all the controls, speeds, and devices of a modern tractor.

Mr. Geissinger, its maker, has been very successful in using it for threshing purposes for himself and farmers in his section. He is able to do plowing with three plows attached to the rear, and it climbs the grades without a balk.

The actual cost of this tractor was $265.

After the car has been jacked up, the rollers on this jack permit the car and jack to be pushed around to any desired position

The Automobile Jack on Wheels

SO that it can be slid in under the front or rear axle of a car, the latest thing in automobile jacks is mounted on wheels. Wheels are all right for moving the jack into place, but what about holding it in position when the strain to lift the axle is put upon it? Will it slide?

These two very important questions have been answered by the manufacturer; for the wheels are mounted on springs, so that they move up inside of the base of the jack, which then has a firm, non-sliding foundation.

The new jack is further characterized by the fact that it is entirely encased, so that the grease inside the standard cannot soil hands or clothes. A long extension handle, which can be folded to fit into the tool-box, also makes the jack easy to operate.

A One-Man Street-Cleaning Truck

OPERATED by only one man, the improved form of motor street-sweeper shown below sprinkles, sweeps, and collects all manner of street dirt in one operation. The machine is not so wide as to restrict traffic, even on busy streets during the day. Furthermore, it is mounted on solid rubber tires in the rear and pneumatics in front, and is not so noisy as to prevent its use in the residential parts of a city during the night.

The sprinkling equipment consists of a water-supply tank with a capacity of one hundred and fifty gallons and a set of spray nozzles set crosswise under the truck-frame just in back of the front wheels. The water is fed by gravity from the tank at the rear to a bronze gear pump operated from the propelling engine of the apparatus. This pump is under the control of the driver.

After the dirt has been softened by the water, it is left undisturbed until it comes into contact with the broom.

After the dirt has been picked up by the broom, it is conveyed by an entirely new means to the hopper or dirt-container located above the vehicle frame directly in back of the driver's seat.

This new method makes use of an inclined flat steel plate around which are operated a series of rubber squeegees mounted on chains and placed longitudinally of the vehicle center-line. At the top of the steel plate, the dirt drops off into the hopper, while the squeegees snap themselves free from the dirt, pass over the upper chain sprockets, and return for another circuit.

Formerly street-cleaning was a complicated job for three pieces of horse-drawn apparatus. This machine, operated by one man, does it in one third the time

Try These Out on Your Automobile

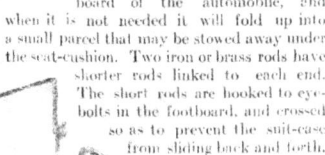

Adjusting Blades of Fan while Running

THE engine of an automobile or an airplane should be kept at a uniform temperature. In hot weather the engine must be cooled, in cold weather warmed—all practically by the same means, a water-jacket and a rotary fan. W. G. Weatherly, of Peola, Washington, has invented a device for changing the pitch of the blades of a fan while it is running.

The fan is surrounded by a frame or rim. The outer ends of the blades are pivoted in bearings in that rim. The inner ends of the blades are fastened to cranks, the handles of which rest between two disks which can be operated by a shaft controlled from the driver's seat.

A Pneumatic Valve-Grinder

A PNEUMATIC valve-grinder has recently been put in the market. The mechanism is contained in a cylindrical casing of die-cast aluminum about four inches in diameter and two and one half inches high. At one end is a swelled handle, to the free end of which an air tube is attached. The air, under pressure of from fifteen to one hundred and fifty pounds, passes through the hollow handle to the valve-chamber, and presses against the packed pistons attached to piston rods.

A Carry-All for Tourists

A CARRY-ALL has been invented by Charles C. Efi, of West Hoboken, N. J. It is a boxlike cabinet with one removable shelf, and is attached to the rear end of the automobile. The cabinet rests upon a bracket, and is large enough to hold blankets, clothing, provisions, etc. The doors of the cabinet are so arranged that they can be turned down to form a table.

A Collapsible Suit-Case Holder

DID you ever ride in an automobile with a suit-case at your feet on the floor of the tonneau? Do you remember how you tried every possible way of wrapping your legs around the suit-case without striking a single position that you could maintain in comfort for more than a minute at a time?

All you needed for your comfort and peace of mind was one of the collapsible suit-case holders shown in the accompanying picture. It will hold a suit-case, lunch-basket, or luggage-carrier on the running-board of the automobile, and when it is not needed it will fold up into a small parcel that may be stowed away under the seat-cushion. Two iron or brass rods have shorter rods linked to each end. The short rods are hooked to eye-bolts in the footboard, and crossed so as to prevent the suit-case from sliding back and forth.

Make Your Automobile Engine Do It

A REEL that can be attached to an automobile has been introduced to facilitate the rolling up or paying out of wire. For take-up use, the reel is attached to the rear wheel of an automobile that has been jacked up. This operation requires no tools, and can be performed quickly.

When the Tractor Strikes a Rock

WHEN a plow runs against any obstacle a horse will stop, but a tractor continues to pull. To prevent the breaking of the plow points in such a case an automatic clevis hitch has been invented, which under great strain disconnects the plow from the tractor.

The device consists of a yoke-shaped casting with two flat parallel arms placed horizontally, with the open end toward the tractor. A bolt through the tractor whiffletree and the open ends of the yoke hold the two parts together. Powerful coiled springs rest against the flanges of the yoke on the outer side of each arm. Through holes in the flanges of the yoke run the arms of a U-shaped clevis hook, which pass through the spring coils and have plates riveted to their ends, which rest against the front ends of the springs.

The clevis of the plow is attached directly to the U-shaped clevis resting upon the springs, but indirectly by means of a clevis arm pivoted to the closed part of the U-shaped hook. When the tractor pulls the plow, the resistance of the plow will cause a compression of the springs. If the pull reaches a certain maximum limit the end of the clevis arm slips out and releases plow.

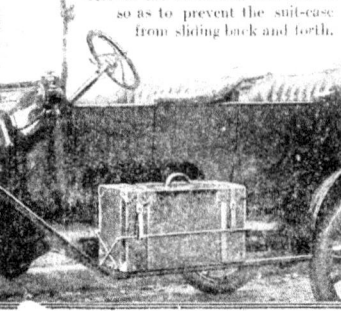

Carrying Milk in Tanks instead of Cans

Perhaps this will put an end to the constantly rising price of milk

By Joseph Brinker

Why carry milk in cans, when it can be carried with less trouble and expense in tanks like these? The tanks are lined with glass, so that the milk never comes in contact with the steel outer shell, and the tanks may be cleaned as thoroughly as the glass bottle

DID it ever occur to you that the cost of milk delivery is approximately twelve per cent of the cost of that milk to the final consumer?

Milk is one of our basic food products. Its price is steadily rising higher and higher, and with it the high cost of living. The farmer blames the distributer and the distributer blames the farmer; meanwhile the price of milk continues to rise.

The cost of milk delivery may be divided into four parts. First, there is the cost of the haul from the farm to the railroad; second, the haul to the city depot by train; third, the haul from the railroad depot at the destination end to the city pasteurizing plant; and, fourth, the final delivery to the housewife. In the final delivery to the housewife, it is, of course, impossible to make the deliveries in bulk because the milk is bought in quarts. But in the other three phases of delivery it is entirely practicable to deliver in bulk instead of in small quantities.

Under the present scheme of milk distribution the farmer pours his milk into ten-gallon cans and hauls it or has it hauled to the nearest railroad to meet the milk train going to the nearest large city. It still remains in the ten-gallon can while in the refrigerator car, and when it is taken off and again hauled to the city pasteurizing plant.

Repeated Handling Adds to Cost

This means four separate handlings from the time the milk leaves the farm until it arrives at the pasteurizing plant. First, the farmer puts it into a ten-gallon can; second, the man who hauls it to the railroad siding unloads it into the milk-car; the third handling is the unloading from the freight-car to the truck to carry it to the pasteurizing plant; and the fourth, the unloading at the plant.

Now, it is a fair law of transportation that any kind of goods can be moved more cheaply in large units than in small ones. Furthermore, this principle has been put into practice to an extent that, though comparatively small, is sufficient to indicate the possibilities of revolutionizing the entire system of milk transportation.

The Tank Can Be Put on a Truck

This change will come about through the use of tanks carrying from five hundred to one thousand gallons of milk at a time. It seems patent that the cost of handling a gallon of milk in thousand-gallon lots would be less than handling it in ten-gallon cans.

The first place where tanks of large capacity can be employed instead of ten-gallon cans, without any considerable change in the existing equipment, is in the collection of milk from the farms and its transference from dairies to condenseries or the like, when no railroad haul is involved. In this class of work it is necessary simply to mount a milk-tank on a motor-truck, and the truck will do the rest. A truck equipped with such a tank is shown in the illustration above.

Milk concerns have not followed this plan before because they were afraid that carrying milk in tanks might lead to contamination caused by the contact of milk with metal. There is no longer any ground for such fears, for the tank shown in the picture is lined with glass and none of the fluid ever comes into contact with the steel outer shell. Furthermore, the tank may be kept scrupulously clean, for it may be flushed with hot water or disinfected with live steam in the same manner as are the glass quart bottles now.

This tank has a capacity of five hundred gallons. It is made of steel, lined throughout with a thick glass enamel. This enamel is fused into the body of the steel itself at a tremendous temperature, so that the finished material combines the strength and resistivity of steel with the easy cleaning qualities of glass. The glass lining extends over the flanges and to the outer edges of the filling man-holes at the top, so that at no time is the milk in contact with the metal. The tank is held firmly in place, so that the glass enamel will not be cracked while the truck is in motion.

Many such tanks as that shown are now in use to haul cream from the dairy to the creamery and to haul milk from the dairy to the bottling plant or condensery.

And Why Not a Tank-Car?

There is no reason why the same idea cannot be applied to railway transportation as well as to motor-trucks. We have oil-tanks and chemical-tanks on railroad cars. Why not milk-tanks? In fact, Mr. F. T. Craft, a milk dealer of New York, in testifying before a commission inquiring into the high cost of milk in that state, predicted that a combination motor-truck and railway-car tank service for handling milk in bulk would revolutionize the entire present scheme of milk transportation and would be one of the big factors in reducing the cost of milk to the final consumer.

By the use of tank trucks in the country, tank-cars on the railroad, and tank-trucks again in the city from the railroad depot to the pasteurizing plant, milk handling could be put on a bulk basis.

There's Always Something New

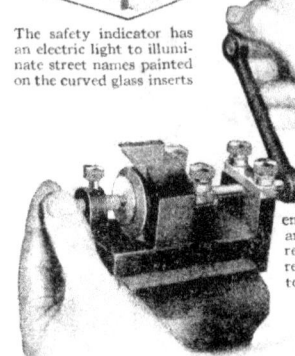

The safety indicator has an electric light to illuminate street names painted on the curved glass inserts

With this simple tool, engine valves with different angles of face are readily refaced and all carbon pits removed; the valve is automatically self-centered when it is put in the holder of the apparatus

Here is a machine that is built like a motorcycle and drives like an automobile. It claims riding qualities equal to the best automobiles

Four of these bar skids are used to each truck wheel to prevent slippage. Each bar is held by a chain covered with a rubber buffer, which prevents the chains from marring the wooden spokes of the truck wheel

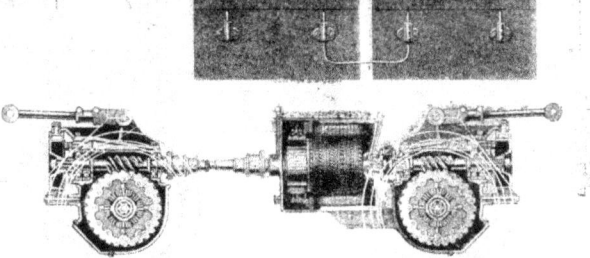

The phantom view of this electric tractor shows the four-wheel drive and axle units. Universal joints and propeller-shaft connect the two with the motor

A new gas-tight piston-ring gets its non-leak properties from the permanent lap joint, which allows for expansion and contraction by reason of the curved surfaces at each end of the ring

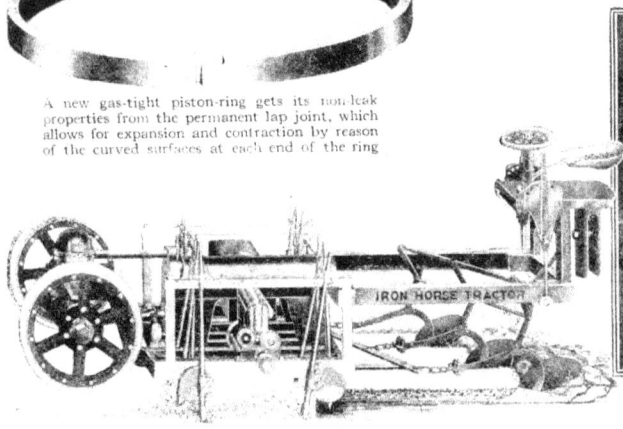

© International Film Service

P. J. Griffen, of Dorchester, Mass., has invented a new air-lock valve which he claims will increase the life of pneumatic tires fifty per cent. It locks the air in the tube for all time unless a blowout or puncture releases it. No more leaky tire valves or flat tires if this device is attached of your tires. It will reduce pneumatic worries to a minimum

By a series of multiple units or legs, this tractor walks on the soil without tearing it by cleats or packing it tight by huge flat wheels. It is also equipped with rubber-tired wheels, which transform it into a farm truck, thus making the tractor serve a dual purpose

in Accessories for the Automobilist

New inside flanges and three collar stud bolts change the ordinary Ford wheel into one that can be put on or removed in a few seconds

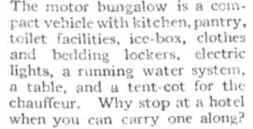

The motor bungalow is a compact vehicle with kitchen, pantry, toilet facilities, ice-box, clothes and bedding lockers, electric lights, a running water system, a table, and a tent-cot for the chauffeur. Why stop at a hotel when you can carry one along?

A leather pad fastened to your automobile door in the manner here shown protects the door top from being marred or scratched

The feature of this new shop vise is quick jaw action, secured by the screw adjustment and the universally pivoted handle

A new spark-plug tester tests the plug inside its chamber. A heavy glass lens at the end opposite the cap permits a view of electrical leaks in the plug

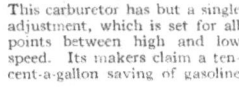

This carburetor has but a single adjustment, which is set for all points between high and low speed. Its makers claim a ten-cent-a-gallon saving of gasoline

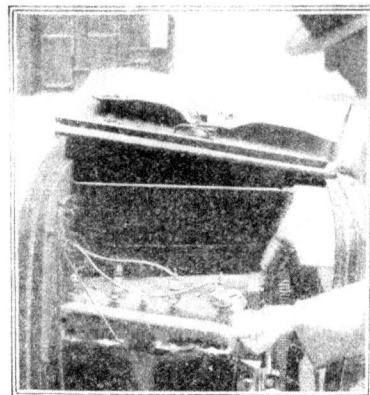

This spark-plug is constructed with a static condenser that fires in a vacuum in the plug. The flame fires through oil

At last—a car for everybody at a price that everybody can pay. True, it accommodates only one passenger and has two cylinders; but then, its maker states that it runs fifty miles to the gallon of gasoline

Newest Automotive Ideas:

The 1920 automobile
new accessories that

Driving at night would be made safer if the motorist would wear this new electric light on the back of his signaling hand

Taking mother on vacation and selling blankets along the way was one automobile owner's idea. He found it profitable and beneficial

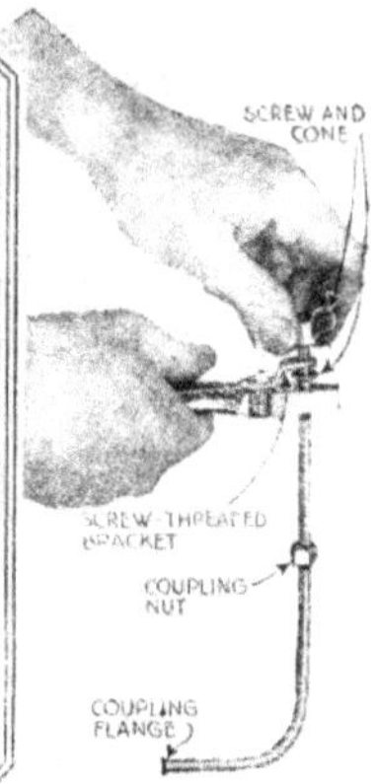

This tractor wheel is really a self-laying track. The wheel rolls on a series of pads

Tube flanging was formerly a difficult job. Now the special pliers shown here hold the tube, while a turn of the screw at the jaws quickly does the flanging

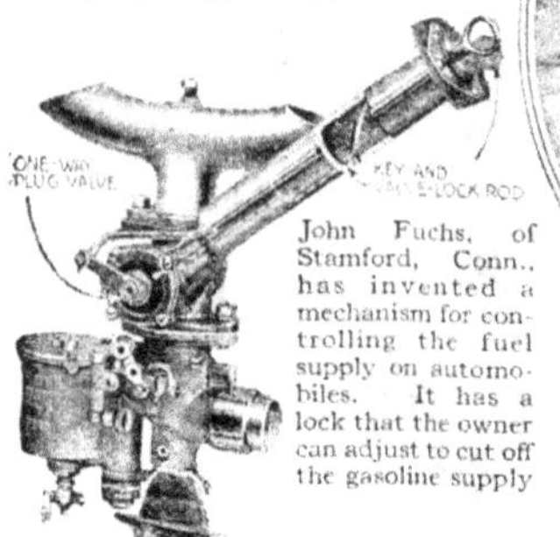

John Fuchs, of Stamford, Conn., has invented a mechanism for controlling the fuel supply on automobiles. It has a lock that the owner can adjust to cut off the gasoline supply

A spider-like device having three outspread wheels over which the tire is mounted permits the operator to rotate the unfinished tire under the grinding wheel and to tone down its tread

For keeping the engine lubricant in its place this invention embodies a specially designed piston and a separating chamber which purifies the oil before it is returned to the crank-case bottom

This small motor-driven cultivator, controlled from the handles, tills corners inaccessible to horse-drawn implements

Pneumatic tires on a motor-truck are not new, but a farm truck so equipped is not only new, but practical and economical. Its owner doubled his running time and reduced running expenses

from Signals to Tractors

shows exhibited many
will promote economy

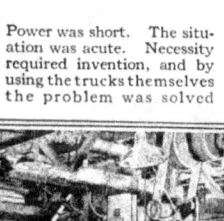

Power was short. The situation was acute. Necessity required invention, and by using the trucks themselves the problem was solved

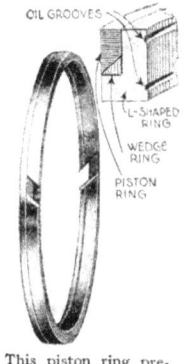

This piston ring prevents leakage by a triangular-shaped wedge ring interposed between the two main parts of the ring

Here is a one-piece honey-comb radiator. No soldered joints enter into its construction, which is effected by electro-deposition

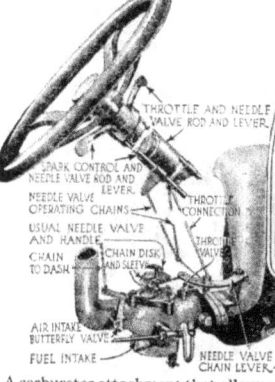

A carburetor attachment that allows adjustment of the needle valve while driving

This metal attachment makes the car driver immune to the headlights of an oncoming car, besides enabling the other fellow to have right road illumination

Springs break in even the best automobiles, and usually the newly fitted spring is stiffer than the others. A block of wood clamped under the old spring in the manner shown brings the car back to an even keel

For hauling milk a specially built glass-lined steel tank of 900-gallon capacity on a five-ton truck chassis has been built. In principle it is simply a giant vacuum bottle and keeps the milk cool even in the hottest weather

He enjoys all the comforts of the inclosed car; the idea embodies a cab built around the driving compartment

179

The electrical test-cart is the first thing of its kind in the way of a complete apparatus for testing any part of the electrical mechanism of automobiles, trucks, and tractors

When the government laid up all private cars, a railroad president wanted something to take him over the company's road. His automobile makes sixty miles an hour upon the rails

Things New in the

Here are a few suggestions comfort: choose the ones

Operated in small spaces this baby tractor was adopted by the government for work in camp gardens. It plows or cultivates, runs a washing machine, pumps water, in fact does a variety of odd jobs

One of the most curious exhibits shown in the recent Paris automobile show was a side-car for use on a motorcycle and serving as a side-car taxi-cab

The rough-surfaced balls move up and down by piston pressure, keeping the spark-plug clean

© International Film Service Co.

"A whale of an automobile," you say. You're right, too. A manufacturer took this means to advertise a certain product. Jonah peers from between the whale's jaws

Here is a truck body made from gas-piping. The rear doors are also of pipes, which slide into each other when opened. The novelty of the scheme should appeal to truck-owners

These metal pads encircle the tires and prevent slipping without the usual unpleasant bumping. The holes in the pads increase the traction

Covered with racing laurels, this aged car came from France to hang up new records for the year 1920. Incidentally it made a fine showing here last year

The slidable trunk-holder includes springs which support the trunk so that its contents are not shaken even when the car travels over rough roads

World of Motors

for your convenience and best suited to your needs

If the windows in your rain curtains tear or break, you can easily purchase others and apply them in a few minutes in the manner shown here

With this simple tool, made of the best cutting steel with a self-centering projection below the cutting edge of the reamer, valve seats may be ground easily, quickly, and cheaply

Instead of a tool-box on the running-board, the 1920 car has a tool compartment in one of the front doors. The flap of the compartment can be locked

Easily transformable into either an ambulance or a limousine, this novel automobile serves a twofold purpose for its renting owner

An electric crane swings a ladle before the mouth of the furnace and the clay stopper is removed. The bucket of molten metal is carried to the melds. Thus the automobile cylinder block is made

© International Film Service Co.

With a suitcase strapped on behind the little red wagon, a week-end can be pleasantly spent, especially if the owner is within a few hours of duck country

A heavy roller, having a long arm by which it is attached to the plow frame drawn by a tractor, utilizes a roller much heavier than the ordinary one. The rolling and plowing can now be accomplished in one operation

Prohibition has not yet reached Bombay, India; nevertheless a wine-cask serves as a water container for street sprinkling. The cask is mounted upon a motor-truck. Formerly ox-carts were used

There's Always Something
Attachments designed to promote

A truck with a water-tight body and special suction pump, used for cleaning sewers and carrying away refuse, is here shown pumping fresh pond water to a water-tank

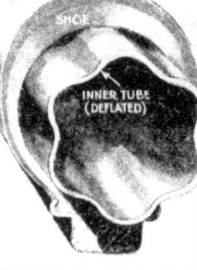

This tube is said to eliminate punctures. Under compression it closes when the nail is withdrawn

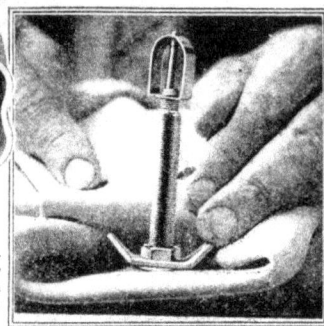

The inner-tube air-releaser is a handy little tool. It holds the air-valve down until all air is exhausted. This tool will pay for itself many times over

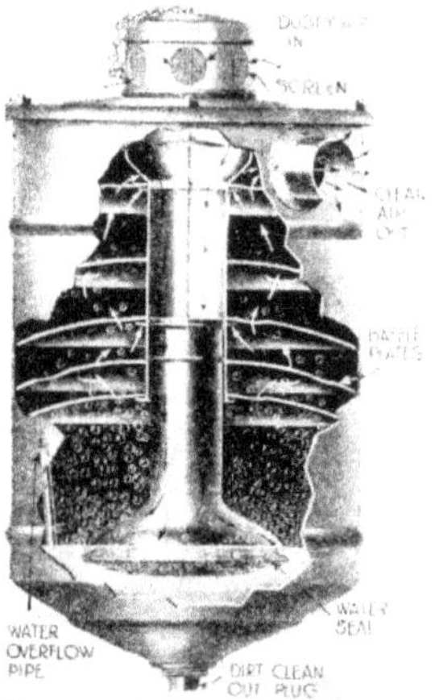

The water-seal air-cleaner allows only clean air to enter the carburetor

The horn button is used the same as a thimble, being kept on the first finger; to blow the horn pressure is applied with the thumb

From coast to coast cars are appearing covered entirely with leather. It eliminates body rattles, and gives a unique though attractive and lasting finish

Now comes the automatic socket wrench, which reduces to a minimum the time for removing and replacing engine bolts. It can be operated from any position

Quickly attached and detached is the power-transmitting contrivance that turns the car into a veritable jack-of-all-trades. It can be belted to almost any machine

New to Help the Motorist
automobile safety and economy

This truck radiator guard is simple and well braced on both sides. The cutaway curved part enables the operator to crank the engine without difficulty

The picnic refrigerator has a special compartment for ice; it fits under the robe-rail

The body is elevated and the ice is thrown into the freight-car refrigerator

Clamped to the under side of the running-board, this luggage-carrier provides ample space for touring accessories. When not in use, it is folded under the seat

The breather-cap slides upward on a bent pin, thus enabling the oil to be easily poured into the crank-case and also preventing the cap from being lost

This camouflaged motor-truck hauls light trains up heavy grades whenever locomotives are unavailable

Where much oxyacetylene welding is done, the small wheeled stand, with its upright sections of pipe and elbows, enables four oxygen-tanks to be moved about with ease

A patent mirror, worn on the back of the hand at night, reflects the light from the car behind and renders the operator's signal easy to see

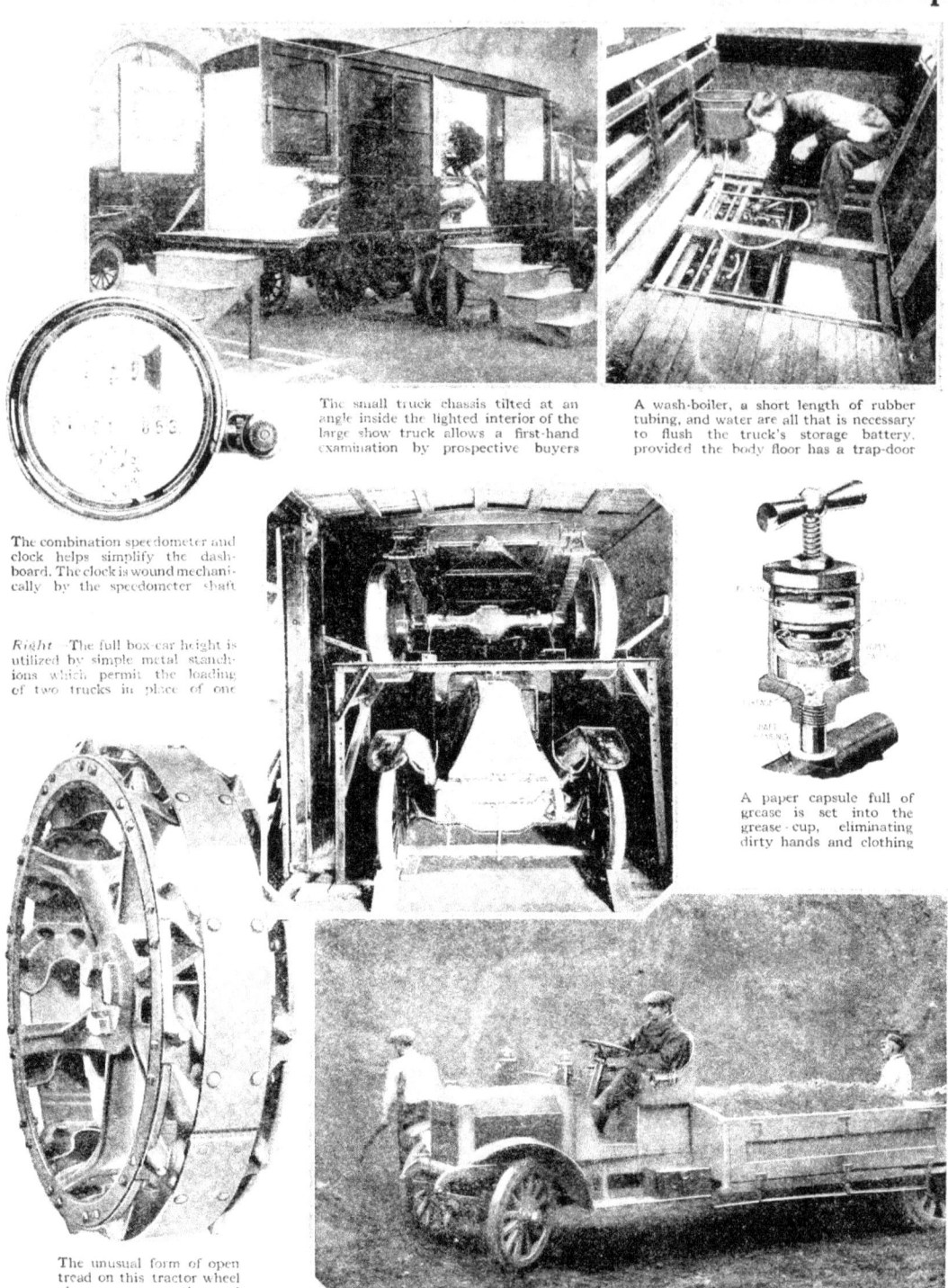

The small truck chassis tilted at an angle inside the lighted interior of the large show truck allows a first-hand examination by prospective buyers

A wash-boiler, a short length of rubber tubing, and water are all that is necessary to flush the truck's storage battery, provided the body floor has a trap-door

The combination speedometer and clock helps simplify the dash-board. The clock is wound mechanically by the speedometer shaft

Right—The full box-car height is utilized by simple metal stanchions which permit the loading of two trucks in place of one

A paper capsule full of grease is set into the grease-cup, eliminating dirty hands and clothing

The unusual form of open tread on this tractor wheel gives greater traction and is self cleaning. The three-piece band may be removed in about ten minutes

Its low-built body dumps the load by gravity, which is an advantage over old-time body methods

Down Automobile Truck Expenses

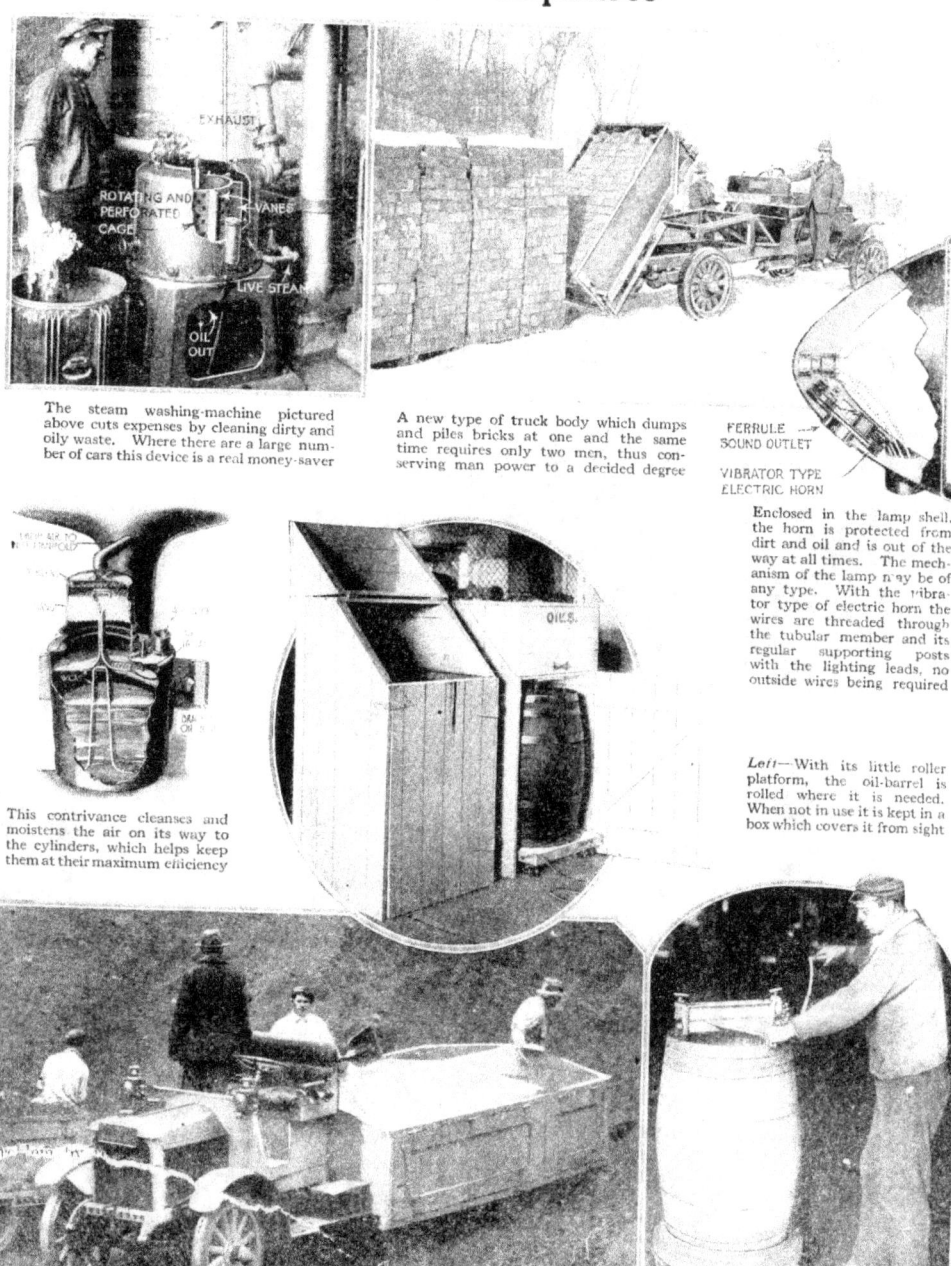

The steam washing-machine pictured above cuts expenses by cleaning dirty and oily waste. Where there are a large number of cars this device is a real money-saver

A new type of truck body which dumps and piles bricks at one and the same time requires only two men, thus conserving man power to a decided degree

FERRULE
SOUND OUTLET

VIBRATOR TYPE
ELECTRIC HORN

Enclosed in the lamp shell, the horn is protected from dirt and oil and is out of the way at all times. The mechanism of the lamp may be of any type. With the vibrator type of electric horn the wires are threaded through the tubular member and its regular supporting posts with the lighting leads, no outside wires being required

This contrivance cleanses and moistens the air on its way to the cylinders, which helps keep them at their maximum efficiency

Left— With its little roller platform, the oil-barrel is rolled where it is needed. When not in use it is kept in a box which covers it from sight

This truck body is so divided into compartments that t measures the concrete or other building material and dumps the desired amount on the construction site

An enterprising garage-owner provided his men with a clothes-wringer to dry the wet chamois cloths used in washing the truck

This carburetor automatically adjusts itself

This truck went out of its way to carry a fourteen-ton engine attempted by anything other than a railroad flat-car and teams

New Inventions for

Proving that peace-time back into their own after

No, she isn't making up, but saying "Home, James," to the chauffeur through her camouflaged telephone

For the driver's convenience the modern motor-truck has a two-step running-board with wooden treads to prevent slipping

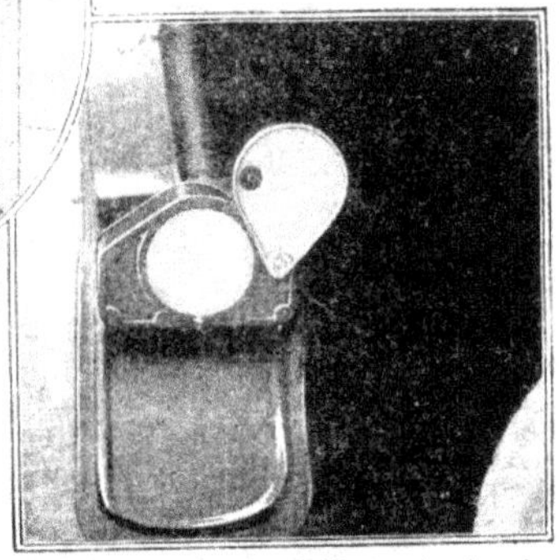

Originally built to meet severe army service, this electric valve-grinder seats a new valve in from six to ten seconds

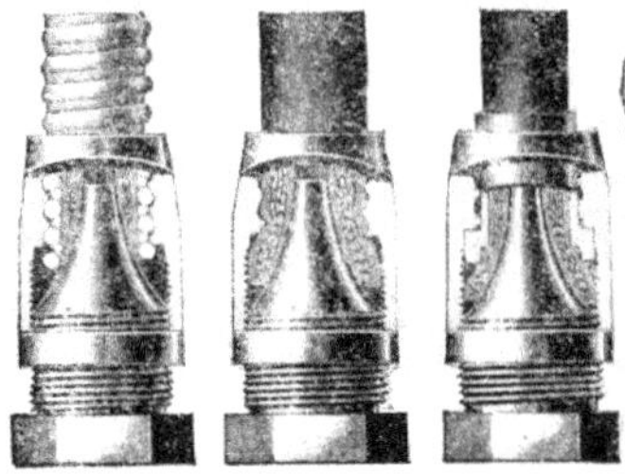

Press the brake-pedal and this stop signal works automatically as long as the brake is down

A new coupling with a cone nozzle fits any 5/16- to ¾-inch plain or armored hose

Forerunner to a living conditions on the road is this portable bath-house ith collapsible tubs and hot running water

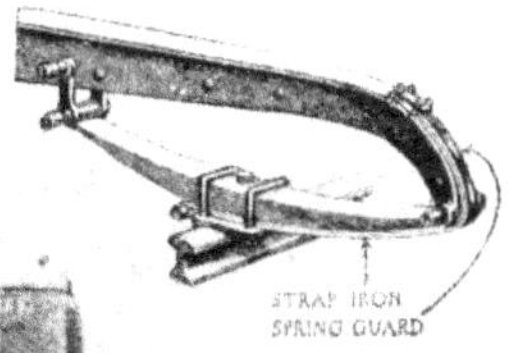

The straps prevent the end of the axle from moving backward, preserving steering control in case of spring breakage

bed over Wyoming's rugged topography, a stunt never before utilizing twenty-four horses. Of course, the truck won its venture

the Automobile Driver

accessories are coming lying dormant during war

An efficient worker is this porch-climbing truck designed for freight terminal work; it turns around in a ten-foot circle

Machines equipped with disk wheels can now use skid-chains with this attachment

RIGHT
LEFT

J. F. Conder's automobile signal, with ruby and green lights, an arrow designating the direction

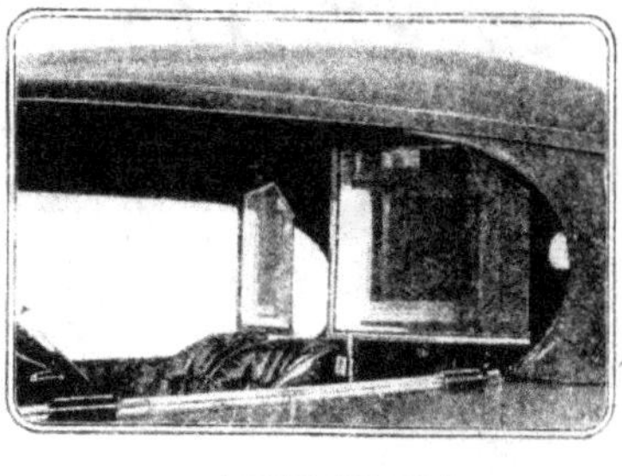

The wind-shield consists simply of two wings attached to the back of the front seats, which are adjustable to any position

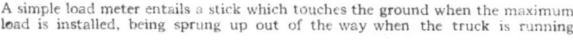

A simple load meter entails a stick which touches the ground when the maximum load is installed, being sprung up out of the way when the truck is running

A comfortable foot-rest for the accelerator that relieves leg strain and the "wabble" so detrimental to gas economy

No chains, slippery road, too much speed —that's the formula for the perfect smash. The results, when followed faithfully, may be seen here—wind-shield and top broken off and two wheels missing. The car turned completely over in the smash

"Always room for one more" is the motto of the moving-van crew, and they usually finish off by lashing your pet bit of furniture on behind, making a topheavy load

How Automobile
The Case of Safety First

Photographs © Publishers' Photo Service

Poetic justice! The big car tried to crowd the little fellow into the ditch, and got the worst of it

Another case of "Forgot my chains." Add a little too much speed and a sharp turn, and you don't need imagination to guess the rest

Both cars had their headlights on full when they met, and both drivers, blinded by the other fellow's glare, misjudged the distance. The crashing head-on collision followed as a matter of course

No, the taxi didn't get tired and try to lean against the tree to rest. This is merely another example of the motorists' great sin of omission—no chains

Usually the center of gravity is kept below the danger line; in this case they got too much stuff atop, and the driver spilled the beans by making too short a turn

Accidents Happen

vs. the Careless Driver

Photographs © Publishers' Photo Service

Tried to beat the train. If he'd won he'd have saved thirty seconds. For this he gambled his car and his life

The driver of this car, which was rammed amidships by the truck in the picture, turned out of a lane into a busy main road without looking to see whether the way was clear

Poor headlights were the cause of which this is the unpleasant effect. The driver simply tried to turn a corner before he got to it

There's Always Something New

Unless one uses the speedometer shaft lubricator shown above, the grease around the shaft hardens and tends to slow it up or perhaps eventually breaks the gear in the swivel joint. The lubricator forestalls forgetfulness on the part of the car-owner

Used to carry moving-picture reels from one country town to another, this truck serves as an advertiser by using the body sides for huge bill-posters

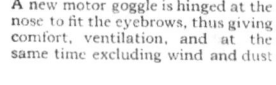

It is claimed that this invention provides a variable speed clutch and control mechanism to replace the friction clutch and variable speed gear used in automobiles

A new motor goggle is hinged at the nose to fit the eyebrows, thus giving comfort, ventilation, and at the same time excluding wind and dust

A skid device for dual or double rear motor-truck tires is made of metal cross-pieces held by a chain in the groove between the tires

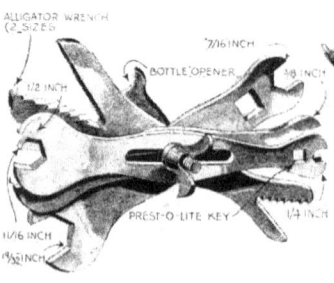

A set of these little bags designed to fit over the Ford operating pedals prevents hot and cold air from entering through the floor-boards of the car

The four-layer automobile wrench fits eight different bolts and nuts, opens bottles, and serves in the capacity of screwdriver when necessary

With the smallest automobile engine, this car comes from France. Five years ago it sold for $950; recently it brought $1,500. Eddie Rickenbacker was its old pilot

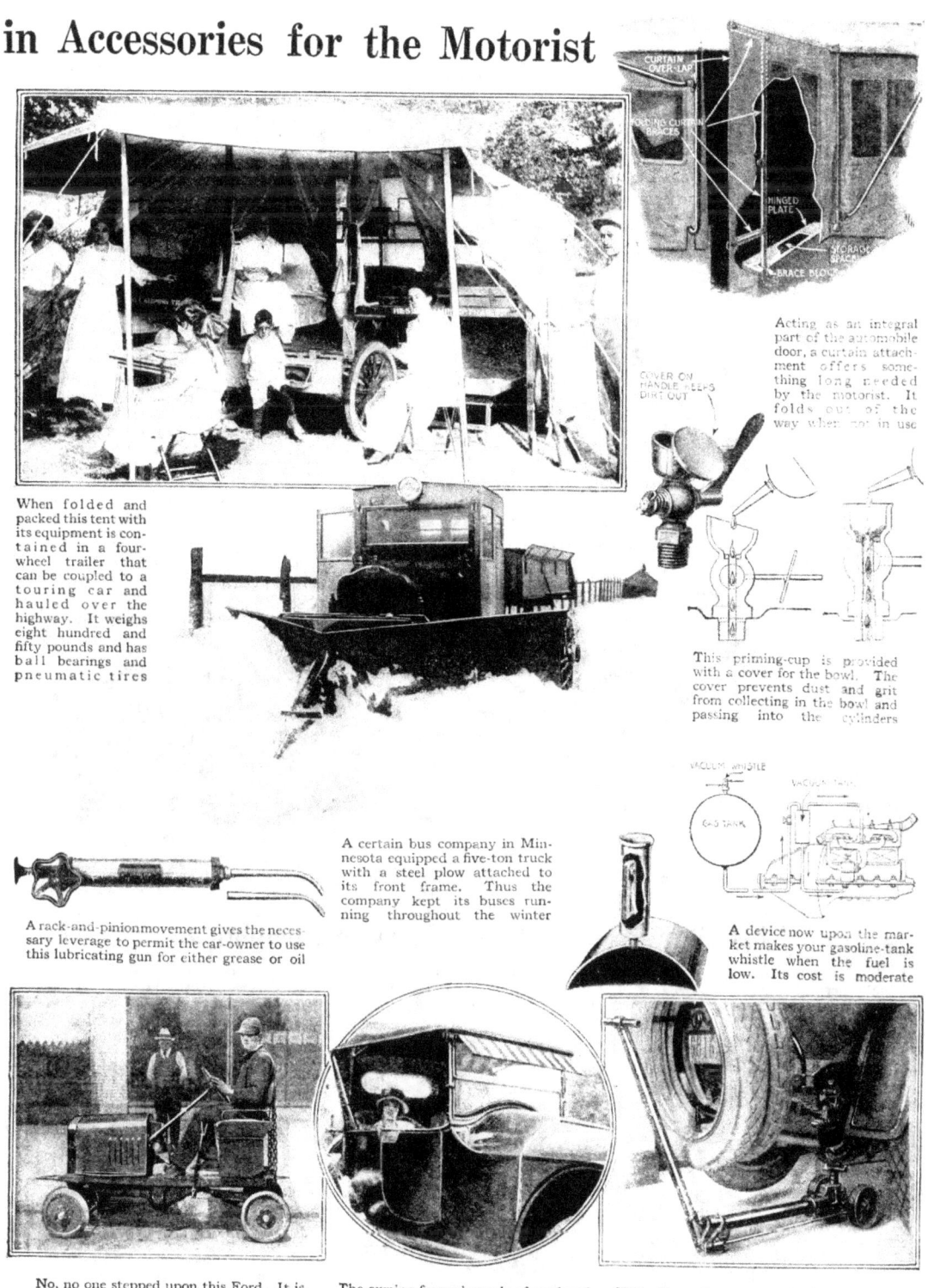

CURTAIN OVER-LAP

FOLDING CURTAIN BRACES

HINGED PLATE

STORAGE SPACE

BRACE BLOCK

COVER ON HANDLE KEEPS DIRT OUT

Acting as an integral part of the automobile door, a curtain attachment offers something long needed by the motorist. It folds out of the way when not in use

When folded and packed this tent with its equipment is contained in a four-wheel trailer that can be coupled to a touring car and hauled over the highway. It weighs eight hundred and fifty pounds and has ball bearings and pneumatic tires

This priming-cup is provided with a cover for the bowl. The cover prevents dust and grit from collecting in the bowl and passing into the cylinders

A certain bus company in Minnesota equipped a five-ton truck with a steel plow attached to its front frame. Thus the company kept its buses running throughout the winter

A rack-and-pinion movement gives the necessary leverage to permit the car-owner to use this lubricating gun for either grease or oil

VACUUM WHISTLE

GAS TANK

VACUUM TANK

A device now upon the market makes your gasoline-tank whistle when the fuel is low. Its cost is moderate

No, no one stepped upon this Ford. It is built this way purposely to do short hauling in and out of factory buildings. It turns around in a very short radius

The awning frame is made of steel and the covering of waterproof canvas. It comes in different sizes to fit any make of car and its position is easily adjusted

With the advantages of short construction, wheels pivot, enormous leverage, and a high handle that cannot strike the car body, this jack is only forty inches over all

An Automobile Ambulance for Injured Cars

POSSIBLY there is nothing in the way of garage equipment more necessary to the man doing a general automobile repair business than the automobile ambulance. There is no work, in connection with the repairing of automobiles, that is so irksome as that of getting a disabled car out of a ditch or off a telegraph-pole and transporting it several miles or blocks to the repair shop, unless the man handling the job is properly equipped for that service.

When attached to the front axle this ambulance is clamped securely to the center of the axle, thus supporting the entire front end of the car and allowing the tongue of the ambulance to control and guide the car. When attached to the rear axle it is ordinarily applied to the brake-drum on whichever side the injured wheel may be located. In extreme cases, however, where both rear wheels are out of commission, the ambulance can be applied to the center by lashing it to the middle of the differential housing or, preferably, to a bar lashed across under the rear system. It will pay for itself in its first job.

The automobile ambulance enables the garageman to save money by handling his "tow-in" jobs with only one man

The very latest thing in commercial steamships is a boat built to carry automobiles and motor-trucks for distribution along the Atlantic seaboard

This Ship Was Built to Carry Automobiles

THOUSANDS of automobiles, and even motor-trucks, have been shipped by boat across the Great Lakes to Buffalo, and then driven overland to their destination in New York or New England. Most of such shipments were made under difficulties, because the vehicles were carried on regular passenger vessels. Now, however, a navigation company of Detroit has designed a boat especially for the service, so that the transporting can be done on a large scale. The vessel, designed by Mr. Frank E. Kirby, is shown in the accompanying broken-away illustrations.

The ship will carry only the officers and crew, and will be operated as the fluctuating production of the two automobile cities direct.

This automobile ship will be three hundred feet long and seventy-five feet beam, and will be capable of carrying between three hundred and seventy-

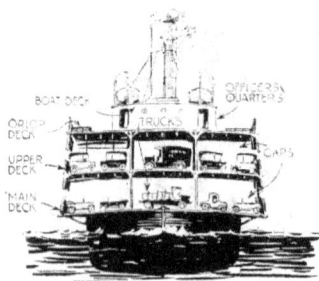

Two elevators located amidships permit even the largest motor-trucks to be taken aboard and stored away in the hold

five and four hundred motor-cars and trucks. Most of these will be carried under cover on the two principal decks, but some may be loaded on the upper deck on the outside.

The ship will have several new features never before included in an ordinary passenger-carrying steamer. These include extra large side hatches wide enough and high enough to load the biggest motor-truck with its body; elevators for transferring the vehicles from the main-deck to the upper decks; and turntables on the two principal decks to make the handling easier.

The two elevators are located on the center line of the ship directly in a line with each of the two large side hatches. The vehicles will be run on to the boat under their own power, and will not be moved by manual labor, except when necessary to place them in position on the ship.

MACHINES, TOOLS, ETC.

Can We Beat Our Swords into Plowshares?

By Hudson Maxim

THERE is none too much war material in the world to-day for the use of the world if a peace-keeping league is formed. The war material should be divided among the nations according to their needs, and then each nation should be permitted, under the league, to produce only a certain limited quantity thereafter. A breach of this understanding would put that particular nation under a ban.

But, assuming that the material is not to be so disposed of, and that there is to be no adjustment between the Peace League nations to distribute the war material among them for future war needs or world police needs, then the question arises, to what peace use can the material best be converted?

From War to Peace

The tanks, especially the smaller or whippet tanks, largely used by the French, could doubtless be converted into tractors. The tank itself is merely a tractor converted from peace to war use, and it would be a very simple matter to convert the tank back again from war to peace use. Few of the farms of France are large enough for the individual farmers to own tractors for their farmwork, but proper arrangements could doubtless be made, either by the government or by private companies, to supply tractors for all-round farm use, on terms that would be acceptable to the small French farmers.

In the Ukraine, however, and other parts of Russia, there would be a wider use for such reconverted tanks, but they would not have very much value even for that purpose. It would be much simpler, with our means of quantity production in this country, to produce tractors specially designed for the work. At best reconverted tractors could never compete with tractors made especially for farm use.

The main value of war material at the present time lies in guns, ammunition, and ships. These guns and the projectiles made for them would have no value except as old metal. The same is true of shoulder rifles. A shoulder gun that cost, say, $50 to make might bring fifty cents or a dollar. Big cannon costing $50,000 to build might be worth from $100 to $200.

As to Airplanes

The cost of transportation and breaking up of the big guns must be taken into account. The same holds true of the big shells. They would bring as old metal only a very small fraction of their original cost. Therefore, if one per cent of the guns and ammunition could be sold for what it is worth for war uses to the smaller nations, it would be a better bargain than it would be to attempt to reconvert or utilize for other purposes all the material there is.

The airplane engine, of which the Liberty motor is a very good example, is a very highly specialized product, made for the purpose of generating the maximum power for a given period. Every part of it is just as strong as every other part, and it has been so tested and timed and tuned that when it goes to pieces it practically goes to pieces after the manner of the one-horse shay. Such a motor is not intended for continuous use. It is intended only for a brief and strenuous use, and the life of the Liberty motor under full load or running at a speed where it is doing its best, I believe, is only about eighty hours.

Some of the airplanes, especially the bomb-carrying airplanes, may have a use as mail-carriers or for similar purposes; but such use is at the present time largely problematical.

There is one war measure that will have the very largest kind of peace use. I refer to the establishment of plants for the fixation of atmospheric nitrogen. Had the war continued a little longer, we should have had some very large and well equipped plants which could in time of peace be utilized for making fertilizer; but most of the work on these plants has now been arrested.

Fixation of Atmospheric Nitrogen

There is no one thing more needed in this country, as an industrial development, than plants for the fixation of atmospheric nitrogen, particularly for the manufacture of fertilizers for agricultural use. The government ban should be taken off waterpower throughout the country, so that individual enterprise may utilize that power

Great conflagrations could be staged as spectacles every sixty days, at a sum far less than that which is represented by Niagara's wasted waterpower

"Whippet" tanks could be converted into tractors. The one pictured here
was used as such in dragging its own fuel supply up to the front in France

for the synthetic manufacture of fertilizers.

All of the water passing over Niagara Falls should be utilized to produce electricity for industrial uses, and a large amount of this could very well be applied to the manufacture of fertilizers. There is practically an unlimited demand for cheap fertilizers for the farms.

Many of these power plants could be shut off in order that the water might be returned to the falls periodically for scenic purposes; but even then it would be a very costly display.

Niagara's Wasted Power

It is estimated that 16,000,000 horsepower is running to waste at Niagara Falls. To produce this power with modern steam boilers would require 32,000,000 pounds or 16,000 tons of bituminous coal an hour. If such coal were to be appraised at $3 a ton laid down at the plant, the 16,000 tons would be worth $1,152,000. That would be the daily cost of the coal supply to produce that amount of power. The value of the water that is wasted is therefore about $300,000,000 a year. This in gold would weigh 450 tons.

A conflagration the size of the great London fire could be provided once every sixty days for public entertainment and amusement for a sum equal to that which is running to waste at Niagara Falls.

The burning of Rome by Nero was a very inexpensive piece of amusement compared with that which is provided for bridal couples who visit Niagara Falls.

Enormous quantities of tri-nitro-toluene—that is to say, T.N.T.—and nitrate of ammonia have been made into high explosives. The best way to employ these materials for peace use would be in the manufacture of high explosives. Stupendous quantities of

Huge cannon, which cost $50,000 to build, are worth only a few hundred dollars when sent to the scrap-heap

high explosives are made in the United States annually for industrial use. To a considerable extent, however, T.N.T. would have perhaps the higher value for industrial purposes. Nitrate of ammonia is one of the best fertilizers in the world, and would be worth for farm use all that it cost to produce it.

The most important use to which war material could be adapted for a period of peace is the machinery that has been constructed during the war for the manufacture of war materials. There are vast quantities of such materials, and large buildings have been erected for the manufacture of war implements. Nearly all of this machinery is very readily adaptable to the manufacture of implements of peace.

Valuable War Machinery

For example, the buildings that have been erected and the machinery that has been installed in them for the manufacture of war tanks can very well and profitably be employed for making farm tractors and automobiles and heavy trucks for industrial use.

The vast equipment for the production of airplane engines can very readily be utilized in major part of automobile and farm tractor engines. One of the largest manufacturers of munitions of war in the vicinity of New York is already planning to launch out in a large way in the production of harvester machinery.

How the Modern Steel-Furnace Does Its Work

What man has learned about handling the metal that gave civilization its start in life

By Ernest Welleck

WHEN primitive man, perhaps prompted by curiosity, picked a formless and heavy mass of a strange substance from the ashes of his camp-fire and, in trying to break it up by pounding it with a stone, found that this strange substance did not break, but could be hammered into any form, he laid the foundation of the framework upon which rests human civilization. For the strange substance was iron, which had been smelted from the iron-bearing rocks around which the fire had been built by chance.

In its various forms, as sword or plowshare, as tool, rail, or wire, nail or screw, engine, boiler, crane, magnet, battleship, or cannon, iron has contributed more to human progress and civilization than any other of the component parts of our planet. Its abundance and properties make it the most valuable—in fact, an indispensable—factor of practically every branch of human activity.

Getting Iron from Its Ore

Iron is a metal, but it is found in its metallic form only occasionally and in small quantities, principally in meteorites. It occurs most frequently in the form of oxides or carbonates—that is, combined with oxygen or with carbon and oxygen. These combinations yield the metal readily when smelted with carbon.

Until about 1860 only three important classes of iron were known—wrought-iron, steel, and cast-iron. Wrought-iron was almost free from carbon; steel contained between 0.30 and 2.2 per cent of carbon, and cast-iron more than 2.2 per cent. To-day a great many more kinds of iron and steel are distinguished. Their characteristic qualities are due principally to the amount of carbon they contain, the method of their production, and the presence or absence of certain other metals combined with the iron to form alloys.

In modern iron-works the iron is obtained from its ore in specially designed furnaces. These furnaces are about ninety feet high, and at their widest part have a diameter of about twenty-five feet. They are built of steel, in form like an elongated barrel, and are lined with fire-brick. The lower parts, in which the heat is

greatest, are lined with hollow bronze bricks through which a rapid stream of water is forced. Under the furnace is the fire-brick basin in which the molten iron is collected, and from which it is drawn off. Another opening permits the slag to run out.

Four Hundred Tons a Day

The furnace holds about 1,300 tons, and is kept filled by adding layer after layer of ore, coke, and limestone as the mass sinks and becomes fused at the bottom. The temperature varies between 300° F. near the top and 3,500° at the bottom. The reduction of the ore takes place in the upper part.

The high temperature required in

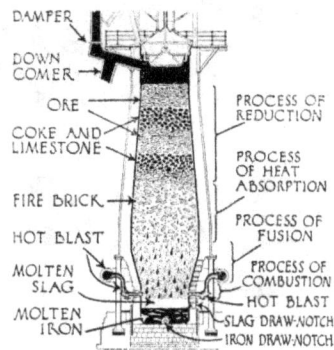

DAMPER
DOWN COMER
ORE
COKE AND LIMESTONE
FIRE BRICK
HOT BLAST
MOLTEN SLAG
MOLTEN IRON
PROCESS OF REDUCTION
PROCESS OF HEAT ABSORPTION
PROCESS OF FUSION
PROCESS OF COMBUSTION
HOT BLAST
SLAG DRAW-NOTCH
IRON DRAW-NOTCH

BLAST FURNACE IS 90 FEET HIGH AND 25 FT. DIAMETER: LOAD IS 1300 TONS

The furnace shown in this sectional view is capable of turning out 400 tons of iron every twenty-four hours

One furnace and four stoves furnish the initial heat for the blast. The stoves are heated red-hot

the production of iron is obtained by blowing into the furnace air previously heated in four stoves accompanying each furnace. These stoves, as big as the furnace, are of steel lined with fire-brick, and are heated to red heat. Only one stove at a time is used to supply the hot blast for the furnace, while the other three are being heated.

The molten iron is run into sand forms and cast into bars, called pigs. Each furnace produces about four hundred tons of pig-iron a day. The pig-iron contains a small percentage of carbon and a number of impurities, which must be removed before the iron is changed to steel.

For changing iron to steel two processes are principally used, the Bessemer and the open-hearth process. In the Bessemer process the iron is changed to steel in a steel vessel, about twelve feet in diameter and twenty feet in height, called a converter.

The Bessemer Process Converter

The vessel, which resembles a big bottle with the neck broken off, is lined with fire-proof material a foot thick, and has a removable bottom, through the fireproof lining of which a number of pipes extend into the converter. The vessel is tilted, and enough molten pig-iron is poured in to cover the bottom to a height of eighteen inches. Then a blast of air at a pressure of twenty pounds to the square inch is forced through the pipes in the bottom, and supplies the oxygen which oxidizes or burns up the impurities. By the addition of a weighed quantity of carbon, the iron is transformed into steel.

In the open-hearth process the iron is converted into steel in a saucer-shaped pan or hearth lined with magnesite, dolomite, and lime, and covered with a roof of silica brick. First a layer of scrap-iron is placed on the hearth, then a layer of pig-iron. Hot gas and air are forced into the hearth and over the iron. The pig-iron melts first, then the scrap-iron. The hot gases, forced over the molten mass, cause the impurities to become oxidized, forming a crust of slag on top of the molten iron. The pure iron thus obtained is then changed to steel by adding carbon, manganese, or whatever metal may be desired.

In the Bessemer process the converter, a steel vessel about 12 feet in diameter and 20 feet high, is tipped over on its side and molten pig-iron is run into its mouth; then it is turned upright and a blast of air is sent through it

In the open-hearth process the molten steel lies 18 inches deep on a bed about 40 feet long by 16 feet wide. Here the current of hot gas and air is shown being forced above and around the molten mass. Ore of less purity than that used in the Bessemer process is successfully dealt with by the open-hearth method

They Make the Wheels Go Round in a Turbine

No other type of steam engine demands such careful workmanship and exactness as the turbine. The thousands of vanes must be absolutely true. Should one come loose it would wreck the whole machine

Superheated steam is forced into the casing, and, directed by the slanting projections of the casing against the vanes of the shaft, causes it to rotate with prodigious speed

The vanes are welded to the shaft, and form rings that fit between similar rings of vanes that project inward from the stationary casing which surrounds the part of the turbine that turns

After the thousands of individual vanes have been welded to the shaft they must be carefully straightened and adjusted to make sure that they fit between the rings of the casing

How Did They Ever Get Along in Father's Day?

Mechanical aids that make life easier for the school-boy

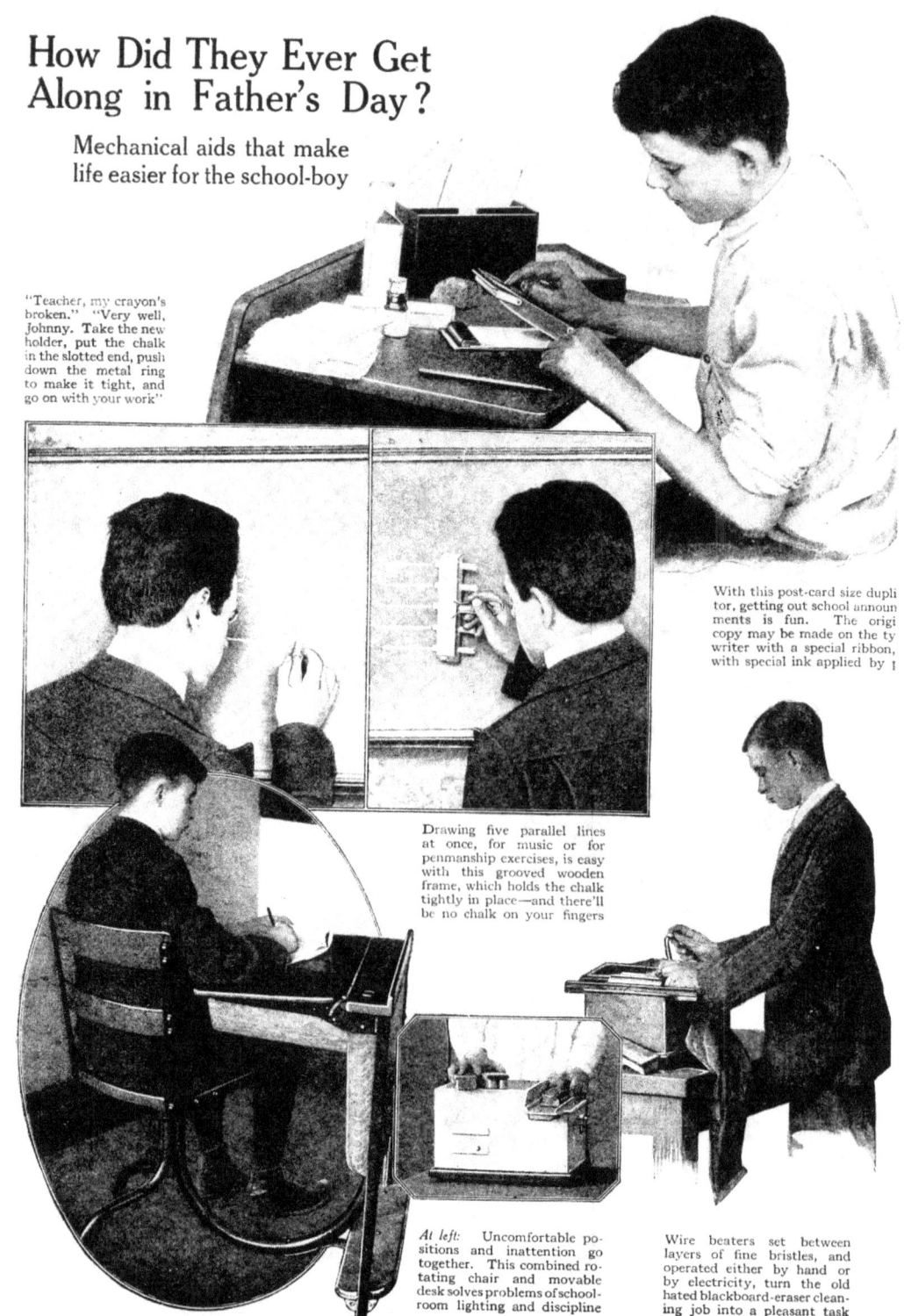

"Teacher, my crayon's broken." "Very well, Johnny. Take the new holder, put the chalk in the slotted end, push down the metal ring to make it tight, and go on with your work"

With this post-card size duplicator, getting out school announcements is fun. The origi copy may be made on the ty writer with a special ribbon, with special ink applied by

Drawing five parallel lines at once, for music or for penmanship exercises, is easy with this grooved wooden frame, which holds the chalk tightly in place—and there'll be no chalk on your fingers

At left: Uncomfortable positions and inattention go together. This combined rotating chair and movable desk solves problems of schoolroom lighting and discipline

Wire beaters set between layers of fine bristles, and operated either by hand or by electricity, turn the old hated blackboard-eraser cleaning job into a pleasant task

Making Piston-Rings by Whirling Them into Shape

And the speed is forty times greater

By E. F. Cone

One man pours the metal into the trough and another operates levers that control the casting

ONE of the many small manufactured products of which millions are used in the United States alone each year are piston-rings. They are thin round objects of cast-iron, made to fit in the cylindrical piston of all automobile and airplane engines. There are many used in steam, marine, and other engines, as well as in pumps. A conservative estimate has placed the number used in this country alone in one year at not less than 200,000,000. The Ford automobile plant uses about 16,000,000 yearly.

It has been customary, and still is in most plants, to make these rings the same as any other small iron casting is made—molding them in sand by means of a molding-machine. By this process about 420 rings a day can be made by each machine.

But a new process has been invented whereby the old way of making these rings is virtually revolutionized. It is called the de Lavaud process, named for the inventor, a Brazilian who is now a resident of New York. By this new process, rings are being made daily in one machine to the number of 18,000 or 20,000. The unusual feature of the process is that the hot molten iron is cast directly into its final ring shape by centrifugal motion.

A Very Simple Process

The process is very simple. It consists in introducing into a rapidly revolving iron mold a definite amount of hot metal of proper composition. This is immediately thrown to the outer surface of the revolving mold, where it quickly cools and assumes its definite shape in the grooves of the mold. It must then be annealed, or softened, for it is very hard, and afterward ground on a special machine, thus rendering the finished product nearly ready for use, only a little machining being necessary. An important consideration is that the rings are uniform in size and thickness of section.

The machine in which the rings are cast consists of three arms protruding from a central revolving standard. On each arm is placed a permanent mold in which the rings are cast. This is a series of alternating rings of cast-iron and of thin boiler-plate separating the rings, these forming a series of grooves, the size and thickness of the piston-ring to be made. By varying these, any diameter ring from 1½ to 8 inches can be made in this machine. By enlarging the machine, rings from 30 to 40 inches in diameter can be made.

These alternating pieces of metal are held together by an ingeniously

He puts the rings into the annealing apparatus and they come out at the bottom softened

arranged set of bolts and pins. A revolving wheel, operating by electricity at the bottom of one side of the central standard, is the motive power that causes the axle carrying the mold to assume its rapid motion. This has a momentum of 1,200 revolutions a minute when the ring is being cast.

How the Rings Are Formed

A movement of the larger hand lever, shown in the picture above, quickly brings the revolving mold over the small trough in which the hot metal is poured. These troughs are replaceable at any moment, and are lined with a refractory. A rapid inversion of this trough while inside the revolving mold empties the metal into the grooves, thus forming the rings. A further slight movement of the lever brings the revolving mold against a brake, which stops the revolution. A reversal of the lever brings the mold back against the central standard. A turn of the machine carries the mold with its rings to the right, and brings a new mold ready for a new charge of metal.

The molds are transferred to the four arms of the pedestal situated at one side of the casting-machine. Here a crew of boys remove the bolts, allowing the rings to fall into baskets. The disassembled molds are kept intact loosely, and placed on the arms of a revolving wheel, by means of which they pass through a tank of cooling liquid. A new machine has nearly been completed for performing the disassembling by electricity.

This new process is unique. The product is not only practically ready for use, but is superior in every way. The metal has been proved to possess striking properties in the way of strength, elasticity, and uniformity, particularly as to its hardness.

Handy Office Devices

Nine aids to efficiency for the use of the office clerk

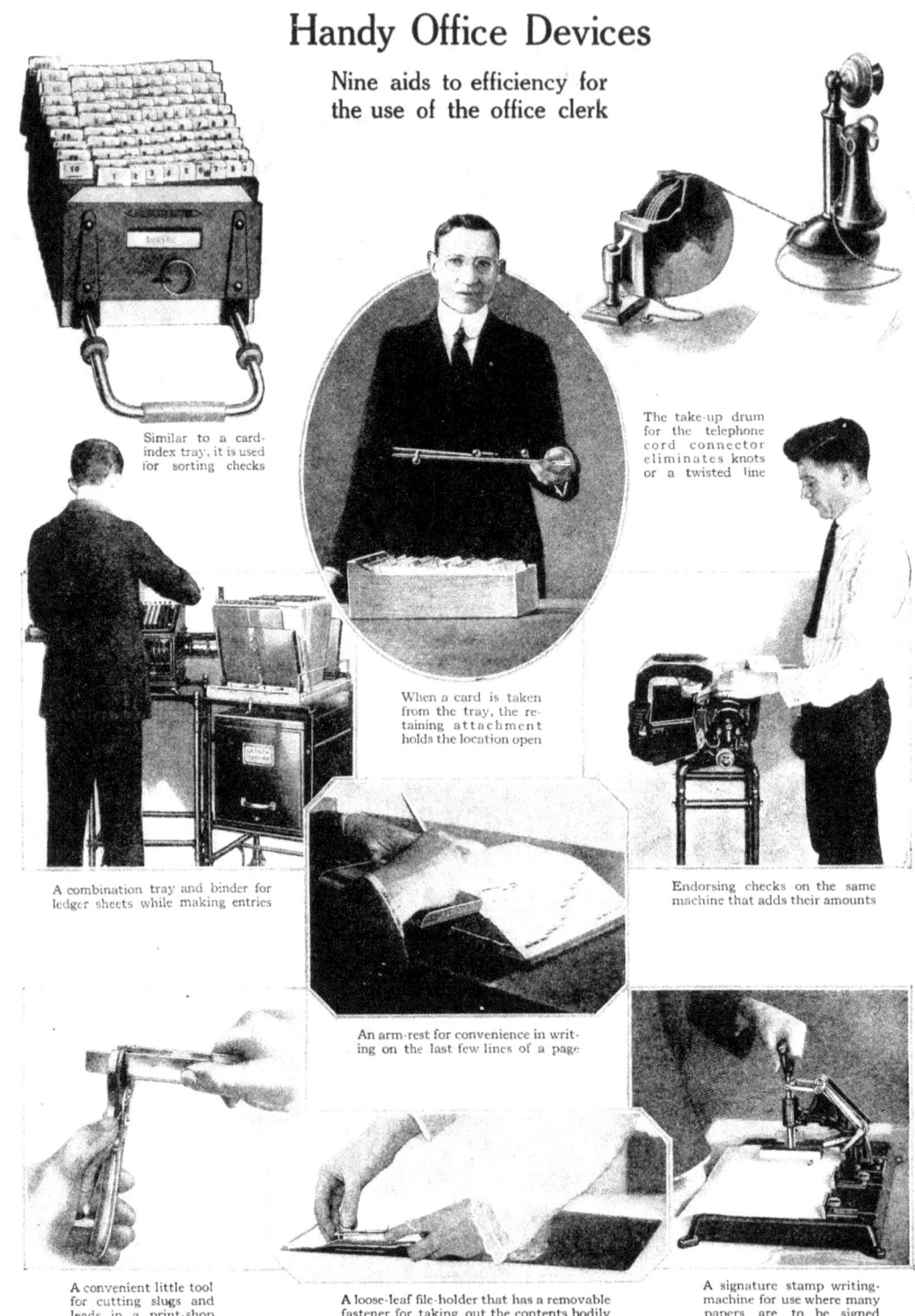

Similar to a card-index tray, it is used for sorting checks

The take-up drum for the telephone cord connector eliminates knots or a twisted line

When a card is taken from the tray, the retaining attachment holds the location open

A combination tray and binder for ledger sheets while making entries

Endorsing checks on the same machine that adds their amounts

An arm-rest for convenience in writing on the last few lines of a page

A convenient little tool for cutting slugs and leads in a print-shop

A loose-leaf file-holder that has a removable fastener for taking out the contents bodily

A signature stamp writing-machine for use where many papers are to be signed

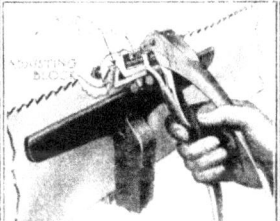

A saw set having operating handles that are so arranged as to enable the operator to assume a comfortable position

A device, easily carried in the pocket, which fits inside jaws of an ordinary vise for clamping pipe

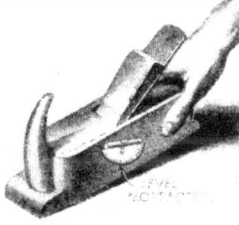

This planer combines a level protractor with a planer, thus producing automatic checking of correct angles while planing

Things Machinists Ought to Know About Their Tools

Rigidity, accurate drilling, front control, and ease of operation were the features kept in mind by the designers of this driller

Making twenty-eight fineness tests of cement in a little more than an hour, this machine is a great time-saver

A high-speed drilling-machine manufactured upon safety-first principles in order to make possible its operation by women

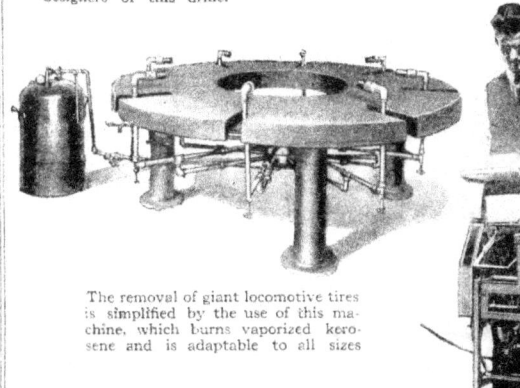

The removal of giant locomotive tires is simplified by the use of this machine, which burns vaporized kerosene and is adaptable to all sizes

This parcel-tying machine ties from twenty-five to thirty packages a minute

A new and compact motor drive for double-spindle shapers, using standard vertical motors that are independently adjustable to suit the length of belting

The appliance is about a foot square and attaches to any electric light socket

Buying Railroads for Junk

This man purchases bankrupt railroads, scraps them, and then sells the material in every part of the world

IMAGINE, if you can, buying railroads just to scrap them. It seems like an impossible proceeding, but it is not. Herman Sonken, of Kansas City, buys railroads, tears them up, and sells them as a sort of glorified junk.

When Mr. Sonken was a boy, his desire was to own a railroad. He has not only owned one, he has owned many. When a railroad gets into the hands of a receiver and has to be sold, Herman Sonken is likely to buy it.

First he finds out how many locomotives, how many cars, how many ties, bridges, and roundhouses belong to the road. He then makes an offer for it. When he has acquired it he scraps it. The cars he uses to transport the other equipments of the railroad to its destination. The roundhouses are torn down, scrapped, and placed on flat-cars and in box-cars, and when he has used these cars for transporting the best of the material, they in turn are scrapped.

How the Rails Are Handled

The wrecking is all done with Mr. Sonken's own machinery. The rails are taken up by a machine built specially for the purpose. A crane mounted on the end of a 30-foot car reaches over a 33-foot car in front of it. The bolts are taken out of the rails, and they are then hoisted on to the car by machinery which grasps them through the bolt-holes.

The rails are loosened on the inside, and the train travels over the loosened rails and picks them up from behind. Since the wrecking train travels slowly, there is practically no danger of the rails spreading and causing a wreck. By this method it is necessary to have a crew of only twenty men to pick up the rails, where ordinarily it would take thirty or forty.

Hoisting rails weighing 700 pounds. The rails are loosened, and the train travels over them, picking them up from behind

The bridges are taken down, and, if in good condition, are sold to other railroads. The cars that are in such condition that they may be repaired are sent to the "hospital," refitted, and sold to other railroads. If they are in a very bad condition they are simply sold as scrap metal. Locomotives are overhauled and resold wherever it is possible.

When a road has been completely dismantled, and nothing remains but the ties, they are sold to the farmers along the way at a rate of $90 a mile. These are treated with creosote and make excellent fuel. The farmers find them even more satisfactory to burn than coal.

A large amount of the railroad equipment that has been scrapped by Mr. Sonken is now over in France. He marks the rails and cars with his name so that they can be identified. Only recently a young man from his office, who was fighting in France, recognized some of the Sonken rails. Parts of Sonken railroads may be found in practically every country in the world.

One that he bought was sent almost bodily to England. With the thirty-six miles of rails went six freight-cars, two combination passenger-cars, three locomotives, and six steel bridges.

Unlike most business men, "live ones" are not what Mr. Sonken wants. He always waits until the courts have written the obituary of a railroad. So, although he has owned many roads,—seven or more,—he has never operated one, and all but one of the seven were long since junk.

The Exception to His Rule

The exception was a short line in Iowa. Mr. Sonken was called in to look at the property.

"As a railroad it's worth nothing to me," said he, "but as junk I'll give $21,250."

He got it. Then, before salvage operations could start, he was offered $250 more than he had paid by a man who wanted to keep the breath of life into the line. It was characteristic of Sonken that he took his $250 profit and let the road live.

Bridges from a railroad scrapped by Herman Sonken. The upper picture shows a bridge before it was dismantled

Box- and flat-cars and other equipment ready for shipment. There are twenty acres of space filled with valuable junk

Machines that Take a Log and Turn Out Billions of Matches

Ever stop to think how much time and labor go into the making of that match you throw so carelessly into the gutter after lighting your cigar? There are at least thirteen distinct processes involved and much ingenious machinery. Below, the splints, or sticks, are coming from the drying-machine

Matches begin life as a three-foot log. First the bark is chopped off; then the log goes through a veneering-machine which cuts it into strips, and these disappear into the chopping-machine, to come out as match-sticks

In closely packed ranks the match-sticks march to the dipping-machine, an arrangement of endless-chain carriers which pass the tips through the fire-making solution. The machine shown below handles 1,000,000 matches at one operation

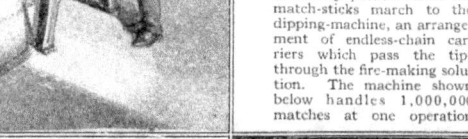

As the match-sticks fall out of the chopping-machine, into the chute, they are sucked up through the large pipe to drying-machines on the floor above

Chaperoned by these two girls, the machine sorts and arranges the match-sticks. The separate cards of splints are then stacked in shallow boxes, ready to go to the dipping-machine, which provides their business ends

Ready to go out to light the world. At the left is shown the packing-room in a modern factory, where the finished matches are placed in their paper boxes

A Mechanical Genius of the Back Woods

Without schooling or training of any kind, he has made himself a cunning artificer in wood and stone

By A. F. Harlow

THOMAS COKE RICE is a Tennessee mountaineer who lives in one of the wildest and most inaccessible portions of the narrow, high, deeply seamed plateau known as Walden's Ridge. An almost unbroken forest extends for several miles in all directions from his home, and a lone cabin three miles away shelters his nearest neighbor. In this lonely spot he has lived for twenty of the twenty-five years of his life. He has never spent an hour of his existence in a school-room, and until the draft caught him in 1918 he had never been more than twenty miles from home. Yet he is a cunning artificer in wood and iron. Had he had the opportunity for mechanical training, he might have been a wonderful craftsman, a manufacturer, an inventor—who knows?

I first saw "Coke" seven years ago. At that time his mother "believed" he was about eighteen years old. The mother is a fine example of the Spartan heroism and stolid endurance common to the Southern mountain women. Her husband left her when Coke and his sister had not yet entered their teens. He rode away on the only horse possessed by the family. The agony of the bitter struggle with poverty during the years that followed can never be told in words. The mother, assisted by the two children, succeeded in raising each year a tiny patch of vegetables near the house. She would walk through all kinds of weather to a settlement six miles away, where she did any sort of work she could get to do.

As the children grew older, the three were able to put half or more of their three-acre clearing into cultivation. Their only implement was an old bull-tongue plow, and, having no horse, they took turns pulling the plow. Then there

"Coke," as he is called, made every part of this forge and blower, excepting the sprocket wheels and chain. And the construction was all his own idea. If he had gone to school, he might have been a great inventor. Perhaps it is not too late yet for the young mountaineer to have his chance

was wood to be cut and brought in from the forest, and in this work the children each "made a hand." The mother has a strong strain of the old Cherokee Indian blood in her, and this, combined with the hardy moun-

Thomas Coke Rice, a Tennessee mountaineer, and his mother and sister have lived in the wilds all their lives. He never went to school, yet he is an expert workman in wood, stone, and, what he likes best of all, iron

tain stock, accounts for the indomitable courage with which she and her children endured their trials.

As Coke grew up, he showed a disposition to tinker with the few tools at his disposal, and began to do repair jobs with remarkable skill. One of his first real creations was a jumping-jack, a crude wooden toy manipulated by a string, the original of which he had seen in the possession of a child at the settlement. When I first ran across him seven years ago, he had already made considerable progress. He had built a little forge out of stones and earth in the yard, and was doing blacksmithing. With a frow, an edged tool used in the country for splitting palings and old-fashioned shingles, he had split out two slabs of oak for the sides of his bellows, and had shaped their edges with his jack-knife. A piece of an old buggy-top served for the leather portion of the bellows. He hacked and whittled out a handle and frame, and soon the forge was in operation. Among his first products were table-knives and butcher-knives, most of which he made out of old wagon-tires. He applied wooden handles, making the tiny rivets with which they were fastened, and sharpened the blades on a grindstone which he had slowly and laboriously fashioned from a chunk of mountain sand-stone. He occasionally found sale for these tediously made tools at the settlement at small prices.

It so happened that I did not see Coke again until the summer of 1919. I found that he had greatly enlarged his list of products. Instead of a bellows he now has a rotary blower, of which he made every part except two sprocket wheels and a chain, which he salvaged from the wreck of two old bicycles, and some scrap tin with which the fan is encased. He got the idea of the machine from seeing a small portable forge in operation at a coal-mine several miles from his home. He has equipped his blower with a treadle as well as a crank, so that he can run it with both hands free to tend his fire and turn his iron. And that blower creates some wind, too!

Coke apologized for the unevenness of the blades on the fan. "I didn't have no square to true 'em up with," said he. He apologizes thus for many of his products. "That ain't made just as it ought to be," he will say; or, "I could do better next time." He seems to have no illusions as to the quality of his handiwork.

He makes practically all the tools that he uses in his farm-work—hoes, mattocks, axes, hammers, hatchets, frows, and so on. It is true that they will not hold an edge as well as steel tools, but he gets much good service out of them. He makes hinges for the gates and doors on the place, as well as latches and bolts for the latter. A razor-back pig comes squealing up to the fence with a small bell clanking at its throat, and you learn that Coke made the bell. He is constantly on the lookout for old iron that he can remelt and use in his work. Parts of old wagons, plows, and other agricultural machinery, discarded tools—all's grist that comes to his mill. Of course he must always carry such junk several miles, either on his shoulder or on the back of the old mule that he now possesses, and of which he is as proud and as careful as if she were a race-horse.

Coke works very skilfully in wood, though he does it, as a rule, only when necessary. He has made some gunstocks that can't be beaten by anybody. He built a kitchen cupboard for his mother, and had no dressed lumber out of which to make it, either—just a few pieces of rough stuff, and the rest he sawed out himself. He made a plane with which to dress it, even making the blade of the plane in his forge. Ordinary kitchen chairs are easy for him, but his best job in the line of furniture is his mother's large, comfortable rocking-chair, which has knobs and curves not easy for a man to achieve when he is absolutely ignorant of geometry or of design, save in a purely intuitive way.

As an example of the expedients to which he has resorted on many occasions, Coke some time ago found in a deserted cabin the burrs or grinding portion of a coffee-mill. He took this home, cleaned and removed the rust from it, built a box for it, forged a crank and attached it, and then, being in a hurry, stuck a piece of corncob on the crank for a handle! The cob does just as well as a polished handle; the calloused palms of the family don't mind its roughness. In this mill the family not only grind their coffee, but grind their corn into meal.

This makeshift handle is an indication of the fact that Coke doesn't care

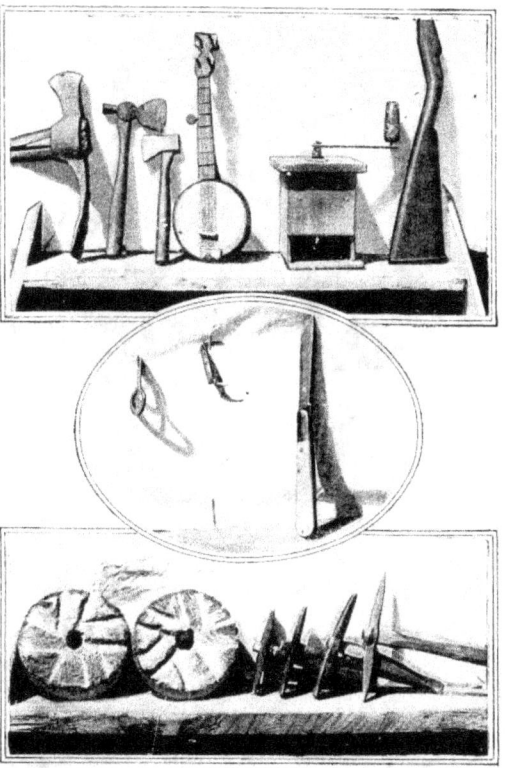

In the top picture are a few of the many crude tools and instruments Rice has made. He can make all kinds of pocket-knives, from the big five-inch-bladed jack-knife suitable for skinning animals to very small ones. Compare those shown in the middle picture. The lowermost picture shows two millstones and four tools. Rice made them all

much for woodworking, anyhow. He says so. Iron is his true love. He will even work in stone at times.

It is twelve miles from his home to the nearest grist-mill. He decided recently that he wanted a little mill of his own. So he has been slowly working out a pair of small millstones, using a variety of mountain sandstone, which is welded together with so much iron that it might almost be called a low grade of iron ore.

But probably his cleverest work is displayed in his pocket-knives. He can make them of all sizes, from the big five-inch-bladed jack-knife suitable for skinning animals or for cutting one's way through a rhododendron thicket, down to the tiny knife shown in the accompanying picture, whose handle is only a trifle over two inches long. The blades shut up easily and accurately into the case, just like a factory-made knife. The handles he makes from cow's horn or from bone. He admits that he has sold two or three of these small knives for as little as twenty-five cents apiece!

Coke still suffers from a lack of adequate equipment. For an anvil he has only a heavy steel axle, fastened across two blocks of wood. The thimble serves as the horn of the anvil. When we first visited him last summer, he expressed a hope that he could some day have an emery wheel on which to sharpen his tools, the ordinary "grindin' rock," as he and all mountaineers call it, being too soft. He has since been supplied with this article.

He digs the coal that he uses in his forge from a vein that outcrops under a cliff two and a half miles from his home, and brings it in, a boxful at a time, on his shoulder.

Coke was taken into the army in the spring of 1918. The mountain soldier did not get any farther than Camp Pike, Ark. No one in the army seems to have comprehended his gifts. He was much interested in the automobiles and trucks he saw, and begged to be allowed to work at something in the motor department, either as driver or repair-man. But he was esteemed too ignorant to be allowed to touch machinery.

There may yet be time to save this latent genius from sinking into the sodden apathy and ineptitude of the ordinary unambitious mountaineer. A few more years in his present surroundings, and his gifts will be lost. He says he would not stay in his present forlorn location an hour if it were not for his mother. She clings to the spot with the unexplainable, feral instinct of the old mountaineer, who sickens and shrivels when away from the big hills. Coke hasn't developed his gifts along the lines that would mean most to himself and others, because he doesn't know how to do so. He doesn't know how to get his wares to the proper market. He doesn't know how to capitalize his talent. Who will give him the chance?

Can You Make a Crate that Won't Break?

Some results of the Forest Products Laboratory's experiments in the testing of boxes, barrels, and crates

By Lloyd E. Darling

SHIPPERS have, within recent years, rather been forced to the conclusion that there is a science and art in the building of shipping containers. Before the war this conclusion was thrust upon them by the mishaps that befell them in attempting to reach South American and other foreign trade.

But during the war the idea crystallized. It culminated in special activity on the part of the Forest Products Laboratory at Madison, Wisconsin—a branch establishment of the United States Department of Agriculture. Here a school was brought into being for the training of men having to do with the packing and crating of quantities of material sent overseas.

The tests made here, primarily instituted to promote war efficiency, have provided a fund of information that will be invaluable to manufacturers and shippers in peace time. The shipping departments of the larger manufacturers have already made tests and experiments of their own. In fact, every package sent out is an experiment. But the subject is still dark and uncertain in a great many of its aspects. So we pass along the results of the Forest Products Laboratory's experience because of their special value in contributing to the general fund of knowledge on the subject.

What methods are followed in this packing-box testing laboratory? We may preface an explanation by quoting D. L. Quinn, in charge of the Madison Laboratory's work:

"A properly designed packing-box is one which has enough strength in each part for the purpose for which it is intended. More than this, it should have no more strength in any part than is necessary to balance the average strength in every other part. The data necessary for designing such a box cannot be obtained from observation of boxes in actual commercial service, because the observer sees the box only after it has completely failed. He does not see the beginning of the failures; and he does not see, and consequently cannot measure, the hazard which completes them.

"A failure frequently bears evidence in itself of the cause of the damage. There is, however, no way of determining from a study of the failure the amount of force exerted by the damaging cause. In cases where several causes have been active, it is impossible to identify each of them."

He concludes, therefore, that the immediate task of a testing laboratory is to go at one thing at a time. That is, the tests applied to containers should be of such a nature that defects

What happens to a box after the freight-handlers have given it their attention. The Forest Products Laboratory hopes to prevent this

Up and down it goes! The block-and-tackle machine finds out whether a box will stand falls

will be shown up singly, and under such conditions that the stress applied, and also the results, may be carefully checked up and kept track of. By thus measuring defects individually and "pooling" the results, so to speak, the tester will be able to gain a comprehensive knowledge of the all-around strength of the box or crate, and its probable capabilities when in actual service.

Initiating Tests

The "compression along an edge" is one of the tests applied. This consists in standing a box upon one edge and applying a steady and constantly increasing pressure to the diagonally opposite edge. The longer the box holds out against such a squeeze, the better box it is. Another box of the same kind will be stood up on one corner, and then pressure applied to the opposite uppermost corner. The longer the box holds out against this test, the better box it is. The illustration at the bottom of this page shows details of the process.

Another test applied to boxes in a laboratory is to drop them. Simple, isn't it? Even a youngster applies that principle. If Amy Brown has a doll with a china head, and she deliberately drops the doll on a cement sidewalk, and the head doesn't break—why, naturally, the doll is a better one than Dorothy Smith's, which does break under such conditions. Simple as A B C.

So in the laboratory they drop boxes. They haul them up off a solid floor with a block and tackle, and then let them come violently down again. They let them come down on an edge, then on a corner, on an end, on a side—every way. If the box

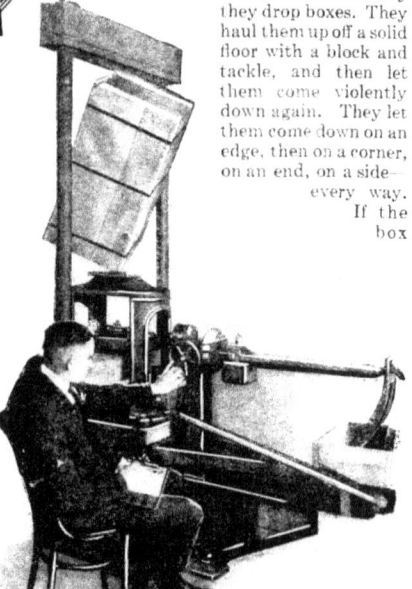

Squeezing a box cornerwise to find out how much persecution it will endure before giving in. Good boxes stand a great deal of this

stands up under all that, it is a good box and can be depended upon to stand up under actual service; because such man-handling as this is what the box actually gets from the freight rustlers and dock employees. The testers reason that it is better to find out in a laboratory whether a given kind of box is no good for its purpose than to find out on the road, and perhaps lose valuable goods in the process. That this is sound reasoning, is evident.

The Revolving Drum

A particularly choice number in the repertoire of the testing men is to run a box through the revolving drum apparatus. This looks a good deal like a concrete-mixer, only its drum is larger. A box is put inside the drum, then the drum proceeds to go round and round in regular concrete-mixer fashion. Meanwhile the box is bumping around among all kinds of pillars and posts inside. When it comes through this test—if it does come through—it is initiated indeed.

The drum revolves so slowly that, as the box drops from one projection to the next, the beginning of breaks and cracks can be noted. The weak points in the construction of the box will, of course, readily make themselves apparent, and in the next box made along this line, the construction can be altered in such a way as to avoid defects of the kind that the test brought out. By continually building and testing in this way, boxes are ultimately developed that are equally

strong in all features. Like the "one-hoss shay," they will go to pieces all at once, if they go to pieces at all. Their failure will not be because of single defects, such as nails pulled through soft wood, or because ends of boards split or broke off, or because the wrong lumber was used for that particular box

Round and round goes the box in the "concrete-mixer machine," receiving bump after bump. If the box has any hidden defect it is sure to be revealed during the process

in the first place. The box constructor will *know* what he is doing.

The laboratory has found that the length of a nail, rather than its gage, seems to be the principal factor in its holding power. Since nails split wood, it has been found that it is desirable that as small gage nail be used as can be driven by a nailing-machine. Mr. Quinn finds that standard cement-coated box nails seem to be most desirable from the standpoint of both efficiency and economy.

"Nails," says Mr. Quinn, "should be driven flush with the wood. They

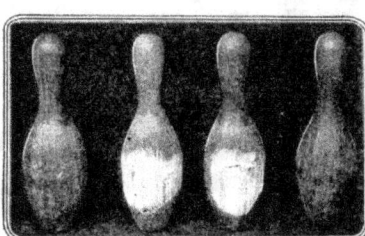

These blocks of wood have been through a stretching-machine to see how much tensile strength they have. The cracks show where the wood broke

should never be over-driven. Over-driving causes the heads to break the fiber of the wood under them, often to such an extent that no holding power is left in the wood. The first strain on the box then causes the wood to pull away from the nails. The evil effect of over-driving nails naturally increases as the material in the boards is made thinner. But *in any box* over-driving is bad practice, and should be guarded against.

"The size of the nail to be used depends upon the species and the thickness of the lumber in which the points of the nails are held. With the woods white elm, red gum, sycamore, pumpkin ash, black ash, black gum, tupelo, and soft or silver maple, as well as hard maple, beech, oak, hackberry, birch, rock elm, and white ash, the penny of the nail should be the same as the thickness of the lumber expressed in eighths of an inch. The following woods take the next penny larger: white pine, Norway pine, aspen, spruce, Southern and Western yellow pine, cottonwood, yellow poplar, balsam fir, chestnut, sugar pine, basswood, cypress, willow, noble fir, magnolia, buckeye, white fir, cedar, redwood, butternut, cucumber, alpine fir, lodgepole pine, hemlock, Virginia and Carolina pine, Douglas fir, and larch.

"Sixpenny nails and smaller should be spaced not more than two inches apart when driven in the side grain of the end, and not more than one and three fourths inches apart when driven in the end grain. The spacing of nails in end construction should be increased from the above, one fourth inch for each penny over six."

More Facts About Nails

One hardly realizes the force it takes to pull a nail out of a wooden board. In experimenting with the holding power of cement-coated, plain, and barbed sevenpenny nails, driven one inch into dry wood, Mr. Quinn obtained the following facts:

Using a cement-coated nail, it takes about a 225-pound pull to get a nail out of one inch of longleaf yellow pine, 133 pounds out of basswood, and 430 pounds out of beech. Plain nails took 140 pounds with longleaf pine, 82 pounds with basswood, and 400 pounds with beech. Barbed nails required only 110 pounds with longleaf pine, 70 pounds with basswood, and 332 pounds with beech.

Effect of Wet Lumber and Storage

The following chart, based on the Forest Products Laboratory's investigations, shows how the strength of a box is altered by moisture and by neglect in general. Keep your boxes dry, but not bone-dry.

	Strength Relation Per cent
Nailed and tested at once at 15 per cent moisture....	100
Nailed and tested at once at 30 per cent moisture.....	90
Nailed at 15 per cent, tested at 5 per cent moisture, 4 months' storage.........	75
Nailed and tested at once at 5 per cent moisture......	50
Nailed at 30 per cent moisture, tested at 5 per cent moisture, one year in storage..................	15
Nailed at 5 per cent moisture, tested at 35 per cent moisture, stored 2 weeks in exhaust steam...........	10

"Tom Thumb" is what the Baltimore and Ohio christened the first locomotive that was ever built in America. Don't sniff too much at this engine. It was a marvel in its day, which was in the thirties

It's strange how the old horse-drawn coach dominated the early railways. Look at this picture of the first American railway train, which made its trip over the Mohawk and Hudson the line between Albany and Schenectady, from which the New York Central system grew) on July 31, 1832. The poor guards had to sit on

Big Oaks from

If you are an inventor don't be
show how some of the men who

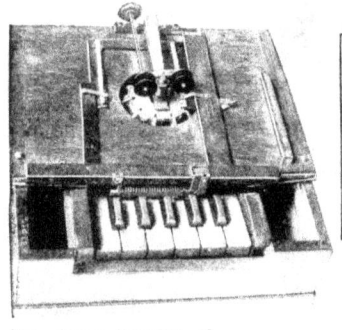

The pioneer inventors of typewriters never thought of writing business letters by machine. They merely wanted to help the blind! This is the invention of Sholes, Glidden and Soule

See this little sketch? It was Edison's first suggestion of the phonograph

Cast your eye to the right and upward and you will behold a picture of the model of a locomotive that Matthias W. Baldwin, of Philadelphia, filed with the patent office 'way back in 1842. He explained how he proposed to convert the wheels into driving wheels by means of coupling rods. His idea has become the standard practice

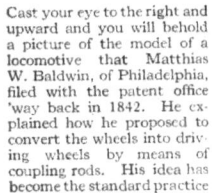

Here is the first telegraph wire that ever carried a message. "What hath God wrought!" were the words that expressed America's wonder in 1835

When next you pick up the telephone to say that you are not coming home for dinner, remember these first crude instruments of

Farmers came for miles around to see Cyrus H. McCormick turn his reaper loose in 1831. This is a reproduction of a contemporary engraving of that historic event. One know-it-all slapped his knee and said: "That invention is worth a hundred thousand dollars!" Too bad he couldn't have lived to see how the industry grew

box-seats, like drivers. Incidentally, the engineers of that day were as sensitive as prima donnas. The first American-built locomotive, the "Best Friend" of Charleston, built in 1830, blew up after a few months because its attendant lashed down the safety-valve. He was annoyed by the sound of escaping steam

When the inventors got over the idea that typewriters were to be made only for the blind, they began to meet the business man's needs, and the first Remington was produced. Notice the lid, that could be swung down over the keyboard and locked. As for the typists of the early days—well, we like them if this is a specimen

Little Acorns Grow

too modest. These illustrations conceived the big inventions began

You still see some hand presses similar to this of Gutenberg's of 1450 on the right. Gutenberg's name is identified with the invention of movable type rather than with presses

Alexander Graham Bell's over which the first message was sent on March 10, 1876. It was telephoned from the attic to the second floor below

It is the outsider that revolutionizes industries. Eli Whitney was a modest school-teacher when he invented the cotton-gin

Selden was a patent lawyer in Rochester when in 1877 he filed an application for a patent on what is the modern automobile. Over $2,000,000 royalties were paid by about eighty automobile manufacturers

This is where the enormous packing industry was born. Here it was that Gustavus Swift, the founder of the business now known as Swift & Company, slaughtered the eighteen-dollar heifer that started him in business

And this is Henry Ford in the first universal car. Ford had thrown up a job as a steam engineer in Detroit to produce this. He fought the Selden patent and beat it

Paper—Made, Used, and Scrapped in a Day

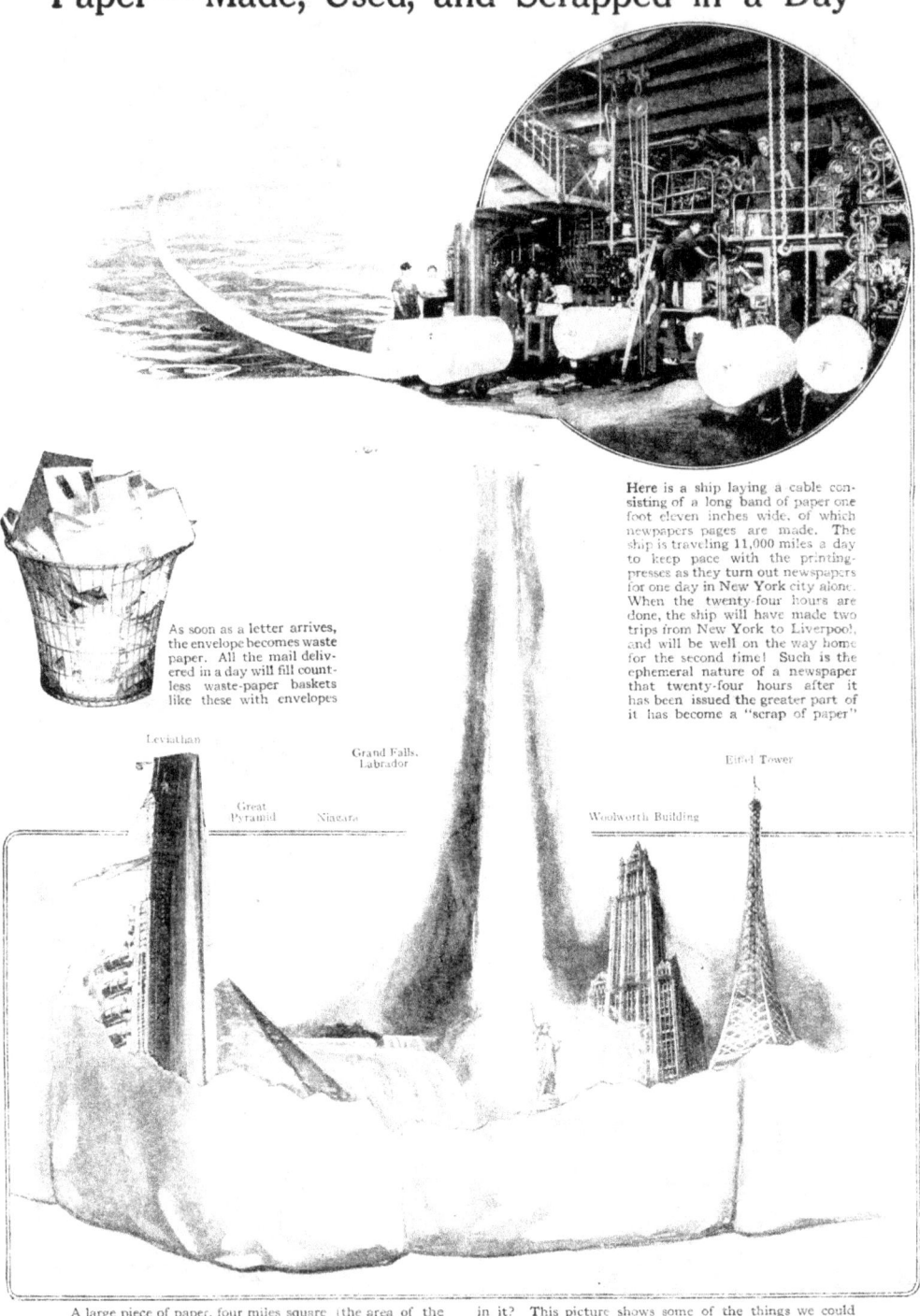

Here is a ship laying a cable consisting of a long band of paper one foot eleven inches wide, of which newspapers pages are made. The ship is traveling 11,000 miles a day to keep pace with the printing-presses as they turn out newspapers for one day in New York city alone. When the twenty-four hours are done, the ship will have made two trips from New York to Liverpool, and will be well on the way home for the second time! Such is the ephemeral nature of a newspaper that twenty-four hours after it has been issued the greater part of it has become a "scrap of paper"

As soon as a letter arrives, the envelope becomes waste paper. All the mail delivered in a day will fill countless waste-paper baskets like these with envelopes

Leviathan

Grand Falls, Labrador

Eiffel Tower

Great Pyramid   Niagara

Woolworth Building

A large piece of paper, four miles square (the area of the daily newspapers of New York City), might come in handy for wrapping up a parcel. What could we put in it? This picture shows some of the things we could wrap up in the bundle, whose side is about 3,400 feet long. But the parcel is not nearly full, as you see

Save Money by Baling Your Clothes
Apply this lesson learned in the war

By Michael Connors

IF you are a manufacturer or an exporter of clothing, it will mean dollars and cents to you to learn how the Quartermaster-General's Department shipped uniforms and shoes to France in a way that saved the Government eighty-five million dollars in a little more than a year.

Few of us realize how desperate was our need for ships at a time when German submarines were sinking six hundred thousand tons a month. "Give us ships, ships, and more ships," was the demand of those who had to equip and feed our Army in France.

In the autumn and winter of 1917-1918 it became apparent that if the war was to last very much longer there would not be shipping space enough to carry the equipment and supplies required by the forces that we intended to send overseas. New methods of shipping equipment had to be invented, and the man who invented them was Major David T. Abercrombie, of New York.

Compressing Twenty Cubic Feet into Four

Clothing is usually shipped in wooden packing-cases. But during the war wood became scarce. Major Abercrombie estimated that 72,762,-300 feet of board lumber, 2,425 tons of nails, and 404 tons of strapping would be needed. The lumber required would have denuded 7,552 acres of timber-land—lumber that could not be grown again for thirty years.

"We won't use packing-cases," decided Major Abercrombie.

"But what will you do?" they asked him.

"Bale everything," was the answer. And that is how it came about that the Government saved eighty-five million dollars.

Major Abercrombie experimented, and found that clothing could be squeezed so that, roughly, twenty cubic feet of material could be made to occupy four cubic feet of space. Consider that ship tonnage is charged for on a cubic foot basis, and that during the war the rate was eight dollars per cubic foot, and you can see at once that by compressing twenty cubic feet to four cubic feet a saving of ninety-six dollars was at once effected. Then

We sent uniforms to France, not in packing-cases, but in bales. The outer burlap was used for the making of two million sand-bags

consider that during the war 1,371,000 bales were shipped, that the net saving on each bale ranged from fifty to sixty dollars, and you can see that the Government saved eighty-five million dollars. Major Abercrombie's project also released 8,180,000 cubic feet of shipping space.

Major Abercrombie discovered that proper baling depends on the proper folding of a uniform. Baling should crease a uniform along the right lines.

Folding the Garments Scientifically

To carry out Major Abercrombie's process, the garment is first carefully folded to a form. A given number of folded garments are then packed in a brick container, corded together so that the bricks cannot fall apart. The folded uniforms in their brick container are then placed in a machine.

Introducing Major David T. Abercrombie, Q. M. C., the man whose baling process saved the Government eighty-five million dollars in a little over a year's time. Instead of shipping uniforms in packing-cases, Major Abercrombie baled them by a process that he invented himself

First of all, hand pressure is applied to form a fairly compact bale. But that bale is not compact enough for Major Abercrombie. It is subjected to the additional squeezing effect of a power baler until it assumes the highly compressed form that he has in mind.

Waterproof paper is used as a lining for each bale. Major Abercrombie had to invent a paper of his own—a creped waterproof paper that would stretch thirty per cent in one direction and ten per cent in the other. Thanks to that paper the bale becomes absolutely waterproof.

In order to save shipping space when the submarine was doing its worst, the Quartermaster-General shipped army shoes abroad in bales, the space they occupied being reduced by about thirty-four per cent. Why not use this system in time of peace?

Not a Bit of Leather in These Shoes

The mechanical cobbler takes on the canvas shoe and reduces the labor to a minimum

The patterns of the uppers are so laid upon the sheet canvas that every available inch of space is utilized. There is practically no material wasted by this method

The assistant at the left of the illustration is shown holding the knives that cut through twenty thicknesses of material at once

Photographs
© Publishers Photo Service

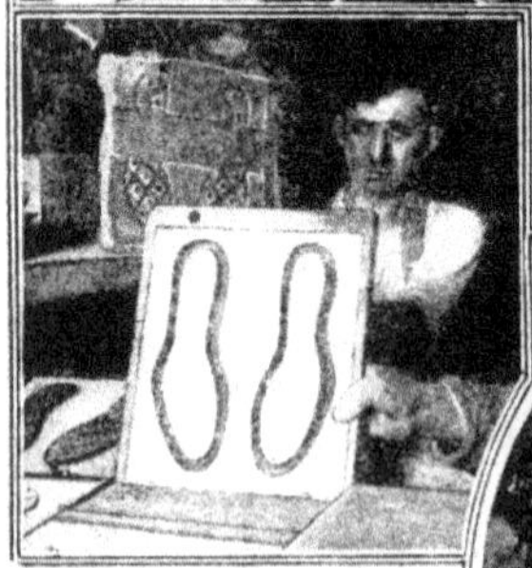

One of a pair of hinged plates between which slabs of rubber are pressed to shape them into soles. The surplus rubber is thrown back into the mixer and melted again

A mechanical Jess Willard that delivers twenty punches to the eyes every ten seconds is an ingenious machine which clinches the hooks and eyelets into the uppers

When the uppers are made this machine pulls them smooth over wooden lasts, turning down the edges under the sole

The shoe finally goes to the ovens, where the vulcanizing process cements the soles, heels, edging, and other parts

Peace-Time Peeps through Periscopes

Useful in many places besides the submarine

Dashing through the waves in a high-powered hydroplane, he watches the path ahead by looking through a periscope; inside a periscope there are two mirrors set at an angle to produce double reflection

© Keystone View Company

© Underwood & Underwood

Perhaps there will be dirty work at the cross-roads to-night, but it won't be at these cross-roads. For the careful gate-keeper has a periscope attached to his house, through which he can see the trains approach around the curve below

In this case the trench is an orchestra-pit, and No Man's Land the stage. By watching the actors through a regular trench periscope, the musical conductor, Herbert E. Hyde, was able to lead his orchestra

© Keystone View Company

This motorman can see people boarding his car in the rear, though he looks straight ahead. Eyes in the back of his head? No; he uses a periscope

"Here comes the bride!" said the woman who had brought her periscope. She saw Princess Pat's wedding procession. The other women could see none of it, and so they stared at the photographer

© Keystone View Company

They alternately glue their eyes to the lower end of a periscope, the top of which overlooks the edge of a fence on the other side of which a thrilling ball game is in progress

The Same Old Job

With power - driven brushes and vacuum cleaners women do their house-cleaning now on railroad cars

Removable parts of doors, frames, and moldings are taken down and given a thorough cleansing

An energetic woman cleaning the wood-work in the confined space of the corridor of a car

YOU know what spring house-cleaning meant before the days of vacuum cleaners; but do you realize what it means to give a thorough cleaning to a whole train of railway cars? It means all the dust and dirt, all the hard work and discomfort of house-cleaning raised to the *n*th power. Where men and women for doing that kind of work are available, the railway companies continue to adhere to the old hand method of cleaning their coaches; but the difficulty of obtaining cleaners is steadily increasing.

Labor-Saving Car-Cleaners

This scarcity of workers has led many railroad companies to the employment of labor-saving machinery with which one man or woman can accomplish as much in an hour as three or more cleaners used to accomplish in the same time. The use of such labor-saving machinery and mechanical appliances has proved a remarkable success economically and highly satisfactory as a hygienic measure. The appliances are of small weight and may be used by persons of average physical strength for many hours at a time without undue fatigue. They are so simple of construction that the cleaners using them need not be specially trained in the proper use of the machinery and auxiliary appliances.

For cleaning the woodwork, rotary brushes and buffing-wheels, driven by compressed air or electric motors, are used. The brushes, which are in the form of disks or cylinders, are mounted on flexible shafts connected with the engine that furnishes the power. These brushes make from five hundred to eight hundred revolutions a minute. They weigh only a few pounds, so that they can be operated without excessive effort by any woman of average strength.

If the cleaning is done in the car-shop, the rotary brush is suspended by a wire from a trolley running alongside the car-track. To facilitate the raising and lowering of the brush while working on the outside of coaches, the suspension wire runs over a pulley and has a counter-weight at the other end.

When the brushing and cleaning is to be done on the inside of a coach, the driving mechanism of the brush is mounted on a stand resting on casters, or is suspended from a bar temporarily placed across the compartment of the coach from one luggage-rack to another.

Disk-shaped brushes are used where it is necessary to clean cracks, corners, and other confined spaces, while cylindrical brushes are used for cleaning large flat surfaces.

The Railway Cleaning Shop

The window-frames and other removable parts of the coaches are taken to a shop especially provided for that purpose, where they are placed on tables in a horizontal position and brushed and cleansed with rotary brushes and buffing-wheels. The re-painting and re-varnishing is done at the same shop.

The cloth of the upholstered seats and back-rests and on the lower part of the doors of the coaches is cleaned by means of a vacuum cleaner. After the dust has been removed, the cloth is sponged with a cleaning fluid and then thoroughly brushed. Next the cloth is washed with water, and then the vacuum cleaner is again used to remove the excess water from the upholstery. How thoroughly this is done by the vacuum method may be inferred from the fact that the complete drying of cloth - covered cushions requires only a day or two. When necessary, disinfectants may be applied to the upholstery during the cleaning process.

Each coach is brushed inside as well as out every day, but only once every fortnight is it given a thorough scrubbing and cleansing. At the same time, necessary repairs are made and missing parts replaced.

Upholstery is freed from dust and dirt by a vacuum cleaner and a rotary electric brush

Women removing the dust and soot from the outside of a coach with an electric brush

Why Small Models Won't Work When Enlarged

The Patent Office abandoned them long ago

By C. A. Briggs

MANY years ago the Patent Office abandoned the practice of requiring working models when issuing patents. This was done in the face of increasing complexities in the devices presented for patenting. Now patents are issued on the basis of drawings and descriptions, except in special cases.

It was found that models were often more misleading than instructive. The full-sized model may prove impractical even when the small one seems to work perfectly; and the reverse may be true—a large design may give results to an extent that was not indicated at all in the small model.

Probably no better device for illustrating the achievements and perfection of modern design could be found than the typewriter of the present day; yet the same arrangement when made on a large scale will serve to illustrate just what a design should not be. A fourteen-ton typewriter was built for the San Francisco Exposition by a typewriter company. When this large typewriter was operated, the arm carrying the type on the end moved ponderously up and struck with a *pung*. It then fell back with a *clank*, and bounced two or three times before coming to rest. These arms were so heavy that a special engine had to be installed in the foundation below the typewriter to operate the parts.

The spring that caused the carriage to move was a relatively weak one. If this spring had been made to scale, its use would soon have wrecked the machine. On the ordinary sized typewriter of this design a speed of one hundred and seventy words a minute has been obtained by the best experts. The huge model was capable of about thirty letters a minute.

There are certain relations between the strength, weight, inertia, size,

This tiny engine is a speed demon; the piston is shot backward and forward by the rapid heating and cooling of the cylinder—the heat being produced by an alcohol flame and the cooling by the shutting of an automatic valve. But would this hold good for a large engine? That is a question

time of action, and deflection that dominate design; and good design is always consistent in this respect.

Some time ago the writer was shown a toy engine. The parts comprised a flywheel, a crank, a piston, a cylinder with an opening at one side near the closed end, and a valve consisting of a piece of thin sheet-iron that could be moved back and forth across the opening by means of a rod engaging a cam mounted on the axis of the flywheel. Just in front of the valve opening there was a tin lamp with a wick for burning alcohol.

When the engine was started the piston moved away from the opening and thus sucked in the hot alcohol flame. Just before the end of the stroke was reached the valve would close the opening, and the cooling of the hot gases would then create a partial vacuum. The result was that the piston was jerked back. The power and speed of this little engine was surprising. When going well it would simply roar. The behavior of the engine was so striking that it occurred to almost everybody, on seeing it run, that it would be desirable to make a large engine on the same principle.

However, one thing on which the action of the engine depended was the very rapid cooling of the hot vapors and gases when they were shut up in the cylinder. Now, as the size increases, the quickness with which the gas cools falls off very rapidly; and if a large design were made the experimenter would find his attempt a failure.

The same thing applies to electrical devices. A small needle can easily be magnetized so that it will support its own weight. When the needle is increased in size until it becomes a bar, the point is soon reached where the bar will not be able to support its own weight.

In electrical machinery the heating of the parts is often the limiting factor in the capacity or use of a device. The heating tends to increase with the volume, and the ability to dispose of this heat increases as the surface is exposed. The volume increases as the cube of the dimensions, and the surface as the square of the dimensions. It is therefore necessary in the larger sizes of machines to alter the design to provide, among other things, the necessary arrangement for getting rid of the heat.

This fourteen-ton typewriter is an exact enlargement of one of the present well known machines. Its operation is difficult and slow. Small models often fail to work after they have been enlarged

Ways to Make the Freight-Car Do More Work

The freight-car has reduced the cost of long-haul shipments; but delays in loading and unloading mean heavy money losses, so that the cost of living goes up. A St. Louis engineer, Kirchner, advocates a car in which a skeleton framework on wheels carries a number of box-like sections which can be loaded and unloaded in a few moments by means of electric cranes

FEW people realize that approximately twenty cents out of every dollar they spend for things to eat, wear, and use goes for freight. Nor do they realize that in the case of perishable goods (milk, ice, etc.) the cost of transportation from the point of origin to the final consumer is often as much as fifty cents for each dollar spent.

But what are we going to do about it?

Government Eliminates Some Evils

The government has taken over the railroads, and has eradicated many of the evils. No longer, for instance, will cars be held as warehouses for excessively long periods at ridiculously low charges; no longer will cars be loaded to only one quarter capacity in order to meet a similar service in next-day delivery by a competing line.

But all these steps toward efficiency do not solve the problem, according to a St. Louis engineer named Kirchner, because they have not utilized freight-cars to the utmost. Whenever a freight-car stops moving, it becomes merely an expensive warehouse; the expensive car framework of the platform, axles, wheels, bumpers, air-brakes, and other devices cease to serve a useful purpose.

All this is true because muscle instead of machine power is used to unload a very large percentage of box-cars.

The yearly freight-car mileage is only eight thousand. This means that a freight-car moves only about twenty-two miles a day, although the average speed of a freight train is thirty miles an hour. Even at an average speed of only twenty miles an hour, and with only twelve hours in motion daily, the possible mileage is 240. Why the difference between the actual mileage of only twenty-two miles and the possible figure of 240? Because the vehicle part of the car is idle while the body is being unloaded.

Suggestion for a New Style Car

In order to overcome the difficulty, when either full carload lots or less than carload lots are shipped, Mr. Kirchner would have the railroads employ a car in which a skeleton framework mounting the wheels, axle, and all movement parts carried a number of rectangular box-like steel sections, which could be loaded and unloaded from the car at each stop in a few moments' time by means of an electric crane.

The sections are made in two main sizes, called the primary and secondary units. Five of the primary units, each of ten tons load capacity, go to make up the equipment of a 100,000-pound capacity freight-car. The secondary units are of two and one half tons capacity each, and twenty of them are carried on one car to make up the same fifty-ton capacity. The primary units are for through freight, while the secondary units are for local freight.

Some of the Advantages

The secondary units may be used to carry still smaller units of from six to eight hundred pounds capacity, the latter units being so designed as to fit exactly inside of the secondary unit in much the same manner as a child's set of wood blocks fits into its outside box. All of the units are made of steel from three sixteenths to a quarter of an inch thick, and are made water-, air-, and weather-proof, so that the goods contained in them cannot spoil. To prove this contention, two of the boxes were filled with two different classes of materials, one with flour and one with sugar, and left to stand in the open air for two years. When opened the contents were in perfect condition.

The advantages claimed for the method of shipping in units are that it would provide for greatly increased car mileage, resulting in increased car profits and decreased rates to the shipper; it would save back hauls of empties; it would practically eliminate demurrage charges; it would save the cost of handling and the cost of crating goods; it would save on insurance to cover breakage, since all the units are of steel; and it would eliminate the shoveling of coal or other bulk materials for loading or unloading.

A Typical Case

Suppose a farmer is threshing wheat. With the unit method he will get some of the small secondary units, haul them to his farm, load them with the wheat, and then haul them back to the nearest rail station, where they will be placed in one of the primary units and sent along to their destination in the first car that stops. They will go at the carload rate, too, since the car will be carrying a full load in the other units. No bags, boxes, or barrels have been used by the farmer, and he has not had to stand any loss for wasted grain, which on the average amounts to eight bushels for every thousand bushels shipped.

Or, in the case of the wholesaler shipping to the retailer, the wholesaler gets an order for a certain quantity of canned goods, tomatoes, soups, peas, pears, etc., or a mixed order for groceries, which are packed into one of the secondary units and the unit sent to the retailer, who shelves the goods and returns the unit to the nearest agent in exactly the same manner

as the Standard Oil Company has its barrels returned.

For shipping wagons, automobiles, or heavy pieces of machinery, the compartment freight-car is obviously impracticable.

The railroad loading and shipping

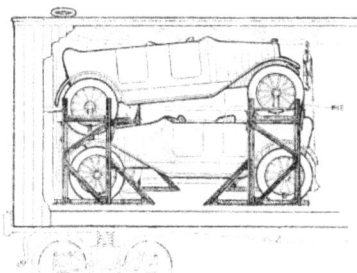

Not more than three automobiles can be shipped in a freight-car. By means of this trestle twice as many can be carried

deck placed on the market by a Western manufacturing concern offers a solution of the economy problem in such cases by providing a method of loading automobiles or other large pieces of freight on two decks, one above the other.

The device consists of a trestle constructed of structural steel framework which can be quickly assembled or taken apart. The trestle

is adjustable, and may be placed and fastened down on or in any freight-car wherever it is required. The loading deck may be used on flat-cars or in box-cars, and makes it possible to increase the capacity of a car from sixty to one hundred per cent.

A Saving in Freight Charges

To illustrate the saving of space made possible by this device, let us consider the case of the average flat-car, which will hold three large automobiles, standing end to end on its platform. With the steel shipping deck three more automobiles of the same size may be placed above the three machines resting on the platform of the car, without exceeding the carrying capacity of the freight-car and without increasing the height of the load beyond the normal height of a box-car. This means a doubling of the loading capacity of the car and a large saving in freight charges for the shipper.

The automobile industry in normal times uses about 90,000 freight-cars every month. Doubling the load of these cars would release one half the number of cars every month for other purposes. It may not always be possible or feasible to use the double-deck method, but it is safe to assume that from 25,000 to 30,000 cars could be diverted for carrying other much needed freight. An increased load of two tons on every car would be the equivalent of adding 200,000 new cars to our rolling stock. These shipping decks would also be highly useful for shipping food.

By means of the steel shipping deck, three more automobiles of the same size may be placed above the three machines resting on the platform of the car without exceeding the carrying capacity, and without increasing the height of the load beyond that of a box-car. The same principle holds good for crated eggs, vegetables, and fruit

Do It with Tools and Machines

The portable belt conveyer has found a wide range of usefulness in loading and unloading cars and trucks and in storing material such as coal, coke, sand, and gravel

It is a portable outdoor lighting apparatus, which weighs only forty pounds, producing a powerful white light at low cost

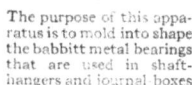

The purpose of this apparatus is to mold into shape the babbitt metal bearings that are used in shaft-hangers and journal boxes

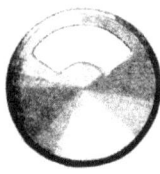 

This ammeter affords the only means of testing the condition of a dry-cell. Contact may be made directly to the battery

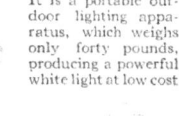

(Right) Emery wheels partly covered with a hood, which catches the dust from the tire casing and then draws it off through a blower

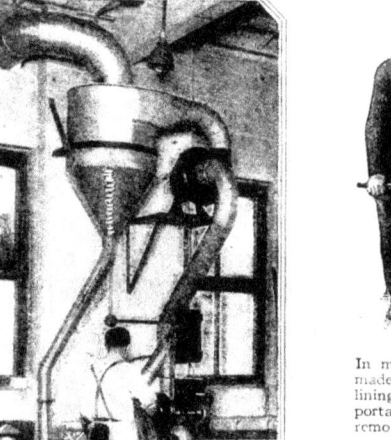

In manufacturing boots, those made with the so-called friction lining stick to the lasts. A portable boot-stripping machine removes them quickly and easily

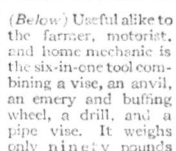

(Below) Useful alike to the farmer, motorist, and home mechanic is the six-in-one tool combining a vise, an anvil, an emery and buffing wheel, a drill, and a pipe vise. It weighs only ninety pounds

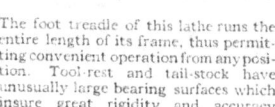

The foot treadle of this lathe runs the entire length of its frame, thus permitting convenient operation from any position. Tool-rest and tail-stock have unusually large bearing surfaces which insure great rigidity and accuracy

Car rails were formerly ground by hand, a long and tedious operation. Now the big traction companies supply their men with an electric motor which drives a flexible shaft attached to a rotary grinder

Do It with Tools and Machinery

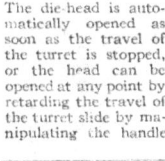

The die-head is automatically opened as soon as the travel of the turret is stopped, or the head can be opened at any point by retarding the travel of the turret slide by manipulating the handle

This kind of "dog" does not need a wrench to operate it: it simply opens, slips over the work, and the cam jaw closes automatically. It is made in five sizes

For metal-working machines comes a simple attachment that automatically opens and closes the work-holder during the feeding and retrograde movements of the tool

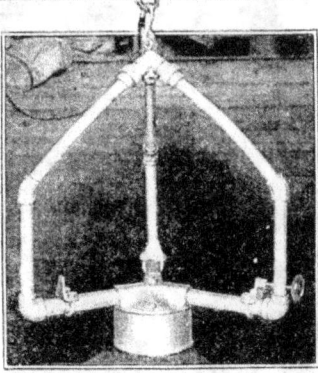

This electrically driven scraper scrapes aluminum and brass castings, automobile crank-cases, etc., and gives the operator a cleaner and more uniform job in one eighth the usual time

Winding armatures of electric motors is a long and tedious job when done by hand. Here is a specially designed motor-driven machine for doing the work easily

About all that could be demanded of a sand-riddle is embodied in this small motor-driven machine. A perfect whirlwind of sand comes out at the bottom when the operator scoops in a supply at the top

This "both-ways" planing tool permits a roughing cut to be taken on the outward or slower stroke and a finishing cut on the return or fast stroke: double efficiency

This apparatus includes an attachment for holding and feeding tool bits against the grinding wheel. The method eliminates the slipshod hand-grinding way, which gives an unsatisfactory edge

Do It with Tools and Machinery

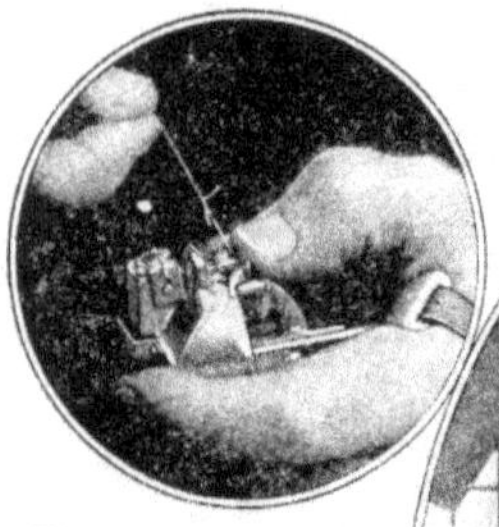

This new device which ties firm knots in threads, leaving the loose ends all very short and of a uniform length, should have a strong appeal to the textile manufacturer

Now comes the saw with an oiler attached to its handle. Simply press down the plunger and obtain enough oil on the fingers to pass over the blade of the saw

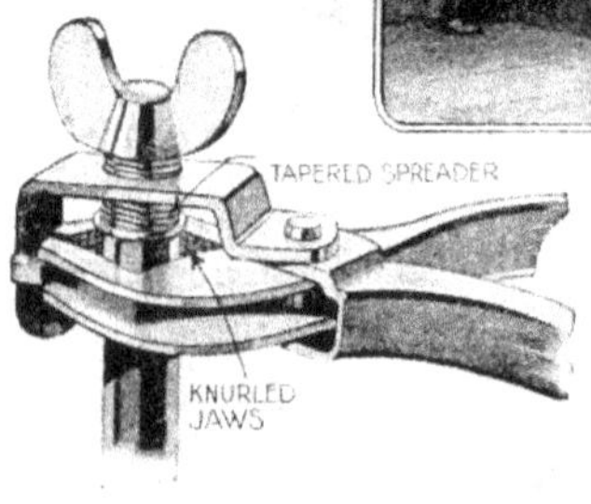

A monster "herringbone" gear weighing 141,210 pounds and capable of transmitting 1,500 horsepower at 40 revolutions a minute. The machine shown cutting the teeth of the gear is said to be the only one of its kind doing work of this magnitude

Capable of flanging a great variety of tubes, a new tube-clamping arrangement is so designed as to enter and expand the end of a copper or brass tube

The constant flow of oil through this drill is a certainty, for there are no loose tubes to loosen, bend or break away from the main body of the drill

Grease-cups attached to the connecting-rods of a machine, for example a locomotive, are a decided improvement over the old method of oiling by hand

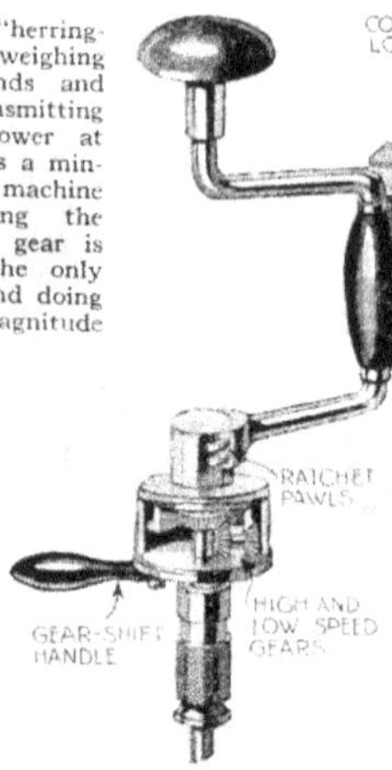

This bit-brace gives the bit various necessary speeds. By releasing one or the other of the upper dogs the high-speed device may be used as a ratchet

It is claimed that the device shown here will clamp all classes of work in almost any conceivable position

A noon and overtime rush in a certain factory restaurant deprived many employees of a considerable portion of their lunch period, hence this portable service stand

The "dope" for coating airplane wings is placed in a special air-tight can to avoid evaporation of the solvents

Machines that Lighten Work and Shorten Hours

Using its rear wheel to steer itself, this self-propelled threshing-machine gets its power from a 40-horse-power kerosene engine

The self-controller on this noiseless automatic electric pump starts and stops by water-pressure

A simple fool-proof magnetic rectifier that charges storage-batteries from alternating current. Its weight is not more than nineteen pounds

This new machine has an electric drive chuck equipped with a balanced hoist

Doing away with mirrors and crawling under machines, this dial indicator locates external and internal repairs to be made

Adjustment of the big guns "over there" was made by a small powerful truck moved by one man, getting the gun quickly into position for firing

A power-driven machine designed for under-cutting commutators without their removal from operating positions or disturbing the brush-holder rigging

A new engine stand, built principally for the Ford engine, allows repairs to be made above and beneath it

Do It with Tools and Machines

The teeth or "grippers" automatically adjust themselves to any shape set into the vise

The sliding jaw of this milling vise is adjusted to the work by clamping the bearing in place

Used as lead screws on lathes for the United States Government, these 55-ft. forgings were easily and safely carried to their destination by a five-ton truck

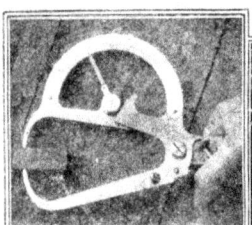

Indicating calipers are something that fills a long-felt want. It also combines both a rule and a magnifier

A handle that stays put is the boast of this vise. It may be moved to any position without varying the jaws

Presses that form and stamp automobile and railroad car parts into shape are ponderous machines, as may be seen from this picture

The amateur carpenter's tools will always be ready if he utilizes cigar-boxes thus

In welding high-speed steel to ordinary steel a saving is effected by using a shank of lower-priced material

Any boy can operate this new type of band-saw; it is practically fool-proof

WIRELESS

The Thermionic Detector

By H. J. van der Bijl, M. A., Ph. D.

An American vacuum tube. A dozen or more different kinds are produced in the United States. Most are distinguished for compactness, a c c u rately worked-out characteristics, and dependable operation

Telephoning to a plane high in the air—one use for thermionic detectors, or vacuum valves. A similar set on the airplane permits the aviator to hear what is said by the speakers below, and also to answer them; thus he can receive and give information and directions verbally

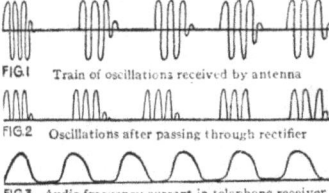

FIG.1 Train of oscillations received by antenna

FIG.2 Oscillations after passing through rectifier

FIG.3 Audio frequency current in telephone receiver

Curves demonstrating fundamental action of crystal detector. Complete story is shown

What the Germans think is high class in a vacuum tube. This one was captured a b o a r d a German submarine

THE art of radio communication owes a great deal of its recent remarkable development to the thermionic vacuum tube. This device is capable of performing a large number of functions. It can amplify telephonic currents with great precision, detect high-frequency electrical oscillations and make them audible, and produce oscillations or "sing" at almost any desired frequency. While it operates with facility over a wide range of frequencies its present singing register ranges from a fraction of a cycle per second to several million cycles per second; the range of power that it can handle is no less remarkable, considering that it detects oscillations the power of which is a small fraction of a watt, and can be made to produce oscillations representing a power of a thousand watts and more.

In practically all radio receiving sets in use at the present time, the crystal detector has been superseded by the thermionic detector, which is more sensitive. As time goes on, the radio experimenter will realize the growing need of a working knowledge of the thermionic tube. This knowledge can be acquired without much trouble by those who know something of the behavior of electrons and are acquainted with some of the fundamental principles underlying electric wave phenomena. The man who understands how and why the thermionic tube operates the way it does will be able to operate his tube under the best conditions and get the most out of it. Let us, therefore, briefly consider some of the fundamental principles involved in the study of this tube.

Crystal Detector Acts as Rectifier

The crystal detector can detect electrical oscillations because it acts as a rectifier; that is, it transmits current readily in one direction, but practically none in the other. Fig. 1 shows a train of electrical waves such as is obtained from a source of continuous oscillations and interrupted at definite intervals.

Suppose these waves be picked up by a wire and passed through the coils of a telephone receiver. If the frequency of a current passing through a receiver lies within the audible range, the receiver gives a tone, because the diaphragm is attracted by the electromagnet so many times a second. But the waves used in radio work are of such a high frequency that they c a n n o t be heard at all. Now suppose we chop off all the lower h a l v e s of the waves shown in Fig. 1, so that we have something like that shown in Fig. 2. There will then be a succession of current rushes through the receiver always in the same direction. The intensity of these current rushes, occurring, say, every hundred thousandth of a second, increases rapidly and then dies out. If this process recurs every thousandth of a second, the effect on the receiver is the same as if a current like that shown in Fig. 3 passes through the receiver a thousand times a second, and the receiver produces an audible tone.

This chopping off of the waves is what the crystal detector does. The early type of thermionic valve of J. A. Fleming can also be made to detect in this way, and in the early days was so used. But it has other characteristics which enable it to be used in a different way. This valve much resembles an incandescent lamp. But, besides the filament, it contains a plate which sometimes takes the form of a metallic cylinder surrounding the filament and sometimes is in the form of a plate or pair of plates.

The discovery was made by Edison in 1883 that if the positive end of the filament of an incandescent lamp is connected through a galvanometer to a plate in the bulb a current is observed to flow through the galvanometer. But when the plate is connected to the negative end of the filament there is little or no current. The same effect can be observed with the circuit shown in Fig. 4. Here the battery E_f heats the filament, and the current between filament and plate is driven by battery E_p. This current is observed to flow when the plate is positive with respect to the filament even in the highest attainable vacuum.

Rear view of right-hand part of radiotelephone set down on page 223. Notice how the valves are mounted

and is carried by electrons emitted from the filament when it is hot. In order to operate a thermionic tube satisfactorily it is necessary to know how, why, and under what condition the electrons can be boiled out of the filament.

Relation Between Electrons and Atoms

The atoms of all substances consist of a number of electrons held together by positive nuclei. The difference in the various elements consists in the number and arrangement of the electrons and positive nuclei. Metals behave as if they have a large number of electrons that are free to move between the atoms, and therefore they are good conductors of electricity, because if a potential difference be applied between the ends of a bar of the metal the electrons move in the direction of the electric field and so constitute a current. Although these electrons are free to move between the atoms of the metal, they cannot escape from it at ordinary temperatures because they are constantly

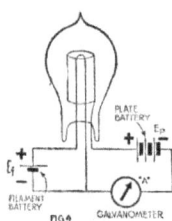

Showing the principle of Edison's vacuum valve discovery

pulled back as soon as they try to reach beyond the surface of the metal. The force tending to pull them back is sufficiently strong to do so, except when the electrons do not shoot out from the surface with a very high velocity. If their velocity exceeds a certain amount the force fails to hold them within the substance and they escape.

Now, how can we give the electrons sufficient velocity to get away from the substance? These electrons, like atoms, are in a state of constant motion, and they move around with a wide range of velocities, which constantly change on account of the collisions of the electrons with one another and with the atoms of the substance. At any instant a few electrons will move with very high velocities, a few others will move with very low velocities, and the rest will possess velocities ranging between these two extreme values. There is one velocity which is shared by more electrons than any other. This is the most probable velocity, and when we talk of the "velocity of the electrons" we shall in general mean this value.

Now, the velocity of the electrons depends upon the temperature of the substance. At ordinary temperatures the velocity is so small that practically no electrons can escape from the substance. But we can easily enable them to do so by heating the substance to a high temperature. Thus a tungsten filament at the temperature it has in an incandescent lamp, or higher, is hot enough to boil out large numbers of electrons. Suppose we heat the filament in Fig. 4 to such a temperature. Then, on account of the potential difference between the filament and the plate, the emitted electrons are pulled over to the plate and a current flows through the circuit.

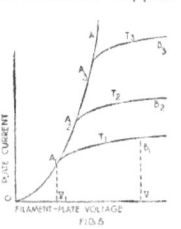

Peculiarities of current conduction through a simple 2-electrode tube

The electrons, of course, flow in a direction that is opposite to the generally assumed direction of current.

The peculiarity of the current conduction through such a tube is that the current does not increase linearly with the applied voltage, as it does in the case of conduction through a wire. While the voltage between filament and plate is small, the current at first

increases slowly, and then gradually more rapidly as the voltage is increased, until finally it does not increase in any marked way, no matter how high the voltage is made. These phenomena are represented by the curve OA_1B_1 in Fig. 5. The current represented by the magnitude VB_1 is therefore very nearly the maximum

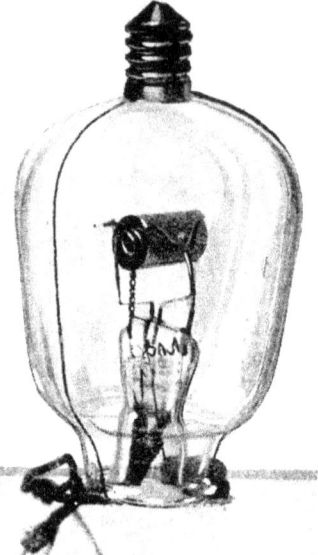

An English vacuum valve. Haste and rush of the war necessitated rough construction. Later valves are improved

current that can be obtained from the tube when the filament is at a temperature T_1. But we can increase this current by raising the filament to a higher temperature, because this would increase the velocity of the electrons and more would escape from the filament. At a filament temperature T_2 we then obtain a characteristic OA_2B_2.

The maximum current that can be passed through the tube is called the saturation current, and is obtained when the electrons are pulled away to the plate just as fast as they are emitted from the filament. From this it follows immediately that when the voltage is not high enough to give the saturation current, then we do not make use of all the electrons that are boiled out of the filament.

Crowding the Space with Electrons

What happens to those that are not pulled over to the plate by the applied potential difference between filament and plate? Well, they go right back to the filament when they are not needed. Of course, they come out again, and so there is a continual movement back and forth, while at the same time some of the electrons are continually being carried away by the applied electric field. It might seem strange that some electrons should suddenly change their minds and return to the filament. The reason for this is that electrons, being negative electric charges, always repel each other, and they get so badly crowded in the space around the filament that some are driven back to it.

This crowding of the space with electrons accounts for the peculiar shape of the curve OA. If just a few electrons came out of the filament they would easily move across to the plate; but when they come out in big crowds it is necessary to apply a voltage between plate and filament to get them across. A few persons running down a quiet street can go at a pretty good speed; but when the street is full of people trying to see a parade, the police can't make them go fast enough. The same illustration applies to electrons. Increasing the number of people would not make them move any faster. This is true also of electrons in the thermionic tube.

Suppose we apply a voltage equal to OV_1 between the plate and filament (Fig. 5), then the current is given by V_1A_1. This is a measure of the num-

Fig. 6—A modern vacuum valve of typical three-electrode construction

ber of electrons that this voltage can pull over to the plate per second. And we see that this number is the same, no matter how many more electrons we boil out of the filament by raising its temperature. The only condition is that the temperature of the filament must not be appreciably less

than that corresponding to T_1. If the temperature is equal to or greater than T_1, the space between the filament and plate is crowded as full of electrons as it can be. Of course, we can drive more electrons across the space by increasing the driving force; that is, we can increase the current by increasing the voltage.

The fact that electrons behave somewhat like a crowd of people is in many respects a great disadvantage. One reason is that it requires an expenditure of energy to get them across the space; another is that the crowding causes the characteristic to be curved. When using the tube as an amplifier it is desirable to have

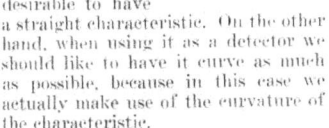

FIG.7

How to obtain the characteristics of the three-electrode vacuum valve. The apparatus used is simple

a straight characteristic. On the other hand, when using it as a detector we should like to have it curve as much as possible, because in this case we actually make use of the curvature of the characteristic.

Controlling the Electron

Before showing how this can be done, let us first explain a much better way of pulling the electrons across, and of controlling the number that can pass to the plate per second. This can be done by inserting a third electrode in the form of a sievelike electrode or grid between the filament and the plate, as was done by De Forest in his "audion." A three-electrode tube

is shown in Fig. 6, in which two flat plates are placed on either side of the filament and two grids intermediate between the filaments and the plates. The pairs of plates and grids are each connected metallically, to act like one plate and one grid.

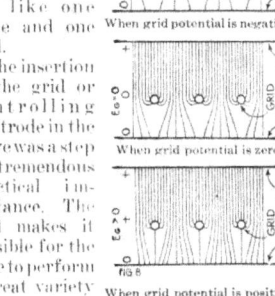

FIG.8

The insertion of the grid or controlling electrode in the valve was a step of tremendous practical importance. The grid makes it possible for the tube to perform a great variety of functions. The characteristics of the tube can be obtained with a circuit like that

shown in Fig. 7. It will be seen that there are really two circuits: the input circuit, connecting the grid to the filament; and the output circuit, connecting the plate to the filament. The battery E_p keeps the plate positive with respect to the filament, and drives the electrons emitted from the filament across the space to the plate.

The number of electrons passing to the plate can be controlled by potential variations applied to the grid. The grid acts like a partial screen to the electron flow. But the screening action depends not so much upon the interposition of the wires in the path of the electrons as upon the attracting and repelling effect of the grid due to the positive or negative potential applied to it. The effect of the grid can be exemplified by the lines of force diagram shown in Fig. 8.

Two Examples

Let us suppose that the negative end of the filament is connected to ground and the plate maintained at a positive potential by means of the battery E_p. Then a certain number of lines of force issue from the plate. Now, if the battery E_g be zero for the present, that is, if the grid be directly connected to the filament, then there can be no lines of force directly between filament and grid. The distribution of the lines of force is then that shown by the central diagram.

It is seen that the electric field, due to the potential on the plate, reaches

through the grid and manifests itself between filament and grid, and that therefore some electrons will be pulled to the plate through the openings of the grid, but not so many as would be the case if the grid were removed. If

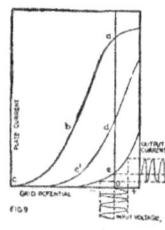

FIG.9

Curves obtained by keeping plate voltage constant and varying grid

now we make the grid somewhat positive with respect to the filament by means of the battery E_g, the electric field between filament and grid will be increased; there are more lines of force in that region, as is shown by the lower diagram Fig. 8. The result is that more electrons will be

pulled over to the plate per second and the current increases.

If, on the other hand, the grid be made negative with respect to the filament, it tends to repel some of the electrons and the current to the plate decreases. It can be still further decreased by making the grid still more negative; and by making the grid sufficiently negative we can stop the flow of electrons to the plate altogether. It will be seen, then, that the current to the plate can be varied by simply varying the potential of the grid, even though the potential of the plate remains constant.

It is important to note that when the grid is negative it tends to repel the electrons that are emitted from the filament, while the plate, being positive, tends to pull the electrons away from the filament. As long as the repulsion of the electrons by the grid is not greater than or equal to the attraction due to the plate, electrons will pass on to the plate and a current will flow through the tube. But, since the grid is negative, the electrons will all go to the plate and none to the grid. This is the same as saying that the resistance of the input circuit is infinite—it is an open circuit. If, on the other hand, the grid has a positive potential with respect to the filament, some of the electrons emitted from the filament will actually be pulled over to the grid, and a current will flow in the input circuit as well as in the output circuit.

When the grid is at the same potential as the filament the number of electrons going to the grid will be very small—in fact, usually negligibly small compared to the number going to the plate. The fact that the plate current can be controlled by potential variations on the grid without the grid taking any current *is an important property of the thermionic tube.* When there is no current in the input circuit there is no power expenditure in that circuit. It therefore means that we can vary the current in the plate circuit without doing any work.

In practice this is not quite true, because there is always a small amount of power wasted in apparatus included in the input circuit, such as induction coils. Of course, this device is not a perpetual-motion machine. Although we can obtain comparatively large current variations in the plate circuit by expending a very small amount of power in the grid circuit, still the force that drives the current through the plate circuit is the plate battery E_p, so that this battery supplies the necessary energy.

Let us now see how the current-voltage characteristic curves of the three-electrode tube compare

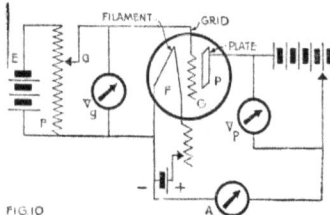

Rack for life tests on vacuum tubes. Thousands of tubes were manufactured during the war

with that of the two-electrode tube shown in Fig. 5. In the first place, we note that the main difference between the two tubes is that one contains a grid between the filament and plate, and the grid acts like a screen to the

electron flow through the tube. If we keep the grid at a constant potential with respect to the filament (zero potential in this case) and measure the current to the plate for increasing potential on the plate; then we obtain curves that look very much like those shown in Fig. 5, except that for the same voltage between filament and plate the current to the plate is smaller. If the grid, instead of being at the same potential as the filament, be maintained at a negative potential by the grid battery, the curves will be shifted to the right, because the repelling effect of the negative grid on the electrons makes it necessary to apply higher plate voltages in order to get the same number of electrons across to the plate.

Further Tube Characteristics

Now, there is another set of curves that can be obtained with this type of tube by keeping the plate battery voltage constant and varying the voltage of the grid battery. See Fig. 9. When the grid is at the same potential as the filament, there is a current flowing through the tube, because the

plate is at a positive potential which drags the electrons through the grid to the plate. This current is represented by the magnitude oa. If the grid be given, by means of the battery E_g (Fig. 7), a negative potential with respect to the filament, the repelling action of the grid counteracts the attraction due to the plate, and the plate current is less than before.

As the negative grid potential is increased the current decreases continuously, as is indicated by the line abc. When the point c is reached the repulsion due to the grid equals the attraction due to the plate, so that the plate current is reduced to nothing. If the plate had a lower potential to start with, the curve de' would be obtained. It looks very much like the curve abc, but lies to the right of it.

For a still smaller plate potential we get the curve c. These curves resemble the one given in Fig. 5. There is, however, this important difference: that a change in the grid potential produces a larger change in the plate current than an equal change in the plate potential. The ratio of the change in the plate potential to the change in the grid potential that give the same change in the plate current is called the amplification constant of the tube. This is a very important constant, and can be measured very easily in the following way:

Suppose the plate battery has a voltage of fifty volts and the grid potential is the same as that of the filament. Then there is a certain current flowing through the tube. Let the plate voltage now be increased to, say, sixty volts. This has the effect of pulling more electrons away from the filament to the plate, so that the current increases. But the current can be brought back to its original value by making the grid sufficiently negative with respect to the filament. If this negative potential is found to be one volt, then the amplification constant of the tube is ten. This can be conveniently done with the circuit arrangement shown in Fig. 10.

The voltage of the plate battery can be measured with the voltmeter V_p, and the current with the current meter A. By connecting the battery E to a slide wire resistance, r, the voltage between the filament and grid, which can be measured with the voltmeter V_g, can be adjusted in very small steps by sliding the contact a until the change in current produced by a change in the plate potential, and indicated by A, is exactly neutralized. In commercially used tubes the amplification constant ranges from about five to about forty,

and sometimes more, depending on the purpose for which the tube is used.

The plate voltages used on commercial tubes cover a very wide range, and so do the currents that pass through them. In the case of tubes used as radio detectors it is not necessary that the current be large, because here we are dealing with very small quantities. Many detector tubes operate satisfactorily with something like twenty volts or less on the plate, and the plate current is usually a small fraction of a milliampere. On the other hand, tubes that are used as amplifiers on telephone lines usually operate on a plate voltage of about one hundred and fifty volts, with a current of several milliamperes. Then again, tubes that are used as radio transmitters need plate voltages of several hundred and sometimes several thousand volts, the plate current being sometimes as high as several amperes. This shows what a tube is capable of.

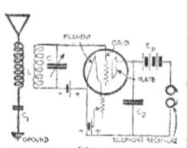

How a three-electrode tube is applied to radio circuits

Let us now see how it is possible to make use of the curvature of a tube's characteristic to detect high-frequency electromagnetic oscillations. A simple type of detector circuit is shown in Fig. 11. The incoming oscillations are picked up by the antenna, and are induced in the circuit LC. The condenser C is variable, so that it can be adjusted until the circuit LC is in resonance with the antenna circuit. (The inductance L can, of course, also be variable if necessary.) We then obtain the highest possible alternating potential difference between the grid and the filament. These potential differences can be superimposed upon the constant voltage of grid battery inserted in grid circuit. But when using the tube as a straight detector the grid battery can usually be omitted.

Suppose, then, that the plate battery voltage is so adjusted that the current for zero grid potential is given by O_e (Fig. 9). The incoming oscillations from the antenna then make the grid alternately increases and decreases, current in the plate circuit therefore alternately increases and decreases. Now suppose the potential variations impressed on the grid by the incoming oscillations be represented by the wavy line shown below O. This causes the plate current to vary in the manner shown by the wavy line to the right of the curve. Now, it can be seen that the difference between those two wave curves is that the voltage wave is symmetrical; that is, the positive peaks are just as large as the negative peaks, but the current wave is lop sided, and the effect produced by the tube is similar to that produced by the crystal detector, as shown in Figs. 1, 2, and 3. In other words, audible current flows through telephone receiver.

Forty Talk Over the Same Line

Although it is done by wireless, wires are used

By John Stuart

THE pendulum swings backward and forward in every line of science: one year the doctors promise to make a new man of one by removing his appendix, and the next authorities are all in favor of removing tonsils.

And now the pendulum is swinging fast in the science of electrical communication. First Morse showed how to make a telegraph system with a battery, a key or switch, and an electromagnet that would "click" when an electrical current passed through it, and a wire between the key and the sounder. In his system the electric current from the battery was allowed to pass along the wire and back through the earth.

What's Wrong with the Telephone and the Cable

Then Bell substituted a transmitter for the key and the sounder by a telephone receiver, which he skilfully patterned after the drum of a human ear. The art of telephony was born. But the telephone lines with a ground return were noisy; current from any trolley line might in part return along the same route as the electrical currents that carried the speech. So Carty, wizard of the present telephone system, gave to each telephone line its own metallic return, and the modern telephone system was created.

Through these early days of the art the swing of the pendulum was always toward more wire.

The wire telephone couldn't cover extreme distances until sensitive little boosters were devised for amplifying its feeble currents. Even then it was not available for trans-oceanic communication. Speak into a telephone transmitter and you set up rather rapid changes in the amount of electric current flowing through it, and hence in the amount that flows along the wire to the distant receiver. The variations of current at this end vary the pull of the electromagnet on the thin plate or diaphragm of the receiver, and so cause it to move in the same way as did the diaphragm of the transmitter that your voice actuated.

On a long line, however, there is a curious effect of capacity. It takes a little time for the current to flow into the wire, and none can flow out at the distant end until there is enough to permit an overflow. Only the variations in current which are caused by the telephone transmitter are useful in giving intelligible speech at the distant end, and they are so rapid that no sooner does the wire start to charge up with more current than the transmitter diaphragm moves so as to supply less current. The result is that very little "juice" flows out of the distant end.

Professor Pupin solved this problem for wire telephony by inserting a coil of wire every so often along the line. More wire, you see: for the pendulum was still swinging. Curiously enough, if the current once starts going in such a coil, it tends to keep on in the same direction; so the electrical inertia of the coils helps to offset the capacity effect of the telephone line.

But even this trick didn't help the telegraph engineers who had to work with trans-oceanic cables. The capacity effect is much worse on cables than it is on open-wire lines, so that the introduction of coils is apparently impracticable except on short lengths, as across a bay. There are no nice places to install the boosters or repeaters, for they must be maintained by operators. The poor oceanic telegrapher had to work at slow speeds, and also learn to read little humps in the ink line traced by his recording apparatus, instead of working with a sounder.

So far, the pendulum swung toward more wire. But meantime it was getting ready for its swing in the opposite direction toward practically no wires at all.

Three Hundred Times Marconi Was Told about his Baby

Wireless was the next swing, and huge stations were erected for transmitting across the Atlantic. But wireless isn't secret. It isn't secret even for its own wizard Marconi. According to the *Manchester Guardian*, Signor Marconi and his wife, having left their baby sick in Rome while they crossed the Atlantic, were to be kept informed by the nurse on the wireless. She radioed, "Baby is better." Eiffel Tower, Carnavon, Poldhu, Clifden, and other stations received the message, and retransmitted it as a matter of courtesy. Three hundred times did the signor receive this message during his voyage.

The man who showed the world how wireless messages can be sent forty at a time over a bare wire, and thus combine wireless with telegraph and cable transmission

Now the pendulum is swinging back. From no wires at all, Major-General Squier is leading engineers to the use of a single bare wire between stations as a guide for the radio waves. Along this wire he proposes to guide, not a single message, but simultaneously as many as forty different messages. Multiplex guided wireless is the next swing of the pendulum.

But the General doesn't care whether the uninsulated wire is above ground or below, or even under water. Hence his method is applicable to trans-oceanic communication —not only telegraph, but telephone. He has tried it out over short distances, between Fort Washington in Maryland and Fort Hunt in Virginia, and also at one of the Signal Corps stations in New Jersey.

It looks like a marvel that one wire can guide forty different telephone conversations without their mutual and destructive interference. To the radio engineer, however, this part of his scheme is commonplace.

In ordinary land-line telephony the transmitter varies a current that comes from a battery, and that would otherwise be perfectly steady. It makes this current increase or decrease in response to the motion of the diaphragm of the transmitter, and hence causes similar variations in the receiver diaphragm. But in radio the current that the transmitter diaphragm varies, that is, modulates in accordance with the voice, is not a steady current at all. It is a specially generated current, which increases and decreases alternately, but at an enormously high rate, thousands and in some cases even a million times a second. Such an alternating current varies too rapidly to affect the receiver diaphragm, and, even if it did, it wouldn't make an audible sound because the human ear can't detect sounds from drums that are vibrating faster than twenty or thirty thousand times a second.

How "Wired Wireless" Works

What counts, in the case of such a high-frequency alternating current, is the "effective value," as it is called; that is, the amount of steady current that would produce the same heating effect. Electric heaters, lamps,

and the like work just as well in an office building having a direct or steady current supply as in a village house where the current is alternating. The thing that counts is the effective value of the current, and not its alternations. In the same way, the rapidly alternating current of radio practice is just as good for being modulated by the voice

Wireless Messages by Wire

Major-General George O. Squier, of the United States Signal Corps, sees no hope in developing cables and telegraph lines. Only one message can be sent each way at a time over a submarine cable, and the insulation costs from $1,500 to $2,000 a mile. He has invented a system combining wireless with telegraphy and cabling. In this article we tell you how he would use bare wires to send wireless messages—forty of them at once over a single line.—EDITOR.

as is the steady current of early wire telephony.

There are also many different devices—or radio detectors, as they are called—that, like electric heaters, respond, not to the rapid alternations in the current, but only to relatively slow changes in its effective value.

The vacuum valve (a kind of electric lamp) is the most justly famous of all the detectors, and is the ultimate prize of every junior wireless amateur, whose pocket-book compels him to use the less efficient and older style "crystal detector," which consists of a metal point resting on a galena crystal. Such a crystal device will pass current most efficiently only in one direction and practically not at all in the other direction. What flows across the contact points, then, from a source of high-frequency current, is merely a one-way current. Any changes in the intensity of the high-frequency cur-

rent will mean corresponding changes in the intensity of the direct or one-way current, that gets through the detector. It is this one-way current that is used to operate the receiver, just as in ordinary telephony.

Such high-frequency currents have one enormous advantage over a direct current to carry the telephone message.

This advantage lies in the fact that each high-frequency current may be separated from all the rest by applying a principle known in the art as "resonance." Just as the trained orchestra leader can attune his ear to any instrument of his orchestra, and apparently be oblivious to all other notes, so a radio-receiving set may be made selectively sensitive to only one of many high-frequency currents. Just as we distinguish notes of musical instruments by their pitch,—that is, by the number of vibrations a second that the instrument sets up in the air about it,—so the "tuned circuits" of radio-receiving apparatus distinguish between different "pitches" or frequencies of ether waves.

Sending Many Messages Along a Single Wire

Over the same wire or through the same ether we may send many different currents with their different frequencies of alternation, and have each selected and received only by its own receiving circuit. Each of these high frequencies may be used to carry a telephone message. A multiplicity of messages is thus sent through the ether without confusion. Of course, where two or more sending stations try to use the same high frequency to carry their individual messages, there will be interference.

General Squier put the case before the National Academy of Science recently as follows:

"In ocean telegraphy the elaborateness of line construction has reached a practical limit. The most promising hope of improving ocean cables is to abandon the present method and to start with the bare wires in water, using high-frequency current."

If that dream is realized, before long we may be able to say to some international "central:" "Give me Paris, East 238,375."

This is a tank in one of the buildings of the Bureau of Standards. General Squier laid bare wires in the water and showed experimentally that it was possible to send wireless messages along the wire under the water. Forty can be sent at one time

"Hello, Is This London?"

The momentous vacuum valve

By Lloyd E. Darling

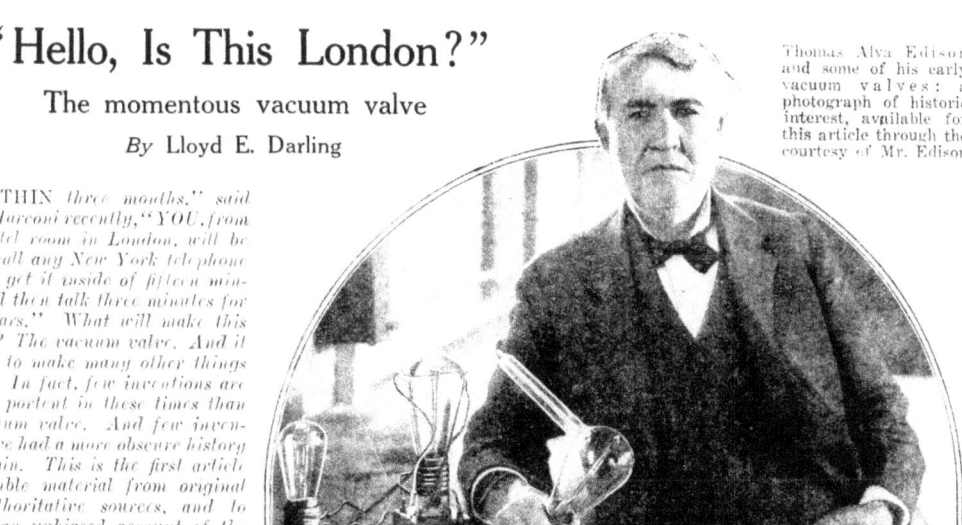

Thomas Alva Edison and some of his early vacuum valves: a photograph of historic interest, available for this article through the courtesy of Mr. Edison

"WITHIN three months," said Marconi recently,"YOU, from your hotel room in London, will be able to call any New York telephone number, get it inside of fifteen minutes, and then talk three minutes for five dollars." What will make this possible? The vacuum valve. And it is going to make many other things possible. In fact, few inventions are of more portent in these times than the vacuum valve. And few inventions have had a more obscure history and origin. This is the first article to assemble material from original and authoritative sources, and to present an unbiased account of the evolution, present status, and future of vacuum valves, and of the men who made them what they are.—EDITOR.

UNLESS you are a radio amateur, or a man who has followed the development of the electron theory on the relations of electricity and matter, you will probably wonder what a vacuum valve is, and why the newspapers haven't had front-page articles about it.

Yet, if a fly takes a notion to buzz around a telephone transmitter in San Francisco, the vacuum valve is the contrivance that will faithfully reproduce the sound-currents at intervals along the great transcontinental telephone line, and, getting them to New York, is capable of amplifying the sound so that the walls of a room will reverberate as if a low-flying airplane had come into the neighborhood. The same thing is true of a man's voice, and they may even do it *wirelessly* from San Francisco to New York, or even from Honolulu, Hawaii, to New York, if they like, though they aren't doing it as a regular practice—yet.

How Fliers Talk to Each Other

Then, too, the vacuum valve is the contrivance that enables the transatlantic naval fliers to talk among themselves and to ship and shore stations on the way over—a feat that may be done regularly in aerial and ship navigation in the near future.

Our huge commercial wireless stations profit through vacuum valves, getting messages with ease from the far corners of the earth. With a vacuum valve, even small stations and indoor aerials can accomplish wonders. The valve enabled the President to stand on the White House lawn, a few months ago, and, in an ordinary voice over a radiotelephone, to direct the flight of aviators thousands of feet in the air and out of sight at times. Secretary Daniels has sent orders to battleships out at sea with the radiotelephone, his voice being heard by the commanders just as clearly as if we were talking over an ordinary nickel-in-the-slot telephone.

Vacuum valves may possibly enable you and me to sit in our homes of an evening and hear a machine on the table reel off the day's news, or the opera that a prima donna in a near-by city may be singing at the moment. We may even save time by hearing it all from a loud-speaking telephone while riding home on the subway or street-car line, the cars being in motion. Major-General Squier goes and

Dr. Lee De Forest's especial contribution to the vacuum valve art was the "third electrode," or "grid." He called his valve an "audion." Here he is manipulating several

drives a nail in a tree, connects apparatus of which vacuum valves are a part, and hears radiotelegraphic messages from all directions, as well as radiotelephonic messages, if there are any.

What Marconi Said

The valve is so sensitive that it makes almost unnecessary the high towers we have come to associate with great wireless stations, though they are not in the way if one has them. At present the valve is only in its infancy. Every month new applications are being discovered. It is possible even to transmit power wirelessly with the vacuum valves, though inefficiently at present.

Marconi himself is credited with saying, a short time ago: "Within three months, you, from your hotel in London, will be able to call any New York telephone number and get it inside of fifteen minutes, and then talk three minutes for five dollars."

Surely, if you haven't heard much about vacuum valves in the newspapers you will soon.

Vacuum valves are not exceptional in appearance. A common form greatly resembles a ten-watt tungsten lamp, the kind the landlady puts in the hall lamp to save current; but it has more wires and plates and one thing and another on the inside than a ten-watt tungsten light. Other forms of vacuum valves are larger than an X-ray tube, and resemble it in general appearance. Details of this are made clear in the illustrations on page 231. The "vacuum" part of the valves' name comes from the fact that the air is exhausted from their interior, as in the ordinary incandescent lamp. The "valve" part of the name results

Vacuum tube gas pressures of a few billionths of an atmosphere being measured under the supervision of E. H. Colpitts and Dr. W. Wilson

(*Right, center*) — Drs. H. J. van der Bijl and C. J. Davisson consider a vacuum valve of unusual kind. This one radiates 100 watts. It will receive and modulate as well

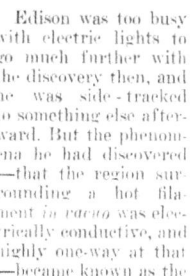

Lieutenant - Colonel F. B. Jewett, chief engineer of the Western Electric Co. He and the heads of his organization, shown above, perfected vacuum valves for trans - continental telephony and the war's uses

Dr. H. D. Arnold has mastered getting the high vacuum necessary for perfected vacuum valves

from the peculiar valvelike action that the bulb and its contents have on some kinds of electrical circuits, a feature that will be made clearer later.

It is interesting to consider the history and inception of the vacuum valve. We want to know the lineage of a contrivance so fraught with wonderful accomplishments and possibilities.

Edison—who else would you suppose?—devised the original vacuum valve. Even he himself did not realize all that he had produced at the time. In October, 1884, Professor Edwin J. Houston gave a talk before one of the first meetings of the American Institute of Electrical Engineers ever held. Said Professor Houston:

I wish to call your attention to a matter which, I suppose, you have all seen and puzzled over. Indeed, I wish to bring it before the society for the purpose of having you puzzle over it. I refer to the peculiar high vacuum valve phenomena observed by Mr. Edison in some of his incandescent lamps. I have in my hand an Edison incandescent lamp. This one, however, has, in addition to the carbon filament, a platinum plate or strip, thoroughly insulated from the filament, and supported between its two vertical parts.

The lamp being lighted by an electrical source, a current will flow in the direction indicated by the arrows. If the filament is just ordinarily bright, no unusual effects are noticed. If, however, the incandescence of the filament is raised from its normal eight candlepower to twenty—, thirty—, forty—, fifty—, or one hundred candlepower, then the needle of the galvanometer is violently deflected by a current passing through its part of the circuit. I have no theory to propound as to the origin of these phenomena. I wish merely to call your attention to the phenomena. At one time Mr. Edison devised other bulbs with long tubular projections at their sides [the kind Mr. Edison is holding in

Dr. J. A. Fleming, F.R.S., of London, England, who applied Edison's vacuum valve to the broad radiotelegraphic field

his hand in the photograph. Ed.] Instead of placing the platinum plate between the legs of the filament as before, he placed it at the far end of this tube, and obtained the same general effects as before, even though the tube was surrounded by a freezing mixture.

There followed much general discussion at the meeting, and the propounding of many explanatory theories, most of which would hardly hold in these days.

Edison has received but very little credit for discovering the valve. He was still working on his electric light at the time he produced it—making those wonderful discoveries that resulted in the forming of a whole new industry, modern electric-lighting and power-distribution.

At the moment of producing the valve, Edison happened to be investigating the deposition of carbon on the inside of the bulbs, and trying to determine the reason for the sharp shadows of the filament sometimes seen. There was also a bluish tinge that aroused his curiosity.

Presently he thought of placing a plate near the filament to see whether currents were being developed simultaneously with these other effects. He discovered that there were. There you have your vacuum valve.

The fact that currents could be carried across a vacuum, and more strongly in one direction than in the other, was startlingly new at the time. Much scientific excitement was created both in America and in Europe.

Edison was too busy with electric lights to go much further with the discovery then, and he was side - tracked to something else afterward. But the phenomena he had discovered—that the region surrounding a hot filament *in vacuo* was electrically conductive, and highly one-way at that—became known as the "Edison effect," after the work he had done. He utilized the vacuum valve practically in 1883, however, in recording changes in electrical pressure in his first New York power-house.

Not much other use was made of vacuum valves until about 1904, when it occurred to Dr. J. A. Fleming, of London, to put them in radiotelegraphic circuits as rectifiers of incoming oscillations. This was a long step. Says Dr. Fleming:

The starting-point for the invention was the interesting observation made by Mr. Edison in the early '80's. In 1885 Sir William Preece read a paper before the Royal Society on his further investigations. I examined the "Edison effect" at still greater length in 1889, 1890, and 1896 in papers read to the Royal Society and to the Physical Society. I came to the conclusion that the phenomenon was due to the emission of negatively charged atoms or molecules from the filament. In those days the conception of an electron, or particle smaller than an atom, had not been reached, and it was not until after Sir Joseph Thomson's epoch-making researches that it was recognized that the particles emitted were not atoms of matter, but atoms of electricity 2,000 times smaller. In 1904 I was endeavoring to use a mirror galvanometer, of the kind used in submarine telegraphy, to detect wireless signals; but this required that the movement of electricity in the galvanometer circuit should always be in the same direction. Receiving circuit oscillations are, however, alternating currents of very high frequency, and with these both the mirror galvanometer and the telephone receiver are inoperative. A rectifier was vitally necessary. I found after careful experiment that the vacuum valve would rectify these oscillations satisfactorily. It then became possible to utilize a receiver or the galvanometer with ease.

As a result of Fleming's work the device came to be known as the "Fleming" valve. Yet, as is evident, it was the old valve of Edison's, even though Edison did not apply it to the radiotelegraphic field, highly important as that application was.

It may be said in passing that the American—Edison—deserves credit in another branch of radiotelegraphy also; that is, for being the "godfather," so to speak, of the whole science of modern practical electric wave propagation, the thing upon which all radiotelegraphy and radiotelephony depends. In the late '70's—even in those early times, electrically speaking—Edison had discovered that you could get electric waves from an electric buzzer—a fact that any youthful wireless experimenter knows now.

At the Paris Exposition in 1881 Edison had a buzzer contrivance rigged up, the waves being sufficient to cause small sparks between the ends of graphite pencils in a darkened box.

Thousand of visitors saw the apparatus, one Heinrich Hertz among them. He went home to Germany, elaborated the idea, rigged up improved apparatus, studied the results, and presently the noteworthy work of Hertz on electric waves was published. Since then you have heard of Hertzian waves, and Hertzian this and Hertzian that, all with

some scientific justification, for the work of Hertz really was invaluable. But as a matter of historic accuracy we should evidently speak of "Edison waves" and such; for Edison was the man whose experimenting, penetrating, everlastingly trying something

Dr. Irving Langmuir and Dr. Saul Dushman, of the General Electric Research Laboratory. This company's men have produced the "kenotron," a two-electrode valve, the "pliatron," a three-electrode valve, and the "dynatron," a noteworthy vacuum tube having negative resistance

mind first made them materialize, though Maxwell, the propounder of the electromagnetic theory of light, thought all along that they must exist.

To Edison, as the first man who ever devised a vacuum valve, and the materializer of electric waves, the whole radio profession owes an immense debt.

As is evident, the vacuum valve has had a good many ups and downs in its history. But the valve would hardly have attained its present-day importance had it not been for one other episode in its career. This was the production, about 1905, by Dr. Lee De Forest of New York, of a vacuum valve containing a third electrode; i.e., a bulb having one other metal part, or "grid," on its interior, in addition to the hot filament and plate of Edison's valve, and the one used by Fleming. What a third electrode does is told by the drawings on Page 232.

Immediately there arose much dispute and patent litigation between the interests owning the Fleming patents and Dr. De Forest as to priority in applying the vacuum valve to the radio field. Only recently did the courts finally decide the case, giving Fleming the credit for having first accomplished the application.

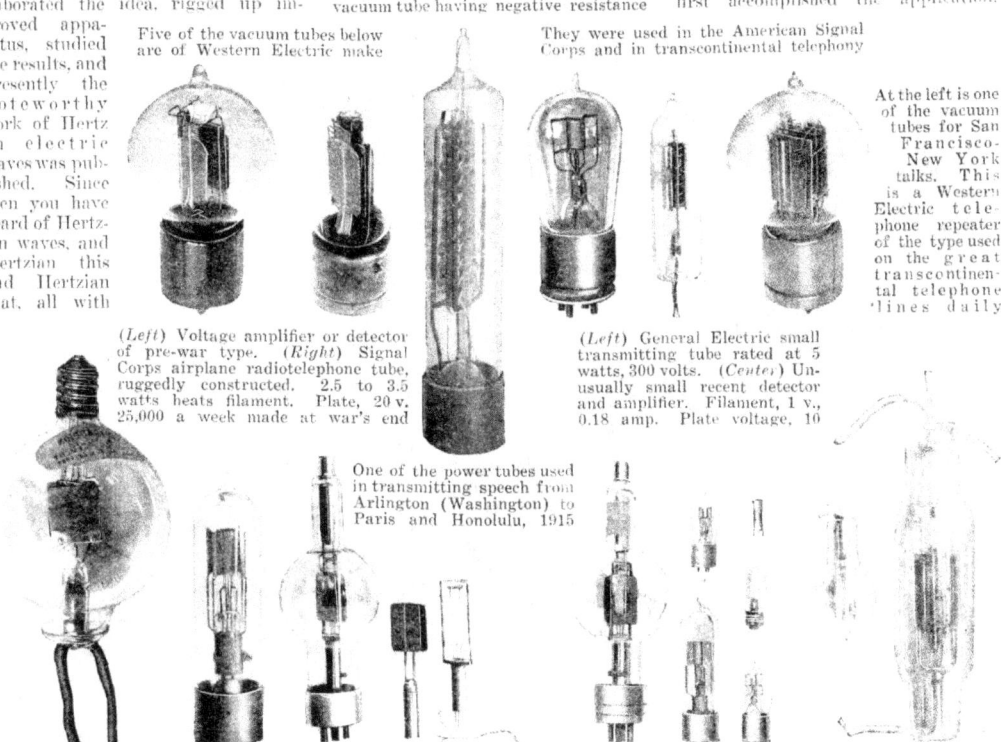

Five of the vacuum tubes below are of Western Electric make

They were used in the American Signal Corps and in transcontinental telephony

At the left is one of the vacuum tubes for San Francisco-New York talks. This is a Western Electric telephone repeater of the type used on the great transcontinental telephone lines daily

(Left) Voltage amplifier or detector of pre-war type. (Right) Signal Corps airplane radiotelephone tube, ruggedly constructed. 2.5 to 3.5 watts heats filament. Plate, 20 v. 25,000 a week made at war's end

(Left) General Electric small transmitting tube rated at 5 watts, 300 volts. (Center) Unusually small recent detector and amplifier. Filament, 1 v., 0.18 amp. Plate voltage, 10

One of the power tubes used in transmitting speech from Arlington (Washington) to Paris and Honolulu, 1915

An audion of before-the-war design—now an old-timer. This was an improved form called an "ultra-audion" by De Forest

General Electric vacuum valves. At left, medium-sized transmitting tube, 50 watts, 750 volts. Center, large-sized transmitting tube and parts, 250 watts, 1,500 volts. At right, comparative size of tubes. D is a receiving tube; E a ballast lamp; and F a grid leak resistance

Large and small De Forest radio oscillion bulbs used for transmission of varied messages—late types evolved from the audion

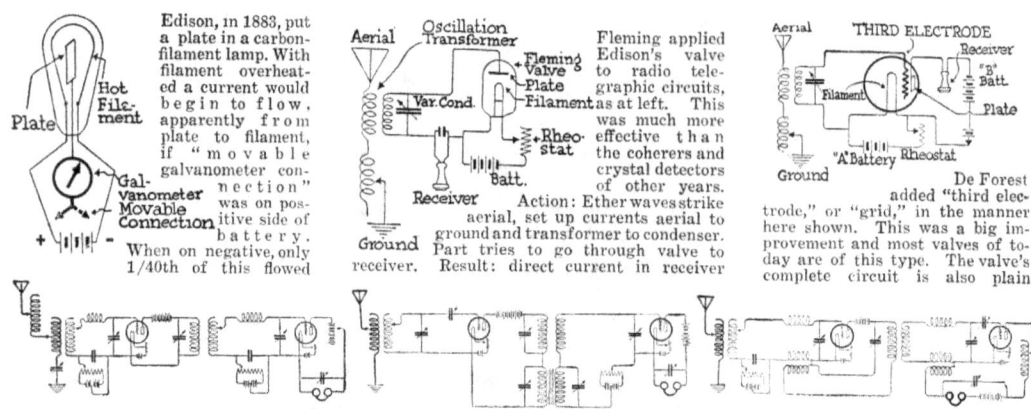

Edison, in 1883, put a plate in a carbon-filament lamp. With filament overheated a current would begin to flow, apparently from plate to filament, if "movable galvanometer connection" was on positive side of battery. When on negative, only 1/40th of this flowed

Fleming applied Edison's valve to radio telegraphic circuits, at left. This was much more effective than the coherers and crystal detectors of other years. Action: Ether waves strike aerial, set up currents aerial to ground and transformer to condenser. Part tries to go through valve to receiver. Result: direct current in receiver

De Forest added "third electrode," or "grid," in the manner here shown. This was a big improvement and most valves of to-day are of this type. The valve's complete circuit is also plain

The three diagrams in this row are after "Bucher, Vacuum Tubes"

This cascade amplification demonstrates complicated circuits attendant to higher applications of vacuum valves. Here output circuit of the left valve is connected to input circuit of the valve at right

Further complication. Here audio and radio frequency oscillations are amplified by a cascade connection. "Audio frequency" oscillations are those under 10,000 cycles per second — "radio" those over

Amplified radio frequency component of plate current of left valve is impressed upon filament of the right valve, where it is further amplified by a second regenerative coupling. Grid condenser then traps it

De Forest, however, is the undisputed discoverer, or inventor, of the third electrode idea and principle, an addition as valuable to radio as the application of the valve itself to that art.

On long-distance telephone lines they use the valve as a "repeater," which application is somewhat different from the valve's service in radiotelephony. Currents on long telephone lines easily become diminished and unrecognizable. So at intervals some device must be installed that will set in motion fresh currents in the line. The currents started by your own voice controlling them. Until De Forest's third electrode valve was evolved, only mechanical repeaters existed — unsatisfactory instruments. Perfecting the valve allowed commercial opening of the transcontinental telephone line in 1914. Several other contrivances go into the making of the transcontinental telephone line practicable, but the valve was the last perfected, and one of the most vital.

The valves of De Forest were excellent detectors of radio signals, but depended for their action as detectors upon the presence of a small amount of gas in the bulb, and their operation varied with variations in the gas pressure. They were inherently low-voltage and hence low-power devices, because of the fact that if voltages appreciably above the ionization voltage of the gas were applied to their electrodes, they would burst

Notes on Valve Working

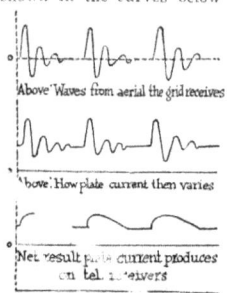

Suppose aerial picks up a signal. Radio frequency alternating voltage is thereupon impressed between filament and grid. What happens is shown in the curves below

Above: Waves from aerial the grid receives

Above: How plate current then varies

Net result plate current produces on tel. receivers

Middle curve shows what top curve does to plate current. Receivers only respond to contour of middle curve. To them middle curve looks like bottom curve. These curves and ones at right from "Radio Communication," a book written by Bureau Standards for the Signal Corps. More advanced discussions are in Proc. Inst. Rad. Engrs.

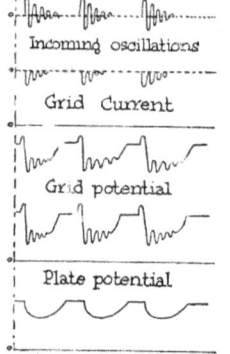

Stopping Condenser

Same circuit as at left, only the grid now has a condenser, making action different. When the grid becomes positive, electrons flow to it, but when negative, none comes. (See curves below)

Incoming oscillations

Grid Current

Grid potential

Plate potential

Current in telephone

Thereupon "grid current" curve results. Grid continues gaining negative charge, see "grid potential" curve. Next should read: "Resulting plate current"

into a blue glow and cease to operate. These peculiarities barred them from uses where an appreciable amount of power is required.

Dr. Irving Langmuir, of the Research Laboratory of the General Electric Company, found means whereby valves can be produced that may be operated with any voltage desired, without the bad effects of gas ionization. Many of the radio sets using these valves have been operated at fifteen hundred volts or more. The former limitations to their use when appreciable amounts of power are required were thus removed, and the wide fields of wire telephony and radio transmission opened up.

Dushman, White, and Hull of this organization did valuable work and university laboratories have contributed materially also. In fact, perfecting the vacuum valve has been the joint work of many.

In the great war organization of the Western Electric Company, and before, very important discoveries and perfections were made. The names of Carty, Jewett, Craft Colpitts, Arnold, van der Bijl, and many others, came prominently to the fore.

This organization's activity raised the war-time production of vacuum tubes from the existing one or two hundred tubes a week of 1917 to a rate of more than a million and a half good tubes a year at the time the armistice was signed. This necessitated establishing factories and training thousands of employees.

Radiotelephonic communication to and from airplanes was brought about; radiotelephonic communication over such long distances as from Arlington, Va., to Paris and Honolulu had already been accomplished as a test. De Forest's laboratory also aided war work materially.

COAL AND HEATING

Locomotives Are Consuming Too Much Coal

By Joseph Brinker

THERE is a tremendous waste of coal in firing locomotives. Cold water is sucked from the tank in the tender by means of an injector and fed into the boiler. The heat generated by the burning coal turns the cold water into hot water and then into steam. The steam is fed to the cylinders, at high pressure, and its power is transformed into motion by reason of its impulse to expand and push on the pistons connected with the driving mechanism of the locomotive.

Heat Goes Up the Smokestack

The exhausted steam, having done its work in the cylinders, is blown away through the smokestack, carrying with it sixty-five per cent of the heat consumed in changing the coal to steam.

To avoid at least part of this waste, a feed-water heater is employed in many modern locomotives. The heater returns to the boiler some of the waste heat of the exhaust steam, and thereby saves at least ten per cent of the coal bill.

In other words, nine pounds of coal, together with the feed-water heater, does the same amount of work as ten pounds of coal in the average locomotive without the heater will do. A saving of ten per cent of a substance as valuable as coal has become is no mean achievement.

How the Feed-Water Heater Works

The feed-water heater, which is shown on this page, takes one sixth of the exhaust steam ordinarily sent up through the smokestack, and utilizes it for heating the water before it is fed into the boiler. Five sixths of the steam is sufficient to provide the necessary draft for the boiler fires.

The one sixth used to preheat the water saves about thirteen per cent of the heat in the exhaust steam. By putting this back into the boiler

ten per cent of the coal burned is saved.

The feed-water heater is a cylindrical tank, which is generally placed just ahead of the front end of the boiler, directly on top of the cowcatcher framework.

By making use of some of the exhaust steam of the locomotive for preheating the water that is fed to the boiler, the cost of coal for an engine can actually be reduced ten per cent

The feed-water pump forces the cold water from the tank in the tender to the intake of the heating tank, where it passes through a series of tubes around which exhaust steam, taken from the valve chests of the engine cylinders on each side of the boiler, is allowed to circulate.

The Water Is Preheated to 250° Fahrenheit

The steam heats the water as it passes through the pipes, so that when the water reaches the boiler it has become so hot that little additional heating is required to change

When the water reaches the boiler it is so hot that it needs only a little additional heating to change it to steam of the adequate pressure

it to steam of adequate pressure. When the exhaust steam gives up its heat to the cold water in the pipes, part of it is condensed to water, which is drained off at the bottom of the tank.

The heater is closed, and will therefore deliver water to the boiler at a temperature proportional to the back pressure in the engine cylinders. In an open vessel water boils at a temperature of 212° Fahrenheit. But when a locomotive is running at high speed the back pressure in the cylinders is sometimes as high as fifteen pounds.

At such times the exhaust steam may have a temperature of 260°. When this is the case, it will heat the water to a temperature of 230 or 250° before it is fed to the boiler.

Since the water passes through the tubes while the exhaust steam circulates around them, it remains free from the vaporized lubricating oil that is invariably contained in exhaust steam. If the oil vapors were pumped back into the boiler they would form a heat-consuming scale on the boiler-tubes.

A Removable Head Provided at Each End

For the sake of greater convenience in cleaning the heater, a removable head is provided at each end.

The tubes are nested in four distinct groups, separated from one another by plates which divide the interior of the heater into four longitudinal sections. On its way to the boiler the water passes through each of the four compartments. This means that it has to pass through sixteen feet of tubing, while the length of each of the four units is only four feet.

To bring each part of the water into contact with the heating surface of the pipes, spirals of brass are provided in each pipe. These rotate at high speed as the water passes through the pipes.

Save the Nation's Energy—Fuel

Learn how to handle your furnace and save coal and doctors' bills

By A. M. Jungmann

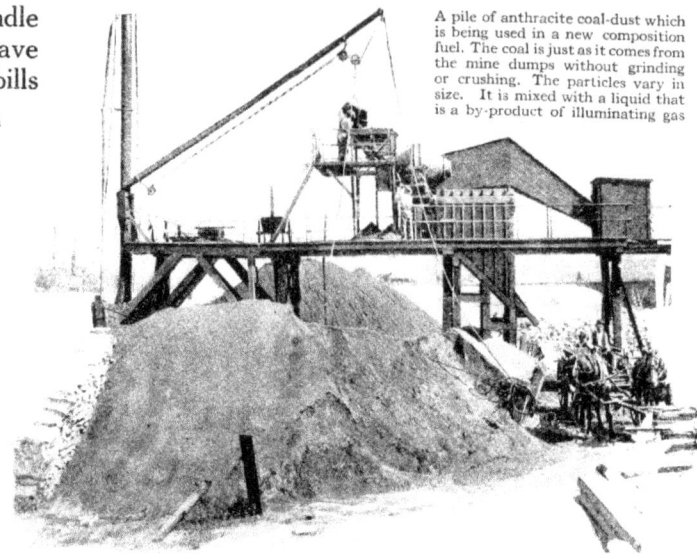

A pile of anthracite coal-dust which is being used in a new composition fuel. The coal is just as it comes from the mine dumps without grinding or crushing. The particles vary in size. It is mixed with a liquid that is a by-product of illuminating gas

IF you want to enjoy better general health and freedom from colds this winter—save coal. A quart of water evaporated in every room of your house every day will mean a saving of one third of your coal bill. What we really need in our houses is humidity rather than a high degree of heat. Moist air feels warmer than dry air of the same temperature, and moist air has the further advantage of retaining heat.

Professor Ellsworth Huntington, of Yale University, recently made a study of some nine million deaths in all parts of the United States and in France, Italy, and Japan. He also made a survey of fifty million deaths in Belgium, Great Britain, Germany, Russia, Rumania, Spain, and other countries, bringing his investigations of deaths up to the staggering number of sixty million. His study has resulted in the conclusion that an average temperature of 64° F. is the best for the maintenance of health. Professor Huntington also found that a uniform temperature is not so healthful as a variable temperature. He concludes that a frequent fall of temperature, followed by a more gradual rise, is an excellent means of preserving health.

Do Not Heat Sleeping-Rooms

Persons who desire mental and physical vigor should sleep during the winter in rooms in which the temperature ranges from freezing to 40° or 50°. It is not necessary for sleeping-rooms to be warmer than 50° at any time of the day. Windows always should be partly open at night and plenty of fresh air admitted to bedrooms.

The reason people keep their houses at such a high temperature in winter is that the outside air, after it is brought into the house and warmed, usually becomes as dry as the desert of Sahara. This dry air feels colder at 70°

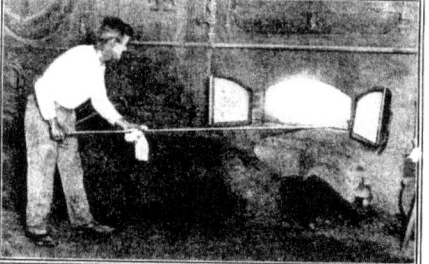

The composition fuel in use. The ash on the floor in front of the boiler shows freedom from clinker and a free-burned granular condition

than moist air would at 60°. By arranging to evaporate water in your house you can save coal because it will not be necessary to keep the air at such high temperatures.

If you are using a steam or hot-water heating apparatus, keep a pan of water on every radiator. These pans should be provided with wicks. There are on the market valves to be attached to steam radiators, which play moisture into the air of the room. If you use a stove, keep a pan of water on top; and if you heat your house by a hot-air furnace, keep a pan of water in the drum, so that the moisture will pass up with the heated air.

Experiments in Ventilation

Professor C. A. E. Winslow, also of Yale University, conducted some very interesting experiments in ventilation for the New York State Ventilation Commission. He found that the latest methods of ventilating rooms by taking air into the cellar, warming it to a temperature of about 67°, and blowing it into the rooms by fans was not so healthful as the ordinary procedure of

The U. S. S. *Gem*, which has been used to test different types of fuel: she has made many successful trips operating on pulverized coal

keeping the rooms at the same average temperature but ventilating them by letting fresh air in at the windows. In the modern system of ventilation the temperature did not vary and there were no drafts. Yet people who worked in rooms under these conditions suffered from colds to a greater extent than those who worked in the rooms where the temperature varied.

Although December and January of last year were unusually cold, the death-rate was low. This probably was caused by the fact that people did not keep their houses heated as much as they have in other years. Shortage of coal, instead of being a hardship, was a blessing in disguise.

Experts in the Bureau of Mines have made some suggestions on as to how to burn coal economically at home. First of all, they lay stress upon the fact that the person who is tending the furnace must acquaint himself very thoroughly with the apparatus. If you want to burn the coal faster, supply more air through the grate. More air can be supplied by opening the damper in the pipe leading to the chimney, by opening the damper in the ash-pit door, or by shaking the ashes down from the grate. If you want the coal to burn more slowly, reduce the draft by closing the chimney damper or the damper in the ash-pit door.

Learn to Manage Your Furnace

To increase the supply of air over the fuel-bed, open the damper to the chimney or the damper in the firing door or both. Air introduced over the fuel-bed helps in burning the gases and the visible smoke rising from the burning fuel. In order to be really economical, you must study your furnace and find out just how much air it requires; for too much air is as wasteful as too little.

A heavy fuel-bed generally gives more satisfactory results than a light fuel-bed, because, during the long periods between firing, a light fuel-bed is apt to burn down too much. When burning bituminous coal it is better to place the fresh coal somewhat in a heap to one side of the grate and leave a small part of the burning fuel uncovered. If the coal contains a large percentage of slack, this method of firing is particularly satisfactory, because the coal is heated gradually and the volatile matter distils slowly and is given a chance to burn. Before the next firing the coked heap of fuel should be broken and spread over the grate. A new charge is then placed on the opposite side of the furnace.

If you will put weather-strips around your windows, the saving in coal will amount to many times their cost

Be sure the glass in your windows is tight. A little trouble in putting the window-panes will save coal

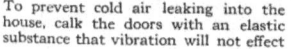

To prevent cold air leaking into the house, calk the doors with an elastic substance that vibration will not effect

If your total coal consumption is fifteen tons a year, you can save at least three tons by insulating the pipes

Mercer P. Moseley, Chief of Conservation, has compiled a list of "don'ts" for coal-users. If you observe these rules you will be able to effect an appreciable saving in coal:

Don't fail to clean furnace before starting fire.

Don't build a fire until necessary.

Don't build a fire larger than is necessary.

Don't fail to make check-draft damper in smoke-pipe do its work.

Don't neglect keeping fresh water in your steam-heater boiler.

Don't fail to keep your kitchen stove clean.

Don't keep your home at over 68° F.

Don't leave your draft open at night.

Don't try to heat all of outdoors.

Don't keep your fire going on pleasant days.

Don't sit in north room when the sun heats the south side.

Don't think it's fur-coat weather when the thermometer is 45° to 50°.

Don't waste water—it takes coal to heat it.

Don't forget that one gas-jet will raise the temperature of a room five degrees.

Don't fail to put up storm doors and windows.

Don't fail to sift ashes.

Don't burn coal when wood is available.

Don't fail to wrap your pipes with asbestos.

Don't fail to keep rooms moist—they heat easier.

Don't forget that moist air retains heat.

Don't forget that dry air causes colds and catarrh.

Don't waste gas—it is made from coal.

How to Bank the Furnace

Do not let your fire burn too low before banking for the night, because it may go out after banking. When banking the fire, leave a part of the surface of the fuel-bed uncovered in order to prevent explosion of the gases rising from the banked fuel. The regulation of dampers after banking

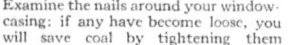

Examine the nails around your window-casing; if any have become loose, you will save coal by tightening them

Take the time to clean your flues out thoroughly at least once a week; soot is a poorer conductor of heat than asbestos

If the air enters from below, in burning soft coal, the oxygen is burned out; the remedy is to let in air from above

Never allow the ashes to get within six inches of the grate; a full ash-pit cuts down the heating efficiency twenty per cent

the fire depends on the amount of air that leaks into the ash-pit, and is something you must determine for yourself. It is generally wise to close the damper in the ash-pit door and leave the damper leading to the chimney partly open in order to prevent coal-gas getting into the house.

Saving Coal in Industry

The best way to start to save coal in boiler plants is to give up guess-work and apply scientific principles. The men who actually burn the coal should be trained in the principles of combustion. No matter how good a man's intentions may be, he cannot save coal for you unless he knows how and knows why it is necessary for him to do certain things in order to accomplish the desired results. Although a very clever fireman may be able to

tell you all about your fire by simply looking into the furnace, you cannot expect every man to do that. The only way you can really know the condition of your fire is by using recording or indicating instruments. Accurate instruments, such as draft-gages, flow-meters, and pyrometers, will help you to handle your coal intelligently and economically.

Some of the things to be guarded against in using coal are clouds of black smoke coming from the stack, which is always a sure indication of waste; the loss due to high percentage of carbon in the ash; and the wasting of exhaust steam. Your steam is a direct product of your coal. There is usually a great deal of economy in using mechanical stokers. Needless to say, it is necessary to see that your entire apparatus is in good condition, that the boilers are clean, that the

baffles are in good order, that the boiler-tubes are not full of soot, and that the flues are airtight, and everything ship-shape. A leaky flue will cause the burning of more coal than is necessary. Wherever possible avoid turns in your flues. The ideal flues are short and straight. Flues that have turns offer resistance to escaping gases and do not afford the free draft necessary to satisfactory combustion of coal.

Pulverized Coal

Pulverized coal is solving many fuel problems. The ideal condition in which to burn coal is pulverized. With pulverized coal there can be no clinkers and no smoke, and the combustion is complete.

Pulverized coal is ground so fine that a pinch of it between the fingers does not feel gritty. A cubic inch of coal is ground into such fine particles that 95 per cent. of it will pass through a sieve having 10,000 openings to the square inch. The advantage of pulverized coal will become apparent when you remember that a cubic inch of coal has a surface or superficial area of six square inches; but when it has been converted into these small particles its superficial area is increased some 700 times. This means that each one of the two million particles is surrounded by air and, when burned, permits perfect instantaneous combustion. There is no waste. Pulverized coal burns like gas.

Not an Untried Fuel

Pulverized coal is not an untried fuel. Nearly ten million tons of it are being used in the United States each year. In the manufacture of cement six million tons are used; in the production of copper one and a half million tons; in the iron and steel industry, two million tons; in the generation of power, some two hundred thousand tons. The cost of pulverizing coal is not at all prohibitive. It can be prepared at a cost of from sixty cents down to twenty cents a ton. The coal is burned by projecting it into the furnace by means of an air-blast, forming a cloud of thoroughly mixed air and coal.

Where scarcity of labor is a factor to be considered, pulverized coal is a very desirable form of fuel, because one man can handle a number of machines; for the process of burning this form of fuel is almost entirely mechanical.

Pulverized coal is not only suited to stationary plants, but is used with gratifying results on locomotives and steamships. Its use on locomotives is especially promising because of the saving of both time and labor. Burning pulverized coal on a locomotive does away with the necessity of a fireman. The coal is supplied to the fire automatically at the discretion of the engineer. There is no smoke, and no cinders are spilled along the road-bed. When it is considered that one railroad in the United States spent in 1916 $375,000 in settling claims for fires started by sparks or burning ashes along its right of way, you can see what a great advantage pulverized coal has over the usual locomotive fuel.

Used on Locomotives

Locomotives operated on pulverized coal do not have to waste time at ash-pits or in cleaning fires and shaking grates. When the fire is extinguished there is no waste of combustible in the ash, for there is no ash. Inspection can be made very easily. A series

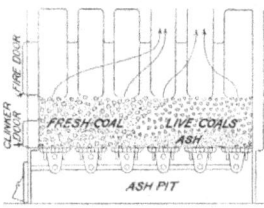

A square fire pot fired by the coking method

of tests show the advantages of pulverized coal for operating locomotives over ordinary coal.

A Mikado type of locomotive in a fast freight service running over a 115-mile division consumed about 29,500 pounds of coal for six trips, as compared with another locomotive of the same type, equipped for burning pulverized coal, and which consumed 22,500 pounds of coal, or a saving of 23.7 per cent in fuel. A locomotive burning pulverized coal has been in constant daily operation near Fullerton, Pa., since January, 1918, and up to the present has not been laid up for repairs.

The Coal Situation This Year

Experiments in burning pulverized coal were conducted within the last few months on the U. S. S. *Gem*, and proved that, where a steamship had storage-room for that fuel, pulverized coal is eminently satisfactory for use in operating ships.

The coal situation this year is particularly interesting. The coal year begins on April 1. The anthracite

This is the apparatus that is used for burning pulverized coal on locomotives

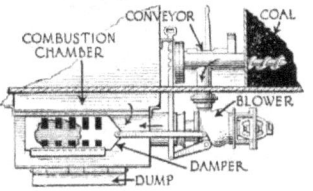

coal consumed for domestic purposes in 1916 was 49,258,000 tons. To this the Fuel Administration has added 2,000,000 tons, bringing the total to 51,258,000 tons, which it has undertaken to supply this year. The first half of the coal year was up on October 1, and the supply for that period was exceeded by 759,136 tons.

Bituminous coal is produced greatly in excess of anthracite; about twelve times as much bituminous as anthracite coal is mined each year in this country.

Shortage of Labor in Mines

This year the United States expects to produce some 600,000,000 tons. In the first six months of this year's coal year 37,000,000 more tons were mined than were produced last year in the same period.

Ever since the war began we have been using an increased amount of bituminous coal—an advance of some 50,000,000 tons for each year.

When you consider the

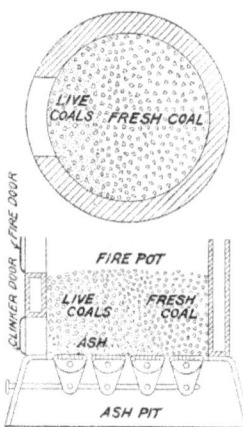

A round fire-pot fired by the coking method

shortage of labor in the mines, the production this year is nothing short of remarkable. If we all do our bit, from turning off the gas or electric light when we leave a room at night, to inspecting the flues in our industrial boiler plants to see that there is no leak, we shall have enough coal this year to keep everybody warm enough for health's sake and to keep the wheels of our reconstruction industries not only turning but humming.

A six-wheel saddle-tank locomotive equipped with a pulverized coal apparatus. It has been in daily operation in Pennsylvania since January 2, 1918. Coal capacity, one and a half tons

Making a New Fuel to Order

How colloidal fuel helps oil and coal to do more work

By Walter Bannard

THE mightiest war-ships burn under their boilers not coal but oil. In 1917 and 1918 it was so difficult to obtain oil that for a time it seemed as if the battleship fleets of England, France, and Italy would be unable to perform their task of bottling up Germany and Austria. The situation was alarming. The submarine was literally turning off the spigot of Europe's oil supply.

In this emergency the engineering committee of the Submarine Defense Association of New York, of which committee Mr. Lindon W. Bates is chairman, determined to begin a series of investigations to meet naval demands. An oil expert of international renown, he decided that a composite of oil and finely powdered coal would meet the demand.

It was not a new idea; but it had come to naught because the heavy coal particles always settled in the oil. Some way had to be found of holding the minute coal particles in suspension for a number of months. When the problem was solved "colloidal fuel" was created.

Particles that Always Dance

What is a colloid? Chemical history answers. Between 1861 and 1864 Thomas Graham discovered that, when certain dissolved substances are separated from a surrounding solvent by a membrane of parchment or fish-bladder, some of them pass through the membrane freely into the surrounding solvent, while others remain behind, only to diffuse very slowly. The particles that do not pass through are evidently larger than the pores of the membrane. Graham called them "colloids."

Victor Henri, a French scientist who has done much to explain colloids, says: "There are no colloids; there is only a colloidal state, just as there is a solid state and a liquid state." When you blow a puff of cigarette smoke from your mouth you see a gaseous colloid of carbon. The finest ruby-glass is a solid colloid of gold. In a colloid there is no chemical union or combination between the particles held and the medium that holds them. A current of electricity passed through a colloid drags the particles with it.

These particles are always in a state of violent agitation, dancing about irregularly because of the collisions that incessantly take place.

Preventing the Powder from Settling

Colloidal fuel is simply a mixture of very finely powdered coal, oil, and a stabilizer ("fixateur," Mr. Bates calls it) to prevent the coal particles from settling in the oil.

The nature of the fixateur is not

The "fixateur" is the magic substance that prevents the finely powdered coal from settling in the oil. You can compound colloidal fuel to suit your requirements. This picture shows a drug-store way; manufacturers will use machinery

disclosed. It produces a state in which the force of gravity is so far overcome that the mixture of coal and oil remain permanent for months. When settling does eventually manifest itself, brief stirring is all that is required to restore the fuel to its previous consistency.

It is possible to sustain for months from thirty to forty per cent of coal in oil. In a colloidal fuel about forty-five per cent oil, twenty per cent tar, and thirty-five per cent pulverized coal can be combined, thus displacing more than half of the oil and securing equal or greater heat values for each barrel, thereby saving much money. At least twenty-five per cent of the fuel oil now burned is conserved and the world's supply of liquid fuel is increased by fifty per cent.

Oil refiners have always wondered how they could dispose of their residues profitably. Colloidal fuel supplies the answer. The "Cinderella products of refining," as Mr. Bates calls them, can be used to prepare a fuel that will command a premium because it is valuable in making the higher grade alloy steels.

It Can Be Pumped Like Oil

Colloidal fuel can be compounded like a prescription. This means that the particular kind of fuel best suited for a particular plant can be made by a central laboratory. One composite in the range of ordinary temperatures is composed of

In the first can is the "fixateur." The second contains wax tailings—a residual waste which has always bothered the oil refiner. In the third can is a binder; in the fourth "anthracite rice." The bottles contain liquid colloidal fuel and pressure still oil used in colloidal mixture about half coal and half oil. Another unctuous semi-liquid is nearly three fourths coal and one fourth oil. The more coal that is added the more paste-like the fuel becomes. But all the pastes are mobile up to sixty per cent coal, and are pumped and atomized as if they were liquid. With the liquid grades the oil-burning equipment of a vessel or a plant need not be changed.

At sea astonishing results have been secured. The net saving of oil on the research vessel with which the Submarine Defense Association experimented amounted to twenty-seven per cent. Moreover, with the same tank or bunker space a longer cruising radius without refueling is obtained. A war-ship or a merchant-ship can increase its steaming radius up to twenty-five per cent.

Cost Savings

The savings of cost involved in the use of colloidal fuel are remarkable. An industrial company in Ohio burns three hundred bushels of oil per day; its oil costs seven cents a gallon; its coal five dollars a ton. A saving of fifty-five cents a barrel, or one hundred and sixty-five dollars a day, is effected by the use of colloidal fuel, quite apart from con-

sidering the conservation of oil and the reduction of transport.

Oil is now extensively used in furnaces of various kinds—brick-kilns, annealing plants, blacksmith shops, brass foundries, and steel works. About seventy per cent of the oil burned by these furnaces can be made to do the present duty of one hundred per cent, and more cheaply at that.

Coal that is useless for ordinary industrial purposes is powdered and mixed with oil to make colloidal fuel. Thus oil is made to go farther and our fuel resources are conserved

A certain steel plant uses three thousand barrels of oil daily, and soon it will require four thousand five hundred barrels. But the three thousand

barrels of oil with coal incorporated and used as colloidal fuel will be ample for the full output, thus effecting a saving of about five hundred thousand barrels a year.

In industrial plants many millions of barrels of oil are consumed annually. If only twenty-five per cent of this oil were displaced by fluidized coal great savings would result. Three million barrels of oil employed in colloidal fuel could do the work of over four million straight barrels. The oil reservoirs of the world are being drained to their very dregs. We must conserve oil, but not at the expense of industry.

Using Low-Grade Coal

Colloidal fuel opens the doors wide to let the world have from the stills all the assistance it needs without starving hungry boilers or flaming furnaces of the grosser fluid fuels which constitute their food. Moreover, great deposits of low-grade coal, as well as large quantities of lignites, brown coals, and waste dusts, are added to the world's fuel supplies.

The world is familiar with three kinds of fuel—solid (coal), liquid (oil), and gas. We now have a fourth fuel which promises to become of world-wide importance.

Farming without the Aid of Laborers

HOW large a farm can one able-bodied man, unassisted, plow, cultivate, plant, and harvest with a fair degree of efficiency?

Not so very long ago it would have been considered extremely presumptuous for any unaided man to undertake the cultivation of a farm of more than a few acres. Now, however, thanks to the formidable array of labor-saving machines invented and placed on the market in recent years, it is by no means beyond the range of possibility for an enterprising worker to cultivate from twenty to a hundred acres, according to the kinds of crops he expects to raise.

The power-driven cultivator shown in the accompanying illustration is one of the recent additions to the labor-saving agricultural machines that have completely revolutionized farming.

The machine, which is shown in the picture in the capacity of a cultivator, may also be used for drilling, planting, and sowing—special attachments quickly

This machine, run by one man, will cultivate, plant, or sow a fair-sized field in a few hours, and, it is said, will do the work better than it could be done by a half dozen men using implements of a more primitive type

adapting it for the particular work that is to be done by it.

This machine is equipped with six rollers for breaking up the clods after plowing. By adding one or more rollers the capacity of the machine may be materially increased. It can also be used for planting three or more rows at a time. With a different attachment the machine can be used for sowing oats or grass, and by still another attachment it may be adapted for drilling the ground or row-planting.

The drive wheels of this versatile agricultural machine are also its steering wheels. The power is supplied by a gasoline engine, and one man can run it easily.

The construction of the machine is simple and substantial. To make the changes for adapting it to the different kinds of work which it is capable of performing takes but little time and does not require the services of an expert machinist.

With a machine of this kind in his possession, the farmer is truly independent of the labor market.

Out of the Garbage-Pail into the Fire

Fuel bricks are now added to the list of things salvaged by science from the nation's waste

By Joseph Brinker

LONG before the world war taught us economy, among other lessons, there was a saying that a French family could live, and live well, on the waste from a typical American table. The waste leak through the garbage-pail was known to be enormous—and, indeed, we were rather proud of it, as an evidence of prosperity.

We've learned better now, and in learning we have unearthed some stupendous facts. For instance, we know that, about the time the war began, there were twenty-nine companies in the United States engaged in rendering garbage for the recovery of grease. They treated 1,200,000 tons of garbage a year, and produced 172,000 tons of tankage—a coarse brown powder which makes fine fertilizer.

How to Save $4,000,000 a Year

Not counting the value of the fats extracted,—a value that was well understood when the war pressed home the need for glycerine from which to make explosives,—the tankage was worth more than $2,000,000 a year; and it has been figured by experts that the installation of rendering plants in all cities of over 30,000 inhabitants would result in a tankage production of more than $4,000,000 annually.

Grease, from which high explosives are manufactured, and tankage were formerly the principal things of use recovered from garbage.

But recently there was announced the discovery of a new method of making fuel from garbage and refuse. This method has been perfected to the point where the government is to conduct exhaustive tests by having one hundred tons of the fuel made in the city of Washington.

Fuel Bricks from Garbage

It's a pleasant thought that, even if we are careless in the kitchen, we may escape the just punishment for our sins, since in wasting what might have been fuel for our bodies we gain fuel for the furnace. Of course, in carrying out the tests, what the government has in mind is to prevent waste of any sort, to conserve the coal supply, and to cut down the number of cars now needed to move coal from the mines to the consumer.

The process by which it is hoped the garbage of yesterday may be the cheerful fire of to-morrow is the invention of E. L. Culver, of Chicago. His plan is to use garbage and coal-dust, the waste from coal-mines, with about seven per cent of tar. For steam purposes no coal-dust is required.

After the garbage is sprayed with creosote for sanitary reasons, it is dumped on a moving belt, where material such as iron, bottles, rags, ashes, and the like are sorted out. The residue is fed to a grinder, where it is pulverized and mixed with the coal-dust. Later it is impregnated with hot water, live steam, and a certain percentage of tar. When the mass becomes pulp-like in form, it is deposited in a brick-making machine, and pressed into fuel bricks ready for firing.

The bricks weigh two pounds apiece, so that one thousand bricks make a ton of fuel. A single brick-machine will turn out twenty-five thousand bricks of fuel in eight hours, or seventy-five thousand in a day of twenty-four hours.

Advantages Claimed for New Fuel

Among the advantages claimed for the new fuel are that it will not slack, no matter how long it may be stored; that it is impervious to water; and that it burns to a cigar-like ash without

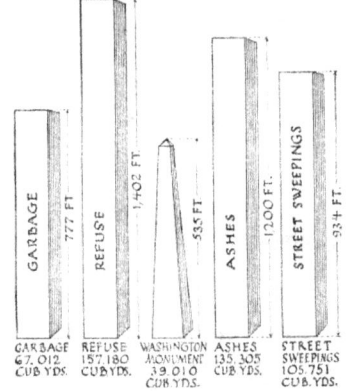

If all of the waste material collected in the city of Washington in one year were sorted out into its component parts, and the parts set up beside the Washington Monument, the monument which is 535 feet in height—would be the smallest in the group

Fuel that is better than bituminous coal is made from garbage by this plant, which sorts and grinds it, mixes it with coal-dust, and turns it out in neat bricks weighing two pounds apiece

leaving the clinkers which are so destructive to fire-grates. Bituminous coal has from 9,000 to 12,500 heat units to the pound, and recent tests of the garbage fuel show it to have from 9,000 to 12,920 heat units. The bricks recently sold in Austin, Texas, where an experimental station was set up, at six dollars and fifty cents a ton.

A second plant has been erected in Washington, D. C., where one hundred tons of brick will be manufactured for government officials for testing purposes. Later a larger plant, somewhat similar to that shown in the cross-sectional illustration below will be erected.

Great Heat Value

While burning, the fuel resembles oak wood, and leaves about the same white ash, though its duration and heat value are much greater. Tests have shown that the garbage fuel contains more heat units than the best bituminous coal. For steam purposes, still better results are obtained if equal parts of garbage and coal-dust and from five to seven per cent of tar is made the mixture. The binder thus created holds together about eighty per cent of the coal-dust while it is burning on the grates.

Daily Average Waste per Person

It may be some time before you will be heating your house with fuel born of the garbage-pail; and in the meantime, or at any time for that matter, it is well to recall how large is the waste to be reclaimed. Definite figures can be worked out for the quantities and contents of garbage, in spite of its variable character and amount. According to a report of the American Chemical Society, the average quantity per person is one half pound a day. It consists of 70 per cent moisture, 20 per cent tankage, 3.5 per cent grease, 1.5 per cent bones, and 5 per cent rubbish. This can all be converted into a fuel product capable of being burned over a grate with less trouble than with the ordinary forms of coal.

The fuel that is made in the plant shown below is called oakoal, obviously made up from the words "oak" and "coal." It is so named because in burning it resembles oak wood, and leaves about the same white ash.

Aside from the advantages of the fuel as compared with coal in respect to its greater heating value per pound and less ash, its brick formation makes it easy to handle and to save in storage space, since it can be stacked in regular rows.

It produces no soot and little smoke in burning, so that smoke-nuisance ordinances can be forgotten when it is used. The possibility of spontaneous combustion is eliminated, and there is practically no loss in transportation.

Don't Let the Coal Set Itself on Fire

Under pressure it may ignite spontaneously

The bigger and higher the pile of coal, the greater will be the danger of spontaneous ignition in the interior of the mass

WHEN coal is stored in large quantities there is always danger of its becoming ignited spontaneously. The larger the quantity of coal stored, the greater the possibility of its catching fire. The higher the pile, the greater the pressure by the overlying mass on the coal near the bottom of the pile, and the more thorough the exclusion of air from its interior.

The organic matter of which all coal consists is constantly undergoing a process of slow oxidation. On the surface, in the absence of pressure, and with unrestricted circulation of air, the heat set free by this chemical reaction does not rise to the ignition point of coal. But under pressure the temperature steadily rises until it reaches the ignition point of coal and combustion begins.

In some of the big yards, wrought-iron pipes are driven into the coal-pile at intervals. Electrical thermometers are placed in these pipes from time to time, and when the temperature reported from any point in the pile reaches the danger point, the coal at that point is thoroughly aired.

Electrical thermometers placed in vertical tubes driven into the coal-pile at regular intervals will give warning of danger

COAL PILE — WIRE TO MAIN CABLE — POINTED IRON PIPE REACHING TO BOTTOM OF COAL PILE — ELECTRICAL THERMOMETER — ROWS OF IRON PIPES TO BOTTOM OF COAL PILE — WIRES FROM ONE ROW OF ELECTRICAL THERMOMETERS — CABLE TO ENGINEERS OFFICE

How to Make the Gas Bill Shrink

Learn to regulate the flow of gas in your range and save money

IF a gas-range does not burn the way it should, its owner should give it attention at once. A neglected burner wastes gas. Popping of burners when lighted should be a signal to see that the burners are cleaned.

Even if the burners are clean, however, there may be a waste of gas. The most frequent cause is too much or too little air. Either defect is very easily remedied with a screw-driver.

The air intake is directly behind the handle, or valve, used to turn the gas off and on. This air intake has a sliding shutter fastened in place by a small screw. Loosen this screw and turn the shutter until the intake is completely closed. Then turn on the gas and light the burner. You will find that it burns with a yellow flame, consuming a great deal of gas and producing little heat. Open the shutter slowly, watching the flame. When it is all blue, the maximum amount of heat per gas unit consumed is being produced. When the shutter is open to this place tighten the screw so as to fasten the shutter firmly in place.

The commonest cause of wasted gas is the rapid boiling of foods. In boiling potatoes, for instance, the gas is usually turned on full force; yet the potatoes are cooked only a little more rapidly than if the gas were turned low so that the water barely simmered. In either case, the water is at the boiling temperature, and no higher.

By turning the gas low after the boiling stage is reached, the potatoes can be cooked almost as quickly as with the gas high, little water evaporated, and almost half the gas saved.

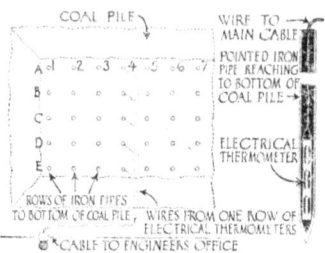

(At right) When the shutter has been adjusted to make the gas just right for the blue flame, the screw is tightened to keep the shutter in that position

(Above) With the flame high, most of the gas is wasted in evaporating the water. The other picture shows the flame as it should be with the water simmering

An Electro-Thermostatic Control for House-Heating Boilers

A home-made electric device to operate the draft doors by the temperature of the rooms

By E. F. Hallock

THE average house-heating steam boiler comes fitted with a highly efficient regulator which automatically opens or closes the draft according to the steam pressure, and tends to maintain that pressure constant without regard to the temperature of the portions of the house being heated. Where temperature conditions are such that a full head of steam is needed at all times in order to make the quarters comfortable, the pressure regulator for all-around efficiency and good service can hardly be improved upon.

Climatic conditions in many parts of the United States are such that steam is needed to keep the place warm one day, while on the next none is needed. These conditions call for a draft regulator that will be responsive to temperature fluctuations in the quarters being heated, with the added precaution that the pressure cannot increase above a certain predetermined maximum.

Pressure Regulator

In other words, what is required is a system that can be set to maintain the temperature at some fixed point, say 65 or 70 deg., and that will supply the heat needed at a pressure not to exceed 2 lbs. per square inch, this being the pressure most house-heating boilers are designed to operate.

In supplying the pressure regulator, the boiler manufacturer has done more than half the work for the man who would pattern his boiler after the foregoing suggestions. He has adequately taken care of the pressure-regulating end of our requirements, and has provided the mechanism for closing the bottom draft at the same time the check draft in the flue is opened, or vice versa. By taking advantage of this linkage, the necessity for making structural changes in the

boiler or the flue system is done away with.

The pressure regulator consists of a very flexible brass bellows in communication with the steam dome of the boiler, so that the slightest pressure causes the bellows to extend. The free end of the bellows is connected to a long arm or lever, which in turn is linked to the check draft in the flue and to the bottom draft door, so that when the bellows extends the check draft is opened, admitting air directly into the flue above the fire, while the bottom draft is closed, cutting off the air from beneath the grate.

The pressure at which the drafts will operate is changed by shifting a

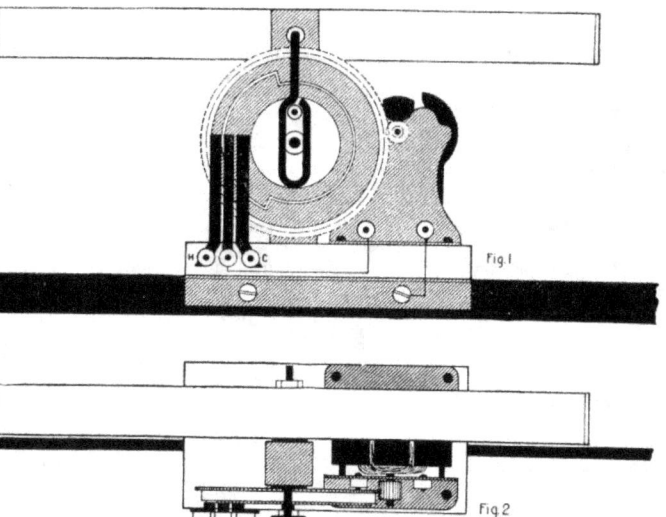

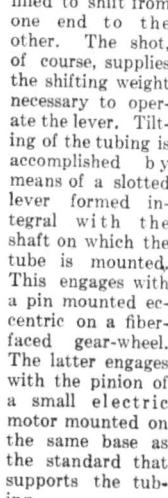

Fig.1

Fig.2

A small electric battery motor tilts the tube in which shot are placed to roll from end to end for operating the draft dampers on the furnace

sliding weight along the arm of the lever.

How the Regulator Works

With the weight placed near its fulcrum, the moment of the force due to the weight is reduced and the pressure required is low. Sliding the weight out toward the end of the lever has the effect of increasing the pressure necessary to operate the drafts. Naturally, when the pressure falls off after the drafts have been closed a short time, the weight pulls the lever down until the fire brightens up sufficiently to raise the pressure again.

If the counter-weight is removed,

and replaced by a device that shifts automatically according to the temperature of the heated rooms, midway it can be arranged either to close or open the drafts at will.

To Make the Apparatus

The simple apparatus shown in the elevation Fig. 1, in plan Fig. 2, and in application to the pressure regulator on the boiler (Fig. 3) accomplishes this purpose. It comprises a closed length of 1-in. square brass tubing, $11\frac{1}{2}$ in. long, pivoted at its midpoint to a wood standard so that it can be tilted in either direction. Tilting the tube causes 2 lb. of BB shot with which it is about one third filled to shift from one end to the other. The shot, of course, supplies the shifting weight necessary to operate the lever. Tilting of the tubing is accomplished by means of a slotted lever formed integral with the shaft on which the tube is mounted. This engages with a pin mounted eccentric on a fiber-faced gear-wheel. The latter engages with the pinion of a small electric motor mounted on the same base as the standard that supports the tubing.

The whole apparatus is mounted by means of two clamp plates screwed to the under side of the baseboard and to the operating lever of the pressure regulator, so that the end of the balancing tube is just flush with the end of the lever itself.

The apparatus is so simple and the sketches so clear that little description is necessary. The apparatus from which the drawings were made was put together from scrap materials, the dimensions being chosen to fit the things at hand.

The baseboard is a piece of cypress $5\frac{3}{8}$ in. long, $2\frac{3}{4}$ in. wide, and $\frac{5}{8}$ in. thick, and the standard is a piece of 1 in. square cypress 5 in. long over all, mortised into the base $\frac{1}{2}$ in. from the front edge and 1⅝ in. from the left edge. The clamp plates are made of two lengths of sheet brass, right-angled and screwed to the under side of the baseboard, with their perpendicular sides 3/16 in. apart to accommodate the lever of the pressure regulator.

The shaft on which the balancing tube is mounted is a length of ⅛-in. brass rod, looped to form a slot, as shown, and elbowed to journal in a babbitt bearing in the standard. Its outer end is threaded to take a thread tapped through both sides of the balancing tube and a lock-nut.

To Reduce Friction

After the whole apparatus has been set up and put in working condition, the lock-nut is screwed tight and soldered both to its shaft and to the balance tube to prevent its working loose and disarranging the apparatus. A washer is interposed between the tube and the standard to prevent binding, and another is soldered to the outer side of the shaft, adjacent to the standard bearing, to keep the shaft from working lengthwise into the standard.

The brass gear wheel 3½ in. in diameter is faced with a disk of fiber 3/16 in. thick and 3¼ in. in diameter, the composite wheel being mounted on a pin anchored in the standard, permitting the wheel to rotate freely. The pinion on the motor-shaft is 7/16 in. in diameter, and the motor is mounted on the right side of the baseboard, so that it engages perfectly with the teeth of the gear wheel.

One important consideration is that the operating lever on the boiler regulator tilts, and the length of the balancing tube should be such that it will be about twice the length to insure the shifting of the shot from one end to the other. In the apparatus under consideration the length of the boiler regulator arm is 18 in. and the tube was made 36 in. long. To bring about the proper results the pin was mounted ⅝ in. off center, so that its total throw is 1¼ in. To reduce friction to a minimum, a roller ¼ in. in diameter is fitted to revolve freely on the pin, and the slot in the operating lever is made 1½ in. long and ¼ in. wide, there being sufficient clearance to keep the boiler from binding at any point in its stroke.

Since it is necessary to have some device to turn off the current from the motor, a commutator is mounted on the face of the fiber disk, and the three brass brushes fastened to the baseboard at the left side make the proper contact when the heat-controlled relay swings from side to side. The commutator consists of two disks of thin sheet copper so shaped that the contact is made between the middle brush and one of the outside brushes, while

the motor is tilting the tube in one direction and between the middle brush and the other one on the reverse motion of the arm. A little experimenting will be necessary, in setting the commutator, to get the operation of the motor so that it will do its work with-

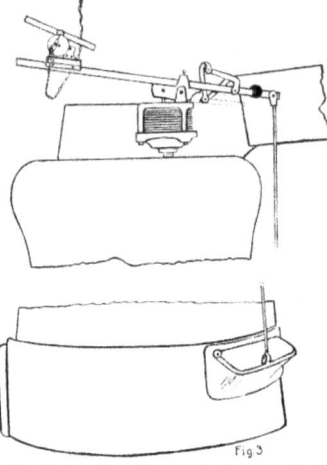

Fig 3

The tilting tube and its mechanism mounted on the draft damper operating the lever

out keeping the arm tilting back and forth when once started.

The thermostat consists of 22 elements, of which 20 are exactly alike and the other two only slightly different. Each element consists of a bar of soft flat steel 4¾ in. long, ½ in. wide, and ⅛ in. thick, and a similarly sized piece of sheet brass—20-gage.

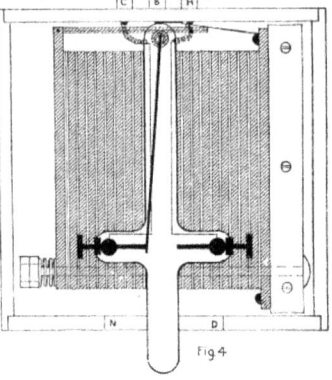

Fig 4

A simple thermostat to control the current by the temperature of the house or room

In a cold temperature not to exceed 30 deg. the steel and brass bars are clamped tightly face to face, and a 1/16-in. hole drilled about ⅛ in. from each end, and then they are riveted tightly, using a copper or brass rivet. The ends of these steel and brass bars are then soldered together for at least ¾ in.

If the brass bars buckle a trifle by the heat, pay no attention to it; this is just what is wanted to make the thermostat. The steel bar for the unit on the right is 5⅜ in. long, each end being drilled for fastening the bar to a wood piece with screws. The brass bar is soldered to it, as with the others, leaving a slight hang-over of the steel bar at each end. The left end unit has a steel bar 5¼ in. long, the upper end of which is forked for the reception of a piece of rack which is pinned in the fork when the thermostat is assembled. All solder projecting is filed, as well as the rivethead, to make the surfaces smooth.

Centering and ⅜ in. from the lower edge, a 3/16-in. hole is drilled through each unit for a bolt. The units are assembled as shown, the bolt passing through the wood base. A good stiff spring is placed over the bolt, and the nut is screwed down to apply plenty of tension. The bowing of the brass bars will cause the uppermost ends of the units to separate fanwise, and the lever action of the units themselves will enable you to take advantage of this extension, which increases as the temperature of the bars increase and decrease in changes of temperature.

The short length of rack shown pinned to the forked member of the thermostat engages with a pinion mounted on a shaft that carries a long spring brass lever. Also mounted on this shaft, so that it can be independently rotated, is a cross-shaped piece of fiber with two terminals in the shape of adjusting screws with lock-nuts, and they are so set that the swinging brass lever makes contact with either one or the other. Holes are drilled in the sides of the box, so that the air may freely circulate around the bars.

Simple Electrical Connections

The electrical connections are very simple. Each contact screw is connected to a binding post located on top of the case: the one on the left side or the low temperature side is connected at C and the other at H (Fig. 4.) The frame of the thermostat is connected to the central post B. These terminals are connected with like lettered parts on the regulator. The terminal B is connected to six dry cells in series, and the other battery terminal grounded to the nearest water-pipe.

The thermostat should be placed out of drafts and quite out of range of a radiator. At night it is necessary only to pull the fiber lever over to the left toward N. This will retard connection until very low temperatures are reached. The setting may be determined to suit requirements. With such an arrangement the basement need be visited only to place fuel on the fire or remove ashes.

Don't Blow Out the Gas

Before you start the Welsh rabbit put on that becoming bungalow apron. Chiffon sleeves are really not safe near a gas stove

Don't search for the leak with a candle—call the plumber

A frequent cause of fires—leaving the gas lighted near an open window

Many a leak has resulted from this unnatural use of the fixture

DON'T blow out the gas is good advice to remember this winter.

Don't allow a draft to blow out the light. A sudden gust of wind from an open window may do the trick, and you will wake up in eternity.

Don't hang clothes on gas fixtures. It is an easy way to start a leak.

Be sure to turn the gas off. Don't forget to see that your gas fixture is in good condition. See that the pin is properly placed so the key won't turn all the way around. If the key does turn all around, you are quite likely to turn the gas on again after you have turned it off. If there is anything the matter with your fixture, the gas company will repair it.

If you smell escaping gas, notify your gas company. Many persons have lost their lives through looking for a leak with a match or a candle.

If you feel you must locate that leak yourself, put a wet cloth over the place where you think gas is escaping. If the gas really is escaping, you will know it by the appearance of air bubbles.

Don't use inferior gas tubing.

Gas mantles are safer to use than is the open flame, because if you turn the gas off in a mantle-equipped fixture and then turn it on again in five seconds the gas will light. The use of gas mantles will prevent an accident through inadvertently turning the gas on again.

Don't allow curtains or draperies anywhere near an open flame.

You Can Make Burning Gasoline Absolutely Safe

IF ANYONE told you that you could pour gasoline into a burning tank of gasoline without any danger to yourself, you would tell that person he was crazy. Yet one of the accompanying illustrations shows a man pouring real gasoline through a real live sheet of flame flaring out from the top of a tube on an automobile gasoline tank, and another shows a young woman doing a similar stunt with a small gasoline tank on a camping stove.

These unusual and almost uncanny performances are made possible by inclosing the tube entering the tank with a fine wire mesh screen which prevents the burning flame at the mouth of the tube from passing down into the fuel in the tank.

But this is only one of three advantages of this safety tube. The other advantages are that the tank cannot explode, and that there is no evaporation of the gasoline.

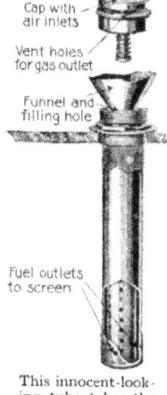

Cap with air inlets
Vent holes for gas outlet
Funnel and filling hole
Fuel outlets to screen

This innocent-looking tube takes the danger out of gasoline in several ways

It looks dangerous, but the safety tube is in place

The tube keeps the flame from getting to the fuel

Since the flame cannot enter the tank, the only other thing that could make the tank explode would be the expansion of gases in the tank. The way to overcome this difficulty is to allow the gases to escape when the pressure gets too near the strength of the tank parts. This the safety tube does automatically.

As soon as the pressure in the tank reaches the danger mark, the gases press up through the small holes in the cap top and lift the tank pressure valve. When sufficient gases have passed off, the spring automatically seats the valve again.

Evaporation is prevented, notwithstanding that air must be admitted to the tank to take the place of the fuel drawn off as it is used.

You put the ash-sifter upon your can; then you move the handle back and forth and vibrate the screen. No dust is scattered, because everything is inclosed

GRAVITY SIFTING SCREENS

Dump the ashes into the hopper. They fall upon the first inclined screen. The heavy unsifted particles pass to the next screen. What coal is left is taken out by way of the door which is shown open

You don't have to shake anything in the apparatus shown at the left. The ashes trickle down on the cone - shaped screen, and this in turn spreads them

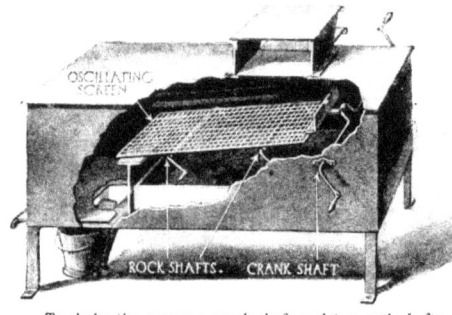

OSCILLATING SCREEN

ROCK SHAFTS. CRANK SHAFT

To shake the screen a crank-shaft and two rock-shafts are depended upon. Turn the crank-shaft and the screen moves back and forth. The ashes fall into a receptacle and pass out through a gate into a can

You shake the screen much as if it were a corn-popper, and then you open a door and tilt the screen so that the ashes drop into the can. The chief merit of the arrangement is that the dust is not scattered

Here we have the rocking-chair principle. Johnnie simply rocks the apparatus back and forth, the ashes drop down into a receptacle at the bottom, and the unburned coal can be withdrawn by opening a gate

COAL DUMPING KNOB

DOUBLE SCREEN

SLIDING GATE FOR DUMPING COAL

Fairy Godmother to Modern Cinderellas

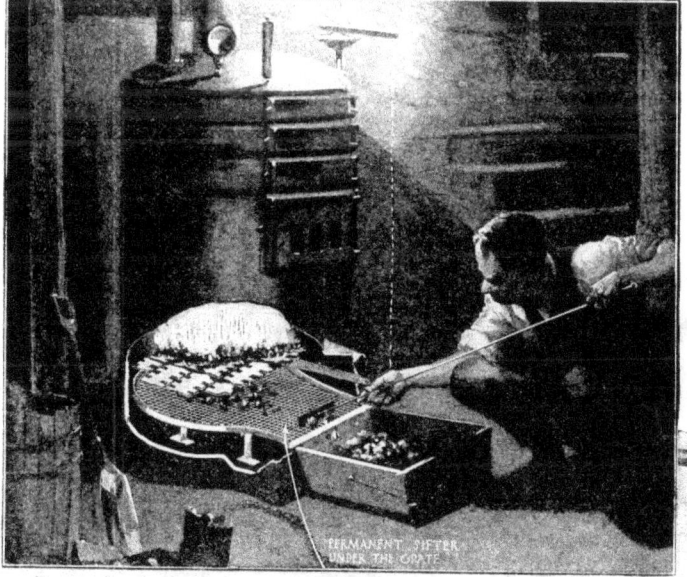

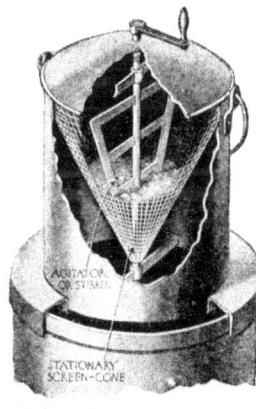

Another dustless ash-sifter. Here we have a stirrer used inside of a stationary screen. Turn the crank-handle at the top and the stirrer forces the ashes through the screen. The whole apparatus is placed on an ordinary ash-can

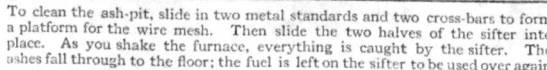

To clean the ash-pit, slide in two metal standards and two cross-bars to form a platform for the wire mesh. Then slide the two halves of the sifter into place. As you shake the furnace, everything is caught by the sifter. The ashes fall through to the floor; the fuel is left on the sifter to be used over again

This is a wire screen shovel. The ashes are scraped up by means of a combined gate and scraper. During the shaking process the scraper becomes a gate, which opens and discharges the coal

Connected with the rotary screen is a fan which blows the ashes into the furnace so that they pass out by way of the natural draft

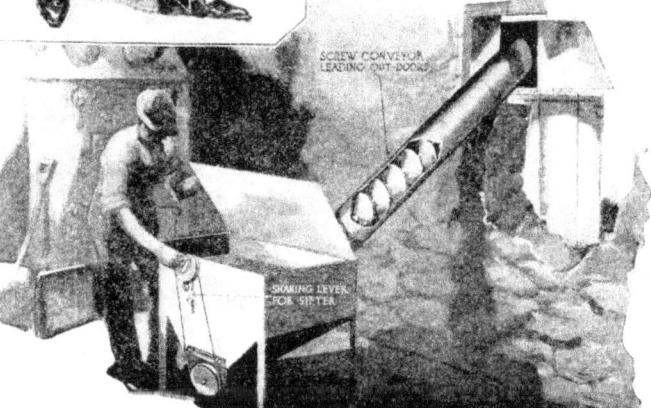

This ash-sifter, built into the stove, shakes the grate. The ashes drop down into a rotary screen, which is rotated in such a way as to shake out the ashes

This saves the trouble of carrying the ashes out of the cellar. A screw conveyor leads up from the ash-sifter. Turn the crank, and you not only sift the ashes, but you convey them out of the cellar at the same time

How Are We to Keep Warm This Winter?

By B. Kaye

BUT one fuel can successfully take the place of hard coal, and that fuel is coke. By coke we do not mean the ordinary gas-house coke that has so often been used, unscreened and unprepared in any way, but a domestic coke, made by a special by-product process, screened and assorted in sizes like hard coal.

What Is Coke?

Coke is the coal that is condensed by a special process to a form approaching, as nearly as possible, pure carbon, and that is further made more combustible by its light and porous structure. It is the solid part of the coal left after the volatile matter has been driven off.

In distilling by the retort-oven method, coal is placed in a closed chamber heated by the combustion of gas in flues, arranged in the wall of the chamber. In such a chamber the highest grade of coke possible is produced, and all the by-products that rise in the form of vapors are recovered.

The heat-producing qualities of coke surpass those of hard coal. A ton of coke can be made to go further than a ton of hard coal, because it is pure fuel, contains less waste, less incombustible material, and about one third as much ashes.

How to Use It

A coke fire will "come up" more quickly in the morning than a hard-coal fire, at the same time holding just as well overnight; it is all a matter of damper adjustment. Then, too, coke can be stored out in the open better than any other fuel, and handled without depreciation. It is clean: for there are no clinkers.

Coke produces only one third of the amount of ash produced by an equivalent quantity of coal

These lumps—coal on the left and coke on the right—are of approximately the same size, but the lump of coke will give you 30 per cent. more heat than the more solid-looking piece of coal—and no clinkers

Although coke has all the advantage in heat-giving qualities, and will "come up" more quickly in the morning than will a hard-coal fire, it is twice as bulky, the coal-scuttle on the right holding the equivalent in coke of the coal contained in the more modest scuttle at the left. So build your bins large when you buy coke

A coke fire must be checked to keep down the heat; a hard-coal fire must be poked into activity. A great deal depends on how the fire is kindled, but once the fire is started, coke burns hotter and faster than coal does. Care should be exercised in regulating the draft, for a beginner often blows out his fire before it is well started.

To start the fire, kindle it as you would hard coal; but, as fast as flames appear, cover them with smaller pieces of coke, thus closing in the heat and deepening the bed of coals. This is important. Coke kindles best from its own heat.

What to Remember

When the fire is once started, shut off all drafts, especially chimney draft, to keep the heat from going up the chimney.

Do not shake or poke a coke fire half as much as you would a coal fire.

Keep the bed of the fire deep, with coke well piled up.

Remember that a large body of coke, burning with a gentle glow, will give more heat than a small quantity burning fiercely.

Use nut size for ranges, egg or stove size for furnaces or heaters, pea coke for self-feeders.

Coke, being lighter in bulk and of porous body, thus admitting air readily, must be fired in larger mass than hard coal, and it has been found best to use as small a size as will readily stay on the grates, stove, nut, and pea being the favorite sizes.

Remember that coke takes more space than its fuel equivalent in coal and arrange your bins accordingly.

FIRE FIGHTING

Fighting Fire from the Sky

The Forest Ranger Becomes an Aviator—Forestry Bureau's New Fire Fighters Airplanes and Dirigibles Now Used

By Robert H. Moulton

VISITORS to our national parks this summer have been considerably puzzled by the frequency with which airplanes and dirigibles have skimmed overhead. No reason for this activity was apparent, now that the war is over.

However, an excellent reason has existed: the flyers were acting as aerial patrols, watching for and reporting any forest fires they discovered. Thus Uncle Sam has found an important use for his aces of the air right here at home.

Heretofore the work of patrolling United States forests to locate and fight fires has been entrusted to men on horses, motorcycles, or railroad speeders, and partly to watchers stationed at lookout points.

The former travel about from place to place, while the watchers, who are located at advantageous

The dirigible has just dropped a bomb—not to blow up the forest below, but to put out the fire there; the bomb is filled with fire-extinguishing chemicals

spots, usually on the top of some high mountain-peak, never leave their post, but remain constantly on the lookout for fires.

The observers at lookout stations are provided with all the latest scientific instruments for locating distant objects, and with carefully prepared maps showing the names of all canyons, streams, roads, ranger stations, and the like.

How the Watchers Do Their Work

These men have powerful field-glasses, and a telephone leads to the nearest ranger station and to the forest supervisor's office. If a fire is discovered, the watcher immediately telephones to headquarters, giving the location of the fire as it is projected on the map of the station; a crew of men is despatched to fight it.

The efficiency of this system has been increased from year to year, until now it is as up-to-date and as successful, in its own field, as that of a municipal fire department. In fact, the same essential factors for success are employed in the forest work that are universally recognized as necessary for fighting fires in cities: namely, first, detection of the blaze before it has gained headway; and, second, prompt attention.

According to reports compiled by the Forest Service, there were in the United States an average of 28,735 fires a year between 1915 and 1917. The ground covered by these fires amounted to 8,052,945 acres. The estimated damage was $9,875,000.

A report covering the period between 1913 and 1917 showed a yearly average of 6,184 forest fires, with 666,663 acres burned and damage to the extent of $599,000.

Last fall it occurred to officials of the Forestry Bureau to supplement the work of the regular forest rangers by the addition of air scouts, and a few experimental flights were made to determine the value of the airplane and

dirigible in this connection. These experiments proved so satisfactory that Secretary Baker issued an order directing the Air Service to cooperate with the Forest Service in the work of forest fire patrol, and, beginning with the first of June, the plan was regularly put into operation.

Air scouts and bombers who saw service in the Great War were set to work in the forests; and dirigibles — filled with the new non-inflammable helium gas instead of dangerous hydrogen—will also be used in patrolling the forests for fire.

Reports received by the Forestry Bureau show that the airmen experience no difficulty in detecting fires in heavy timber at an elevation of ten thousand feet and even more. From this it will at once be evident what an advantage the air scout enjoys over the watcher in a lookout station. Even when these stations are located on mountain-tops, the range of the observer is more or less limited by reason of the character of the country cut by deep canyons and broken by mountain-ridges. Furthermore, the area of country that he can see at any time is always the same, whereas the airplane scout, flying swiftly at a much higher altitude, is enabled to cover a wide and ever-changing stretch of country within the space of a few minutes.

The aerial observer has still another advantage, of the greatest importance, over the watcher in the lookout station. The latter, when he has discovered a fire, can do nothing more than tele-

Airplanes and dirigibles will cooperate with the watchers at lookout stations; this watcher has before him a map of the forest, and near by there is a telephone, for use should he detect smoke

phone its location to the nearest fire guard; he depends upon others to go to the scene and combat it. On the other hand, the air scout, when he has located a blaze, can send in the alarm, and then, swooping down over the spot, drop on it bombs charged with suitable chemicals for extinguishing fires. By this means he is frequently enabled to put out the fire unaided, or to check its growth until the regular fire-fighters have time to reach the place before it has gained any considerable headway. In this connection, a plan that is soon to be tried out proposes transporting fire-fighters in dirigibles from which ladders can be lowered to the ground.

Warnings Flashed by Wireless

The observer in a lookout station locates fires by means of triangulation, reports being telephoned from separate observation points. The air scout locates them by coordinates in the same way that gun-fire is directed to a particular spot or object—that is, by means of squares drawn on duplicate maps, one being in the possession of each airplane observer and another in the office of the forest supervisor.

The tentative plan is to equip the patrolling airplanes with wireless sets, with which the pilots may flash the first signals of fire. Roads and pathways where a carelessly thrown cigarette may be the cause of thousands of dollars' worth of damage will, of course, be watched particularly.

Regular war-time bombing planes are used; it is planned also to use dirigibles filled with the non-inflammable helium gas

For years fires have been fought by crews of men alone; they chop down trees at the edge of the fire in order to check it

Fighting an Oil-Well Fire

It's a case where "Water, water, quench fire," doesn't work — Suffocation by Steam and by Foam

By P. Schwarzbach

SUDDENLY, without warning. there was a loud crash, and a flaming automobile shot up into the air. This happened as the automobile was crossing a bridge over a creek in Drumright, Oklahoma.

What caused the accident? Near the creek was a flowing oil-well. Gases from the well had collected along the edge of the creek, and the headlights of the onrushing car had exploded them. Gas and oil will burst into

flames at the toss of a spark, and even, as this case proves, from the heat of a headlight. And these fires are the hardest of all fires to fight. Water will not quench them—they must be smothered.

Imagine trying to smother 100-foot flames shooting out of an immense gas-tank. This is one of the problems that harrow the souls of the owners of oil and gas wells and tanks.

In combating gas and oil fires, the

fighters have tried suffocation by steam and by foam. The foam is made by bringing together two chemical solutions that produce tough, non-inflammable bubbles. They have also tried cutting off the source of supply by digging a tunnel up to the well and then forcing the oil to flow into it.

Oil was flowing from a well in Humble Field, Texas. It was pouring out of two pipes—one twenty feet

A group of men advanced toward the burning well, shielded from the flames by a piece of corrugated iron; in their hands they carried a long pipe with an asbestos-lined cap at the end

Fighting Fire Under Water

Water, water everywhere, but never a drop to put out the fire that threatened the lives of two divers

By Francis Arnold Collins

FEW firemen have ever faced so appalling a danger as did the deep-sea divers who found themselves trapped far below the surface of the water in a burning wreck. The submerged hull in which they worked was filled with compressed air and the abundance of oxygen made the flames spread with abnormal rapidity. Though they were literally buried under water, there was no time to wrench open the hull and let it in. Should they abandon the job?

From the first the work of salving the wreck had proved extremely difficult. The ship had gone down in quiet but deep water, and every effort to raise her with pontoons had failed. It was finally decided to close her hold and some of the superstructure and to blow out the water by compressed air.

The ship had been injured in a lower section of the hull and the work of plugging her hold was less difficult than might be imagined. Like a bottle upside down in the water, it didn't matter so much if there were a small hole or two in the lower side—the water would enter only a little way. But the holes in the top would let out any air forced in and cause the trouble.

Several days had been spent inserting strips of oakum in the openings to stop the leaks and make the upper sides air-tight. Finally the air-pumps were started. Several hours of pumping, however, failed to bring up the sunken ship.

"She leaks," decided old Bill Andrews, veteran at the pumps. "The water's gone out of them upper cabins all right, but when the air gets to pushin' the water on down into the hold, and out the bottom of the ship— why, one of them cabins is leakin' air just enough that the pressure keeps goin' down and the water don't back out only so far. We can't raise the ship unless all the water goes out. Them leaks is got to be fixed."

This statement, though ungrammatical, was entirely correct. Two experienced divers were selected to go down into the air-chamber and investigate. They wore full diving-suits.

The divers reached the hull in safety, made their way through the hole in the bottom and up into the superstructure to the point where the air had driven the water out. Candles were lighted and they started on their tour of inspection. The tiny flame was needed not so much for illumination as to show the presence of a current of air. If a leak had sprung in one of the cracks, then when the candle was held near it the flame would blow the way the air was going—out the leak, and would indicate at a glance the location of the leak.

Clumping laboriously around in their leaden-weighted shoes and heavy suits, the men moved forward, holding the candles to the cracks. Slowly and cautiously they worked, for no one knew better than they that the air-chamber, far below the surface of the water, was ripe for a bad fire. The compressed air had dried out the inflammable materials. An enormous amount of oakum had been used in filling the seams, and the ragged ends of oakum are highly inflammable. Once started, a fire under such conditions would spread with rapidity. Even under ordinary air-pressure the oakum would quickly fan the merest spark into a blaze.

Suddenly a draft of escaping air caught the flame of the candle ahead. Before the diver could draw back, it had leaped to a frayed end of oakum several inches away.

"Fire!"

The worst had happened. There were tons of water above, below, and all around, but no time to wrench open some part of the roof or sides of the cabin and let it in. The flames were spreading with terrifying speed. Without thought of self, the men, driven to the most primitive method of defense, tried to beat out the fire with their heavy mits. Had they forgotten in their excitement another danger that threatened? The smallest hole burned through the air-tight gloves meant certain death—suffocation from the smoke that would pour into the inflated suits, or death by drowning should they attempt to leave the air-chamber. And still they fought, inviting death at every blow.

The dark chamber was lighted only by the fitful flames of the burning oakum. The heavy diving-suits made the work extremely difficult. The

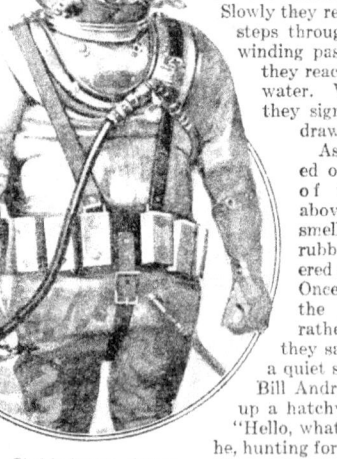

Suddenly a draft of escaping air caught the flame of the foremost candle, and before the diver could draw back, it had leaped to a frayed end of oakum several inches away

Clad in heavy, clumsy diving-suits, the men were sent down to search for the air leak in the sunken hull

movements of hand and foot were slowed down by the weight. A blow that would beat out a flame in an ordinary atmosphere would not suffice here. Desperately they pounded and stamped, working like men in a nightmare who try to escape from peril but are held back by an invisible force. Bit by bit they gained on the swift, fierce flames, which were fed by the light oily oakum and the excess of oxygen in the air. At last pluck won.

Under the frantic slapping and stamping the last spark smoldered out.

Then only did the divers think of retreat. Their gloves, though badly singed, were fortunately intact. Slowly they retraced their steps through the dark winding passages, until they reached the free water. With a jerk, they signaled to be drawn up.

As they crawled over the rail of the ship above, a strong smell of burnt rubber still lingered about them. Once safe above the ground, or rather the water, they sat down for a quiet smoke. Old Bill Andrews labored up a hatchway.

"Hello, what's up?" says he, hunting for his favorite oilcan.

"We are," was the laconic reply, "and if you don't mind we'll stay up."

How to Beat Our Worst Enemy

Properly enforced laws would save the United States millions of dollars that now go up in smoke — Fire losses equal cost of maintenance

By F. W. Fitzpatrick

The United States has the best fire-fighters in the world. But it has been estimated that about two thirds of the fires they have to fight could be avoided by more stringent laws properly enforced

IN normal times Europe's fire losses average but one eighth or one tenth the loss suffered by the United States. One reason for this is that, under the laws of most European countries, a man cannot collect insurance for a fire originating upon his premises which the authorities can trace to his own neglect or carelessness. More than that, if the fire spreads to his neighbors' property, he is assessed, along with the insurance companies, for a proportion of their losses. All of this, of course, encourages carefulness in property-owners.

Result of Carelessness

Half of the fires in the United States are the result of carelessness, and a great part of the remainder are nearly always traceable to arson; and the cost is more than $250,000,000 a year. But this sum represents only the fire losses in cities and towns where fire records are kept. Every year millions of dollars' worth of property of which there is no record goes up in smoke, for instance, in fires on farms or in other remote districts. Then, there is the cost of maintaining fire departments, the cost of high-pressure water supply, and the cost of maintaining insurance. In most of the larger cities the cost of maintaining fire departments is almost as great as the fire losses.

Our Terrible Conflagrations

All of these charges and losses might almost be called fixed or overhead, for they remain about the same year after year, in spite of the splendid efforts and actual accomplishments the United States has achieved in fire prevention. Yet one cannot call those losses "constant," for every few years there is a conflagration that boosts the

One plan for fire prevention would provide for the inspection of all buildings large enough to house twenty or more persons, and their labeling as to fire safety. The sign below needs no explanation

A landlord who could truthfully label his building "Fire Resisting" would have no trouble in getting all the first-class tenants he could accommodate

"Ordinary Construction" would mean reasonably safe; that is, not fireproof, but having the requisite sprinkling or other fire-fighting apparatus installed

From their fire-escapes they fearfully watch the blaze roaring in a paper warehouse only a few houses away

average figures enormously. For instance, in the year of the San Francisco fire the fire losses were $580,000,000 instead of $250,000,000.

In the regular order of things this country is about due for another extraordinary blaze, the chances being intensified in cities where, in connection with war work, there is a great mass of temporary construction.

The terrible fire in Chicago in 1871 broke out in its lumber district, and, because of the fact that two thirds of Chicago's buildings were wooden, spread with great rapidity over the town. The flames even leaped across the river, destroying in all 17,450 buildings. The loss was $196,000,000, a fabulous sum in those days.

The fire that broke out in Baltimore in 1904 ate up the entire business section of the city.

The Biggest Fire

Right on its heels came the worst and largest fire in the history of the United States—the San Francisco fire of 1906. It broke out as a result of a series of California earthquakes, and raged for three days, tearing through the entire business section. The earthquake broke the water-mains, thus draining the reservoirs and making it almost impossible to fight the fire.

It was checked, finally, by dynamiting several houses. The insurance companies bore half the loss, which reached the enormous sum of $500,000,000.

These huge losses could have been avoided if the buildings had been truly "fireproof." Still people will build only as well as they are made to by law, so that every step of the way toward fire protection or fire prevention has been taken after a fight with the very people who should be most interested—the owners of property.

Since it is only by force that we can accomplish anything in that line, let's get still another law or ordinance through: the *labeling* of buildings. We have laws now as to just how buildings shall be constructed in this or that district, and it is the building commissioner or inspector who passes on the matter, interprets the ordinance, and is the authority in each city. Let us give him still another function: the inspection of all buildings of a public or semi-public nature—churches, halls, stores, hotels, apartment-houses, any building that houses more than twenty people under one roof. After these buildings are inspected they should be put in one of three classes and labeled.

Labels for Buildings

"Fire Resisting" would mean that the building is constructed according to all the requirements of fireproof construction; "Ordinary Construction" would mean that the building, though not fireproof, is reasonably safe and has the requisite sprinkling apparatus or other protection installed; "Dangerous" would point out the fire-traps. No law, no police, no terrible lessons would have half the effect upon property-owners as would this labeling. With "Dangerous" marked over his door, what chance would an owner have of renting his building?

Balloons that Can't Catch Fire

We discover how to substitute helium for hydrogen and thus we revolutionize the dirigible airship

By Albert Whiting Fox

The balloon was behaving badly when this picture was taken at Fort Sill, Oklahoma, and so the bag was deflated. Then, as the other picture shows, something happened

IN a recent test of marksmanship at one of the army aircraft schools, two small balloons were provided as targets for two respective groups of marksmen. At the first shot from the first group the target burst into a mass of flame and smoke—the effect of an explosive bullet. The second group fired once, twice, a dozen times. Observers were sure that some of the shots had scored hits—but nothing happened. The firing continued, with the same negative result. Something was wrong. The explosive ammunition was pronounced defective. Report to this effect was made formally to the proper authorities.

Why It Didn't Catch Fire

Only a few officers present knew the secret. The test was camouflaged, in the sense that the second balloon could not have been set afire by any known agency. It contained the new non-inflammable gas developed by the government and guarded as an all-important war secret. This gas—helium—is now being produced in large volume. Its value for peace and war usage is regarded as tremendous. The story of its development from the inception of the idea to its practical usage has all the elements of romance of the middle ages.

It might truthfully have been said to the disappointed army marksmen: The reason you don't explode that balloon is because a Professor Cady, of the University of Kansas, some ten years ago made a survey of the gas and oil resources of that region, and published a scientific paper, which happened to be in the files of an old British scientist in London at the time one of the Zeppelins was brought down.

"Suppose the gas would not burn?" the professor thought. He recalled having seen the report of an American professor, illustrated with so-called iso-helium lines to outline the region where he thought helium existed. The

A company of terror-stricken men fled for their lives when a great hydrogen-filled observation-balloon caught fire at Fort Sill, Oklahoma, during inflation. How did the fire start? An electric spark passed when the folds of the bag were rubbed together. It was the same kind of spark that you obtain when you rub your feet on the carpet and light the gas by touching your knuckle to the burner. When non-inflammable helium is used, instead of hydrogen, there need be no fear of such accidents

British scientist knew, of course, that helium was non-inflammable, and light enough to lift a balloon or Zeppelin. He put two and two together in the quiet of his study, and wrote a note to the British Admiralty. That was the beginning.

Fascination of Helium for Scientists

The Admiralty turned the note over to the proper scientific branch as a matter of routine. Ideas from outsiders came by the thousands, but this one attracted attention, chiefly because helium had a fascination for the British scientists. Its characteristic line had long been a feature of the sun's spectrum. But the gas itself had never been isolated on the earth.

It seemed for a time as if helium would never be obtained. Then Ramsay and Rayleigh got hold of the residue of several tons of liquid air that had gone through the Claude plant in Paris, and fractionated it, discovering unknown constituents of the air (argon, xenon, krypton) and—helium. There on the spectroscope was flashed the long-lost helium line! More light was shed on helium when it was found to be a constituent of radium emanation. Radium and helium are elements. Radium was being transformed into helium! Were the old alchemists right, after all?

Professor Satterlee, of the University of Toronto, made some investigations that led him to suggest the possibility of helium's being the product of radioactive transformations under the ground.

English and Americans Collaborate

The Admiralty took the position that, if there was one chance in a thousand of getting helium in quantities, it was worth taking. Professor Satterlee was requested to consult the American authorities and to investigate Professor Cady's report.

About this time the United States came into the war, and the British sent over experts to collaborate with Colonel Burrill, technical head of our Chemical Warfare division. He, in turn, took the matter up with General Squier, then head of Army Aeronautics, Director Manning of the Bureau of Mines, and Admiral Griffin, Chief of the Bureau of Steam Engineering in the Navy.

Soon after the matter reached Admiral Griffin, there appeared in his office a Mr. Carter, who had given up civilian work to enter the Navy. It

The hydrogen with which balloons are inflated is contained in steel bottles under high pressure. It is thus that helium will be bottled

companies to liquefy the natural gases in the areas indicated by Professor Cady. Congress was asked to appropriate money for the production of

another company about to begin, all located in Texas. Natural gas is collected and compressed and cooled down in the same kind of apparatus that is used to liquefy air. As the cooling process continues everything is liquefied out long before the helium. The regular hydrocarbon series, making up approximately 70 per cent of the natural gas, liquefies first, and then the nitrogen, constituting about 30 per cent, is condensed at a temperature of 195° C. below zero. It requires 268 degrees below ordinary zero (Centigrade) to liquefy helium, which is then collected for use in balloons. The amount of helium in natural gas is .95 per cent.

Had the Germans Known Helium

In a military sense, helium for balloons is the answer to the high-explosive bullet, just as armor in the old Navy was the answer to the guns that riddled wood. Had the Germans filled their Zeppelins with helium instead of explosive hydrogen, they might not have proved such failures as city bombers.

The entire art of designing and equipping dirigibles will be revolutionized. Electrical apparatus, heretofore barred because of the ever-present dread of sparks, may be installed, new types of motors may be used. A new field is opened up for development.

Why is there so much helium in Kansas, Oklahoma, and Texas? Probably because of the quantities of radio-active minerals. Helium, as we have seen, is a degradation product of radium.

Here we have the third stage of the terrible balloon conflagration at Fort Sill, Oklahoma, of which the first two stages are illustrated on the preceding page. The picture tells its own story of men driven to flight and overcome by fatigue. Helium instead of hydrogen, as the inflating gas, would obviate such accidents

developed that he probably knew more about the liquefaction of air and gases than almost anyone else in the country. It seemed as if fate had pushed the right man at the right moment into Admiral Griffin's office.

The Army and the Navy engaged

argon. It would never do to let the truth leak out in the midst of a war.

Professor H. N. Davis took hold of the Army end, Commander Dyer represented the Navy.

Now there are two big plants producing helium in large quantities and

All that is left of the balloon is the basket. As long as hydrogen—highly inflammable and, when mixed with air, violently explosive—is used to inflate bags, such accidents are inevitable. And now the government announces that it had made preparations before the armistice was signed to fill the bags of balloons and dirigibles with helium, which is almost as light as hydrogen, but which is not inflammable. It sounds like an alchemist's dream

What They Do When There Are No Fires to Fight

Six men with but a single thought—promotion; and so they are reading books on building construction, fire prevention, and anything else that will help them

These seven serious souls are proudly holding a framed shawl which they all conspired to make. They have generously decided they will give it to the Red Cross

If no fire inconsiderately interrupts him he turns out one porch swing a day, increasing his income by selling them

Intrigued by the shape of the base of the base-burner stove, he uses it as a model for flower-pots. They are made of cement and painted

When he isn't playing with fire he plays barber and cuts the hair of his comrades, even unto his chief

His side issue is cobbling. When he has mended and patched the boots of his fellow firemen, he fixes the shoes of his family, so that no one can say, "Who is worse shod than a shoemaker's wife?"

They take turns rescuing each other from imaginary fires to keep in trim. The rescuers carry the rescued over one shoulder, flinging an arm around the victim's leg and holding hands with him

Fighting Flaming Oil with Foam

"A MASS of twisted sheet-iron and pipes from which rivulets of oil, gasoline, and naphtha continue to burn is all that remains of the great oil plant. The blaze gave the fire department the hardest twenty-five-hour fight in its history, but there is no longer danger of the flames spreading, as all the damage that could be done has been accomplished. The loss is estimated at $1,500,000."

That is the report of one New York newspaper, written on the fourth day of the great oil fire in Brooklyn last autumn. The firemen had to fight the blaze with water and sand. These were inadequate for the purpose, and the only thing to do was to let the fire burn itself out, at a tremendous loss.

Non-Inflammable Bubbles

Chief among oil-fire quenchers is suffocation by foam. The great non-inflammable bubbles are poured over the surface of the oil, and the fire quickly subsides. The bubbles are made by combining sulphuric acid with bicarbonate of soda. Carbonic-acid gas is thus formed, and it quickly suffocates the flame. Many oil-storage plants are now equipped with these bubble tanks, and some of the tanks work automatically. Thus, should a fire start when no one is near, the tank will see that it is put out—perhaps before any one arrives.

A large oil-storage tank was being built at Midland, Pa. The first ring, about five feet high, had been erected on the foundation, when the owners suddenly decided to try out their new automatic extinguisher. They poured two feet of water in the ring then eighty barrels of crude oil, and on top one hundred gallons of gasoline. They soaked some cotton waste in kerosene, set it on fire, and threw it into the tank.

The flames burst forth, and two minutes later the foam began to flow. Two minutes more, and the fire was out!

Foam-Pipes Solve Problem

The pictures of this fire test show the two standpipes, one on each side of the tank. Both of them contain the foam-making liquids. The one on the right was used for this particular fire. Bicarbonate of soda and soapbark dissolved in water are kept in the bottom of the pipe. An acid-tank for the sulphuric acid is mounted above the level of the soda solution. A pipe leads down from the tank to the solution below.

At the place where the acid-tank and the pipe meet there is a glass plate to keep the acid from rushing down the pipe under normal conditions. A plunger is mounted directly over the glass, and above the plunger there is a hammer, held in place by chains which have links that will melt when heated to 212° F. Thus when the fire was started the links melted, the hammer dropped on the plunger, the plunger broke through the glass, the acid ran down the pipe, mixed with the soda—and the foam came bubbling out of the spout of the stand-pipe.

There are many other ways in which foam-pipes may be made to work automatically. In one instance, the valves that let loose the liquids were worked by electricity; the heat melted a fusible connection that started the current.

To test a new automatic oil-fire extinguisher, flaming cotton waste was deliberately thrown into this tank; when the heat of the fire reached the pipe at the right, things began to happen

The pipe contains two tanks, one holding bicarbonate of soda and the other holding sulphuric acid; when these are mixed a foam results that contains carbonic-acid gas. This gas suffocates the flames

Two minutes after the foam started to flow over the burning oil the fire was out. Since the foam method has proved to be so successful it is being installed in many oil plants

Fighting a Fire with Brains

The cub electrician remembers a lesson and saves the power plant

By Charles Magee Adams

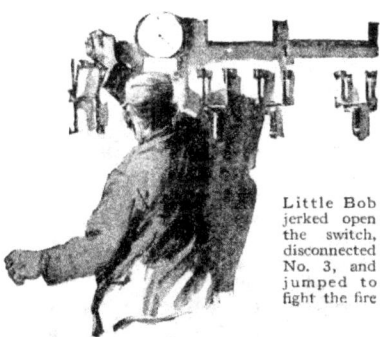

Little Bob jerked open the switch, disconnected No. 3, and jumped to fight the fire

OVERHEAD an arriving flash of lightning crackled sharply across the horn-gap arresters. But down on the floor of the power-house, in front of No. 3 unit's panel in the main switchboard, little Bob Fitzgerald, the chief engineer, continued to frown up at his new attendant.

"No, no," he protested impatiently. "Think! What's the field circuit for, anyhow?"

Johnny Gibson pondered a moment. "Why, it's to excite the alternator," he replied.

Outside, thunder, in the wake of the first lightning flash, rumbled away in the distance, and overhead a second flash crackled more sharply across the horn-gaps. But little Bob continued to frown up at his green assistant.

"Then what do you suppose happens when you open and close this?" he demanded, jabbing a thumb toward a piece of equipment on the slate panel before him.

It was the field switch, and Johnny eyed its heavy polished blades and considered.

"Why, I reckon if you'd open it that'd kill the alternator," he decided at length. "'Cause then there wouldn't be any field."

The thunder boomed closer as little Bob nodded jerkily.

"Of course," he retorted. "When any alternator's running, it's live as long as this switch's closed and dead as long as it's open. Remember that now, and don't go—"

Then the Big Crash Came

A crash drowned out his words. Overhead, in the row of horn-gaps, a huge ball of fire leaped out, blazed for a moment in blinding brilliance, and then broke into a score of crackling streamers.

Little Bob turned quickly toward the switchboard. But scarcely had he done so when, from behind the bulk of unit No. 3, appeared the tall figure of Mike Collins, the oiler, his long arms waving excitedly.

"Fire! Fire!" he yelled. "Number three's a-fire!"

And No. 3 was.

As little Bob, followed by Johnny Gibson, rounded the smoothly swaying Corliss engine and came upon the big generator, a dense cloud of heavy black smoke was pouring out of the pit beneath it, and through this could be seen the flicker of flame.

"Get some extinguishers!" little Bob commanded.

Mike obeyed. The chief turned and raced back around the engine; and Johnny, thinking he too was in search of extinguishers, followed. But little Bob simply jerked open the big oil switch, disconnected No. 3 from the bus-bars, and dashed back once more.

"Want me to shut her down?" Johnny called.

Little Bob shook his head.

"No! Her breeze'll drive the blaze away from her coils. Come on!"

Chemical Extinguishers to the Rescue

Mike had hurriedly assembled a half dozen chemical extinguishers, ready for use. But the fire likewise had been taking steps in the way of preparedness, and even more rapidly.

In the pit under the generator was a large quantity of oil, accumulated there because of disabled overflow equipment. This had rotted the insulation on the three main leads where they emerged from the stator frame, and the lightning had of course flashed across this weakened spot. Now, as the three men turned the streams of liquid into the smoke cloud, it lifted, disclosing a flare of arc sputtering from one cable to another and igniting both oil and insulation.

"Hurry! Hurry!" little Bob yelled excitedly. "Stop it before it spreads!"

Mike and Johnny obeyed. In spite of the choking gas that rose the instant the liquid came in contact with the heat, they sent their streams hissing into the blaze.

The blazing oil was extinguished. The smoking insulation was extinguished. The spitting tongue of arc was snuffed out—but only as long as the streams were held directly upon it.

"What's the matter with these extinguishers?" little Bob demanded. "Here, give me one!"

But nothing was wrong with the extinguishers. The gas that rose when the liquid struck the heat had the familiar suffocating odor. Yet, as soon as the streams were deflected, the

blazing arc flashed up, driving the line of fire nearer and nearer the stator windings.

The supply of reserve extinguishers at hand was exhausted. "Get some more! Hurry!" little Bob ordered.

Mike handed his still charged tank to the chief, and dashed away.

For another minute the two men who remained fought on side by side in the cloud of gas and smoke. But still the line of fire, fed by the crackling arc, crept relentlessly nearer the vitals of the big alternator. Then, while his extinguisher still threw a steady stream, Johnny Gibson dropped it, turned, and darted out of sight behind the still throbbing unit.

Little Bob was frantic.

"Come back! Come back here!" he shouted. But Johnny had gone.

Left alone, the chief engineer turned against the fire with feverish desperation. The pulsing flicker danced between the cables, closer and closer to the stator windings. And then, suddenly, like a match flame caught in a puff of wind, it disappeared.

What Johnny Had Done

Little Bob deflected his extinguisher stream; but it did not reappear. He shot the liquid at the insulation, and its glow died to charred blackness. He sprayed the flickering oil, and it subsided. A moment later Johnny Gibson slid breathlessly into sight around the big generator.

Little Bob swung on him vehemently.

"Quitter! You quitter!" he snapped. "Do you think you can run off and leave me to put out this fire by myself?"

Johnny stared at him, nonplussed.

"Why, I—I just ran around and killed the generator," he explained.

Little Bob shook a blackened forefinger at him. "I disconnected her myself, the first thing," he retorted.

Johnny nodded.

"Yes," he answered. "But you didn't open the field switch, and that left her live. Don't you remember what you told me a while ago? That's why she kept on arcing. I opened the field switch just now."

For a moment little Bob blinked stupidly. Then a smile of comprehension overspread his smoke-streaked face.

"By gum, Johnny, you win! If you keep on learning like that, I'll be darned if I won't have to hunt a new job!"

Just in Time, He Remembered the Field Switch

Warned by the mutter of the approaching thunder-storm, the men in the power-house were ready; but when the lightning flashed across the weakened insulation on the main leads of the generator, it kindled a blaze that they could not control. Nearer and nearer the flames crept toward the stator windings, and trouble loomed big, when Johnny Gibson, the cub electrician, thought of his latest lesson from the chief. He dropped his extinguisher and ran, but not in retreat; for a moment later the flames winked out—Johnny had thrown open the field switch and "killed" the generator

MOVING PICTURES

They Brave Death  for a Picture

THE life of an aviator in the British Royal Flying Corps is hazardous at all times; but there is one task that he often has to perform in which the danger incurred is perhaps the greatest of all. That task is the photographing of enemy positions from the air.

The pilot of a camera-plane must be a man who is not afraid to take any chance, no matter how desperate, in order to secure the desired photographs. His airplane is the finest of its type— generally a two - seater, equipped with a 160-horse-power engine. While not very speedy, this plane is easy to maneuver and very steady in the air. Three telephoto cameras are arranged so that they secure a triple panoramic view of the country below. The pictures are generally taken by the observer and not by the pilot. When he wants to make an exposure, the observer looks through a glass panel between his feet, which acts as a finder. The range of the cameras is really remarkable. I have seen photographs taken from a height of fifteen thousand feet which showed many details to the naked eye.

A Flight Over the German Lines

There are numerous thrills to be had on a picture-taking flight. If you will draw slightly on your imagination I will take you over the German lines with a camera-plane. Imagine that the aerodrome is somewhere on the British front in northern France, and that the camera-plane is already outside of its hangar. Attached to every squadron on active service in the Royal Flying Corps is a photographic section, which is supplied with a dark-room on a motor-

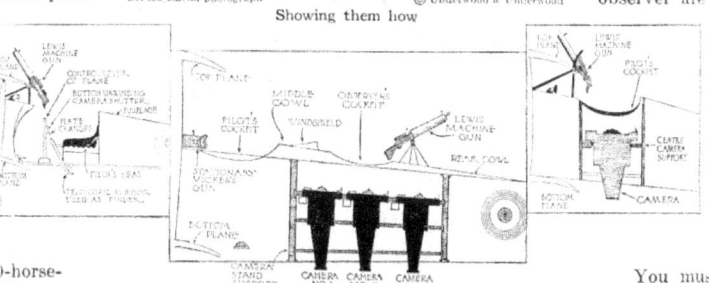

British official photograph   © Underwood & Underwood

Showing them how

Both single-seated and double-seated camera-planes are used. In the single-seater the camera is operated by the "joystick"; the plates are changed by the same means. A reflecting mirror constitutes the finder. Both planes are armed

© Kadel & Herbert

Gaze upon the upper picture, and then upon the lower. Both are photographs taken from an airplane. They show the same village before and after bombardment

truck. This dark-room has all the necessary accessories for developing, printing, and enlarging, and is in charge of a sergeant and five men, two of them master photographers.

The cameras receive a final inspection by the sergeant. In the meantime, the pilot and observer are marking on their maps the area to be photographed. This is going to be a long flight, far into the enemy's territory, and the camera - plane's escort of ten fast fighters are being tuned up for the trip. You must remember that the camera-plane would be a wonderful prize for the Germans. It is the duty of these fighters to see that the precious plates are brought back safely. The weather conditions for the flight are ideal. The sky is absolutely clear.

Protected by Fighting-Planes

It does not take the observer long to decide on the altitude for taking the photographs. He knows, of course, that he has to fly high in order to avoid shrapnel and enemy planes on patrol. The cameras are pronounced "O.K." and orders arrive from G.H.Q. (General Headquarters) to proceed at once to the desired objective. The flight commander climbs into his fighting-plane. One after another, the escort of fighters float off into the air. The last to leave the earth, the camera-plane ascends slowly and takes its place in the "V" formation.

The flight commander leads. Behind him are four more fighters. In the center of the V is the camera-plane, and bringing up the rear are the fighting pilots. Soon the planes are over No Man's Land and then over German territory. Don't think for a

The man at the left is not a pilot, but an observer. When he wants to make an exposure he looks between his feet at a glass panel that acts as a finder

When the prints are ready, a staff of experts reduce them to scale, determine where they overlap, and paste them together to form a photographic map

moment that this group of machines flies level. As the shrapnel screams skyward, they commence to swoop and dodge, never for a moment, however, losing their compact formation. The pilot of the camera-plane sees by his instruments that he is up fourteen thousand feet; but even at this height the shrapnel from below tries to search him out.

Undisturbed, the planes wing their way toward their objective. It is cold at the height at which they are flying. A glance at his map tells the observer that he is almost over the territory to be photographed. Now he concentrates his attention entirely on the glass panel between his feet.

The Germans Try to Head Off the Camera-Plane

Far below are several rivers and canals, gleaming in the sun like tiny silver threads. The pilot is watching the flight commander's plane. As he looks the signal comes, as arranged, that the squadron has arrived. Throttling down his engine, the pilot puts the camera-plane into a quick-turning, rapid glide. Down it drops into the region where the shrapnel is bursting thick and fast. Up above, where the camera-plane has left them, the fighters circle around and around, protecting the camera-plane from attack.

Arriving at the proper level, and utterly oblivious to the shrapnel, the pilot straightens the plane out on an even keel for several seconds. As he does so the observer presses a button which unwinds the shutters of the cameras. As each photograph is taken, and before he pushes the handle that inserts new plates, he writes the number of the slide on the map of the territory which he has just snapped. Back and forth over its objective flies the camera-plane, until the observer is satisfied that he has photographed every inch of the ground below. Then a signal to the pilot, and the big plane ascends again to the level of its escort. Once more in formation, the fliers start on their way home.

Word of the evident success of the

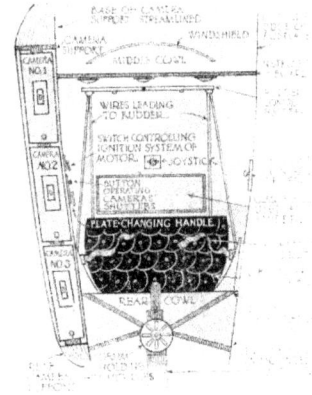

As soon as the photographer touches ground again, he dispatches his camera to the developing-room installed on a motor-truck

Looking down on a camera-plane

raiders has been flashed to the nearest enemy aerodrome. The result is shown when a German squadron appears with the evident intention of cutting our squadron off. There is no question of

Don't think for a moment that photographing positions is as safe or as pleasant as yachting. The bullet hole tells why

dashing forward into the fight. The camera-plane, with its all-important information, must be seen safely home. The enemy squadron separates into groups of three, in an effort to lure one or two of our fighters to attack. But the formation must not be broken, and the tempting bait must be overlooked.

Nearer and nearer to the Allied lines flies the squadron, and the Germans, fearful lest their prey will escape them, swoop to the attack. Four of our planes immediately detach themselves from the convoy and engage the enemy, while the remainder of the convoy flies home with the camera-plane safely in its midst. The four planes left behind will keep the Germans busy. It is not long before the remainder of the squadron breaks formation and lands safely back at our own aerodrome.

Developing Photographs and Making Maps

As the camera-plane comes to earth, the sergeant and his assistants dismount the cameras and hurry away to the developing-room on the truck. Here the pictures are developed. Meanwhile the observer fills in a form, which is then sent to the sergeant to be forwarded to G.H.Q. with the finished prints. These prints, by the way, are finished in a little more than an hour after the plates are received. The moment they are ready, they are dispatched to G.H.Q., where a staff of experts reduce them to scale, determine where they overlap, and paste them together on the photographic map.

This great map has to be correct to the minutest detail. In order to keep it so, several small squadrons go out during the day, watching for changes over the enemy lines. The camera-plane of this small squadron is only a single-seater, equipped with one camera. An interesting feature of this plane is that the camera is operated from the "joy-stick" (control-lever) of the plane, even the changing of the plates is accomplished by the same means.

Noise for the Silent Drama

Parrots shriek, lions roar, airplane motors hum — all from the orchestra pit

Photographs © International Film Service

He's got his hands full of instruments for imitating birds. When Mary's pet gets a close-up, be it a parrot or a canary, Mr. Manne reaches over, picks up an instrument, and soon the theater is filled with the singing or chattering of birds, as the case may be

They look like frying-pans and saucepan lids, but they are only some of the gongs that hang on the wall; Mr. Manne sits with his back to them and simply reaches up when Chi Chu, the Chinese dope dealer on the screen above, pounds for his servants

Poor Nell, the outcast, sank to the ground exhausted; the rain poured down, the lightning flashed, and the thunder growled ominously. And this is what happened behind the screen: one man made the thunder, another flashed the lightning, and a third imitated the rain

And the villain still pursues her—on horseback; whereupon Mr. Manne gets out his two imitation hoofs and plays a tune with them on a board, keeping time with the horse on the screen

The musical director can order the "Anvil Chorus" any time he wants to; Mr. Manne keeps a hammer and a good sized anvil down in the pit with him all the time

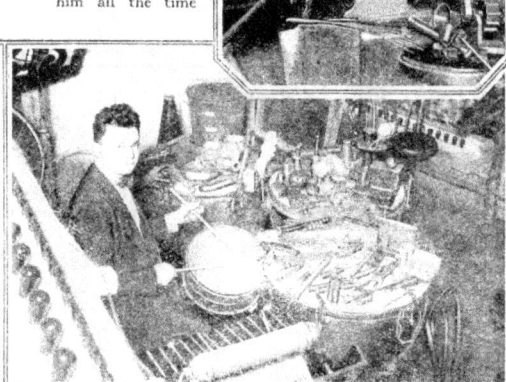

Train whistles and automobile horns are not easy to imitate; an electric circuit is involved and two men are usually needed to run it; the man to the left is getting ready to let off a shrill one

You might think this a hardware store; in reality it is the drummer's end of the orchestra pit of one of New York's largest moving-picture houses; Mr. Max H. Manne, the man in the case, performs on over three hundred instruments

Want to Be a Movie Star, Fido?

Says Joe Martin, orang-outang: "Follow master—that's my motto; I copy everything he does. As you see, today, instead of acting, I'm the camera man; by being able to help out in a pinch, I have made myself quite valuable—and really these men's jobs are simple"

Murphy, Edwin Earle's pet, tells us this: "I'm not an actor and don't want to be one, yet my master insists on featuring me—so sometimes I nip better than trousers"

Imperial Cæsar's tale: "This picture shows—that's what the public think. I'm the confess I think my success is due not to my

"Ours is the ideal life," these four will tell you. "We never fight and we aren't at all jealous of each other—that's why we succeed"

Here's Joe Martin again; he's going to give the baby pig a bath under a hydrant spout. Joe says, "Curly Stecker has really taught me nearly all I know; he is a very good trainer: but then, you know, I film well"

Says Kazan: "I am a full-blooded Russian wolf-hound and thank my parents for my beauty—I might have been a cur! Good looks do help in the movies, but I have incidentally cultivated a most fascinating smile"

We'll Tell You How to Do It

me struggling for life in a swirling torrent
hero in 'The Eternal Triangle,' and I
histrionic ability but to my good looks"

Here's baby tiger's tale: "We tigers have fierce repu-
tations and nobody loves us; but that's no way to
make a success in life. Consequently when Enid
Bennett pets me,
I purr sweetly!"

Says Charlie, the elephant: "Had I stayed in the old country I would have been nothing but an ordinary laborer: here in America I am an actor of no small fame. I have a
fine memory and so this movie game is all very simple—one rehearsal is enough for me"

The extras speak: "All morning we were starved; then
they let us loose in a grocery store and we made for
the pieces of raw meat that we smelled. All the while the
camera man was grinding away. Foolish, don't you think?"

Ethel says: "Once I lived in a Zoo; but that easy life kills
ambition, so I got out before it was too late. I went into
the movies and have starred in 'The Lion's Claws,'
'Lost in the Jungle,' and 'The Lure of the Circus'"

Behind the Scenes in a "Movie" Set

The scenario reads, "Scene on Stairs," and this is what happens in the studio

With clasped hands outstretched in pleading, penitent Helene slowly approaches her wronged husband. And all the while the camera clicks merrily. It is a touching scene, but the directors and the camera-man are too busy to appreciate it

First, the draftsman draws up plans for the setting. He reads the scenario, and finds that Helene, the heroine, married for money; and, since the staircase scene takes place in her house, the setting must indicate wealth

Windows, doors, and staircase are made according to plan in the carpenter shop; they may never be used again, yet they are carefully made and beautifully finished

The staircase is shoved over against the wall of the studio, and the rest of the setting is put into place by experienced stage-hands. From the moment Helene starts to descend the stairs, she will have plenty of time to register penitence and display her beautiful negligée

The painters and paper-hangers paint and decorate until that corner of the studio looks as if it had been transplanted from Fifth Avenue

Tacking decorations over doorways. These small details register wealth—incidentally, also, a decided change from the humble home of Helene's youth

The stage is set. Now for the lights. A great many scenes are taken indoors and powerful lights are needed for the purpose. While these men put on the finishing touches, so does Helene in her dressing-room upstairs

Furnishing Helene's home. This man from the studio has been haunting the furniture houses all day in order to rent appropriate chairs

If You Have Tears, Prepare to Shed Them Now

Like Niobe, all tears: these are real ones

It is plainly seen that when the picture above was taken, June Elvidge, movie star, was in no mood for tear registration; so her manager is applying small drops of ice water. This is but one of the many ways in which movie tears are manufactured

"Hearts and Flowers"—the most popular movie form of what the Greeks called sob stuff—played on a violin by a long-haired, sad-eyed musician while the star is gathering force for a great emotional spill

The movie director's order for a strong, steady flow, accompanied by red eyes, is best filled by applying onion. Onion is not so pleasant as ice water, and so the accompanying expression is very apt to be more appropriate

At the highest, saddest note she will almost always burst into real, wet, unaffected tears. If "Hearts and Flowers" won't do it, the next best choice is "The Curse of an Aching Heart"

Glycerine tears are administered with a medicine-dropper. They glisten in globules most convincingly, and are used entirely in many studios. What is more, these tears will last until the star is told to weep no more

Relying on a star's emotions is a risky business. Here the manager is reading his star a letter from her absent sweetheart. To-day she weeps: tomorrow she may meet an eligible millionaire and smile instead

Thrillers by the Foot

Before the war at least two railroads made from $30,000 to $50,000 a year staging train wrecks for the "movies." The picture in the circle shows a fast freight being wrecked for the "silent drama"

Cleaning up after a moving-picture wreck is a tough job, because, while a train may get off easily in a collision in real life, it is certain to be all "busted up" in reel life

The villain (curse him) responsible for this terrible head-on smash listens in while Si Perkins is telling the conductor of his suspicions

When there's a young woman at the throttle, the train is pretty apt to be safe—in "movie"-land

One film company, since its organization, has purchased and demolished in this spectacular way enough rolling stock to outfit a prosperous branch line

On a blind siding one motion-picture company has set out twenty box-cars salvaged from wrecks like this that it has staged for the "movies"

But no "movie" director could beat this wreck, which happens to be the real thing. The engine fell twenty feet through a trestle

DEVICES FOR USE IN THE HOME

Housekeeping Made Easy

Machines that Lighten Work and Keep the Housewife Happy

With this arrangement you don't have to worry about the plants dying of thirst while you are away on a visit. A wet sponge inside the covering feeds pieces of felt, which in turn pass on the moisture to the plant in small and well regulated amounts

It is no longer necessary to have a "clothes-horse" encumbering the laundry when small linen pieces are being ironed. This ironing-board has wooden bars attached at one end to hold the linen after it is ironed, thus saving many steps

Clamping preserve jars one above the other with this device permits the use of a deep container for steaming operations

With this little rotary grinder in your kitchen there is no excuse for dull knives

This bowl is weighted at the bottom to keep it steady when beating cream

A pinless curtain-holder that makes curtains hang both evenly and gracefully

By holding the embroidery cloth taut, this pinless frame facilitates work

An unexpected guest? In two seconds you can turn this chair into a bed and ask him to stay all night

This combination bath-chair and boot-blacking outfit is hard to beat

Old Duties Done in New Ways

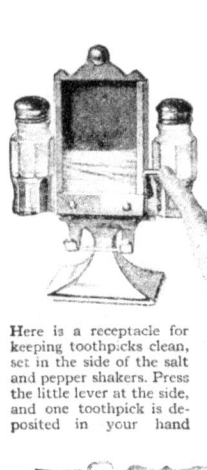

Here is a receptacle for keeping toothpicks clean, set in the side of the salt and pepper shakers. Press the little lever at the side, and one toothpick is deposited in your hand

The little thermometer inside your fruit-jars tells you whether the cover is air-tight or not. In case the jar is faulty, liquid creeps up the thermometer and you must do it all over again

A vacuum device permits the used water to be drawn from the tub and ejected into the sink. It is easily attached to any water faucet and is economical in operation

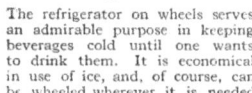

CONCAVE RECEPTACLE TO HOLD AN EGG

No more stringy eggs if you use the poaching-spoon to lower them into the kettle. The bowl of the spoon is pierced with tiny holes

The refrigerator on wheels serves an admirable purpose in keeping beverages cold until one wants to drink them. It is economical in use of ice, and, of course, can be wheeled wherever it is needed

Ice-cream can be kept in this container for six or seven hours without melting, for the container embodies the same principle as a vacuum bottle. The retail price is slight compared with the ordinary freezer

RUBBER-FACED SQUEEGEE

PERFORATED NOZZLE

HANDLE

PUMP

DRAIN TANK

SUPPLY TANK

A small hand-operated device recently placed upon the market enables even a child to make the most beautiful and intricate embroidery. It is easily carried in a hand-bag

This window-cleaning contrivance makes it possible to clean either the outside or the inside of a window without danger of falling

Soak a porous stone in a pan of kerosene, then place it among the furnace coals and ignite it. It starts the fire easily and can be used over and over again

Little Things that Help

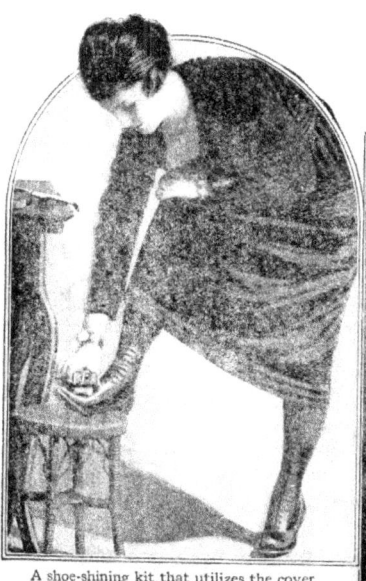

A shoe-shining kit that utilizes the cover of its container as a polisher. The box holds the cleaning materials compactly and is easily stowed away in a drawer

Hanging ten skirts, this adjustable folding rack is made to fit the inside of a closet door. When the skirts are in place the rack folds against the door, occupying very little space

A combination dressing-stool and step-ladder for the bath-room has steps covered with rubber, while the wood is finished in a pleasing white enamel

This medicine-spoon has a spring cover easily lifted by pressure of the thumb when in use

 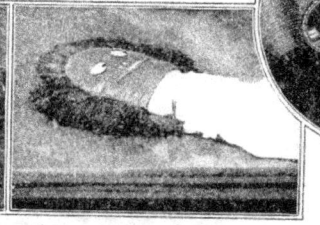

The revolving thief-proof milk-container has as many locked compartments as there are tenants, and is attached to the dumb-waiter

An ingenious dust-cloth, painted to resemble a "mammy's face," is a mitt that fits the hand; the cleaning is done by the woolly hair

A velvet-covered phonograph-record cleaner which enters into the record's tiny grooves

Perforations in the hose permit the pump operated by the faucet to suck the water from the tub into the washstand

First as a handsome library table and then as a comfortable bed, this contrivance economizes space in apartment-houses

Some Devices that Will Save Energy in the Home

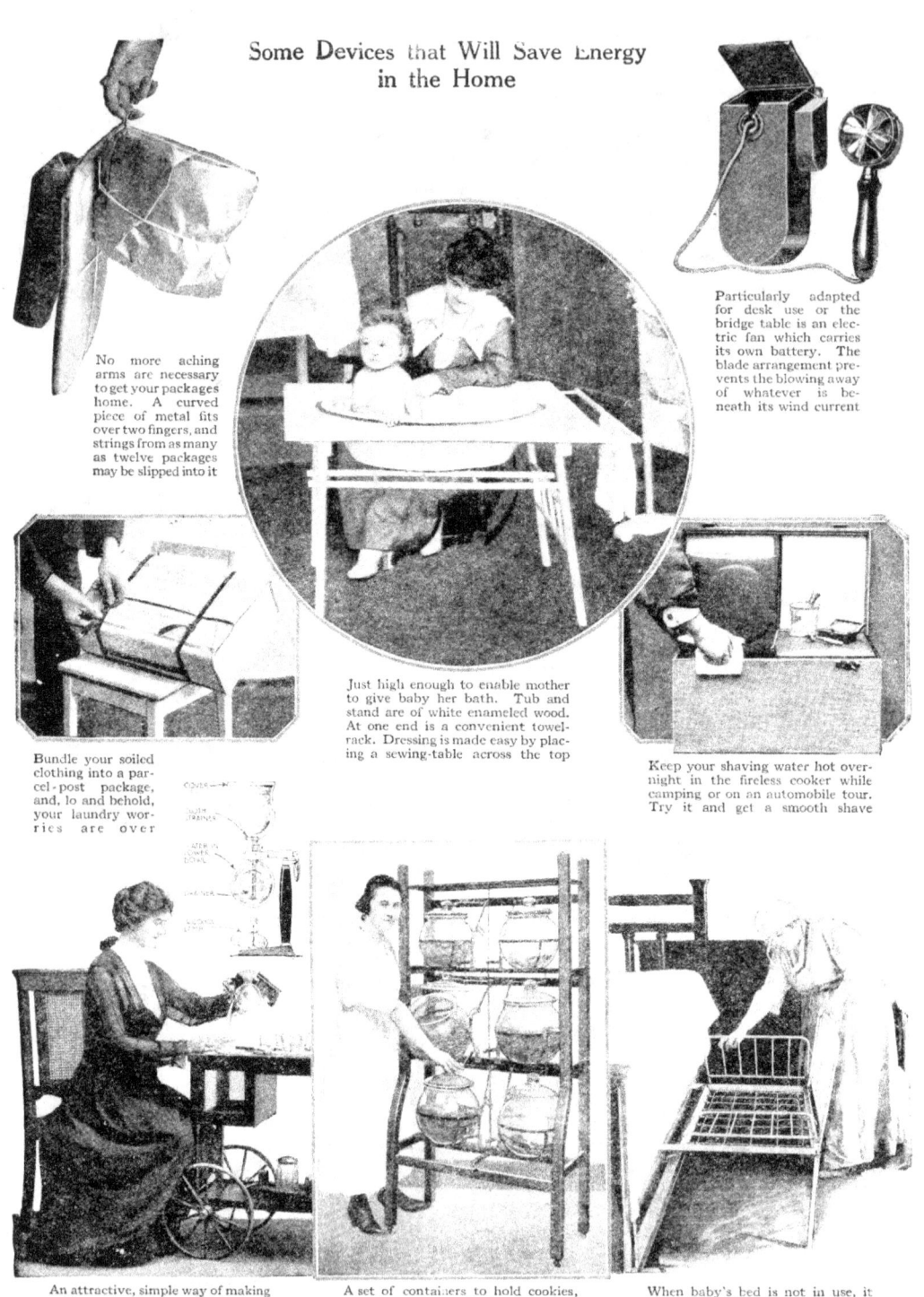

No more aching arms are necessary to get your packages home. A curved piece of metal fits over two fingers, and strings from as many as twelve packages may be slipped into it

Particularly adapted for desk use or the bridge table is an electric fan which carries its own battery. The blade arrangement prevents the blowing away of whatever is beneath its wind current

Just high enough to enable mother to give baby her bath. Tub and stand are of white enameled wood. At one end is a convenient towel-rack. Dressing is made easy by placing a sewing-table across the top

Bundle your soiled clothing into a parcel-post package, and, lo and behold, your laundry worries are over

Keep your shaving water hot overnight in the fireless cooker while camping or on an automobile tour. Try it and get a smooth shave

An attractive, simple way of making coffee on the dining-room table is accomplished with this glass percolator; the coffee does not touch metal

A set of containers to hold cookies, etc., in a store. Push a lever and the glass jar tilts, at the same time lifting its cover automatically

When baby's bed is not in use, it is quickly folded and slipped between the spring and the frame of the bed to which it is attached

Here Are Some New Machines that Help Promote Household Efficiency

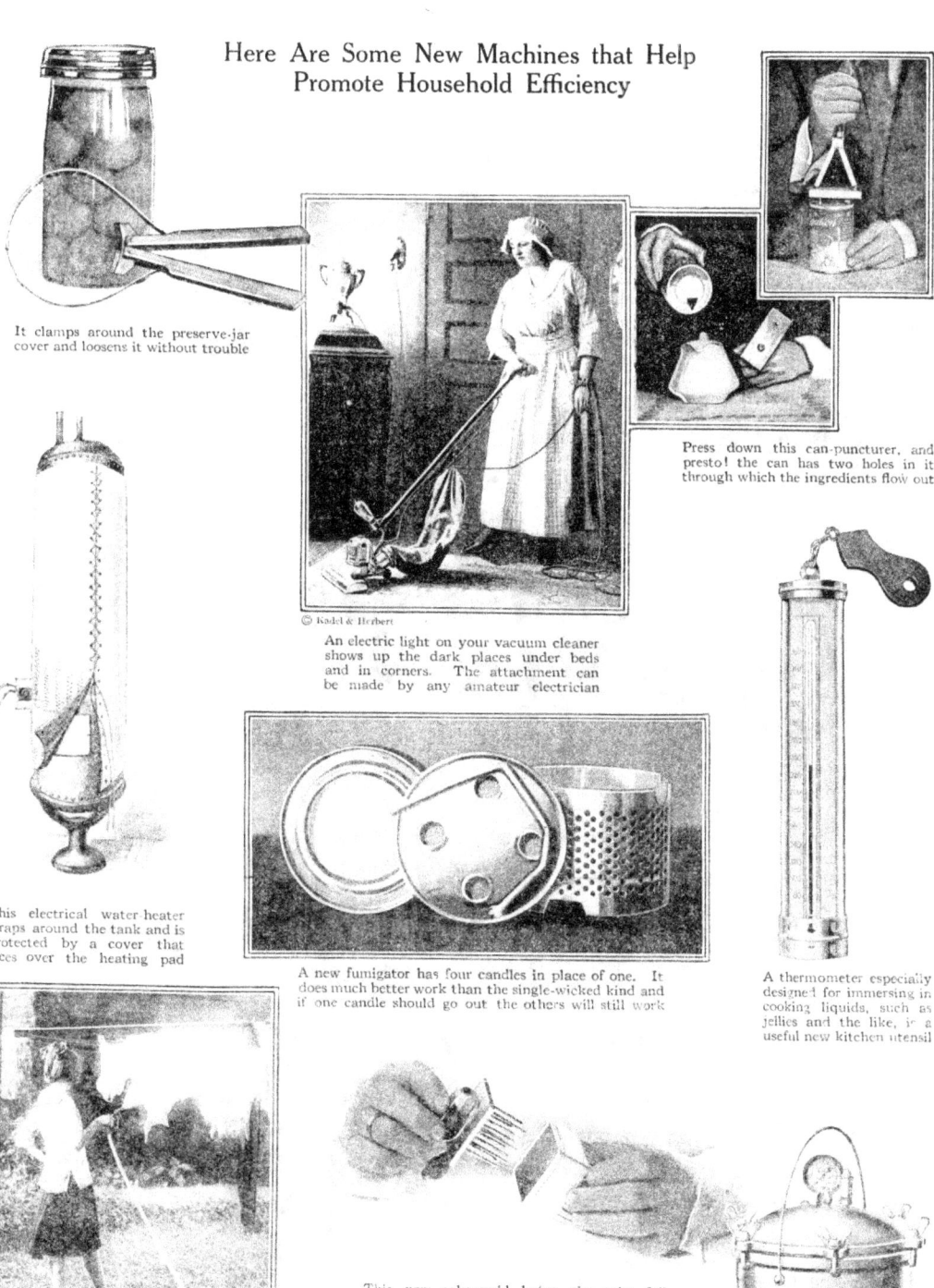

It clamps around the preserve-jar cover and loosens it without trouble

Press down this can-puncturer, and presto! the can has two holes in it through which the ingredients flow out

© Kadel & Herbert

An electric light on your vacuum cleaner shows up the dark places under beds and in corners. The attachment can be made by any amateur electrician

This electrical water-heater straps around the tank and is protected by a cover that laces over the heating pad

A new fumigator has four candles in place of one. It does much better work than the single-wicked kind and if one candle should go out the others will still work

A thermometer especially designed for immersing in cooking liquids, such as jellies and the like, is a useful new kitchen utensil

Sprinkling the clothes on the line by means of a hose is a much better way than the old hand method, and they can be placed in the basket at once, all ready for ironing

This new cake-mold bakes the cake full of holes: do holes render cake more edible?

Here is a cooker that steam-cooks the food. It cannot burn out, and no basting is required. A tight cover, which carries a pressure-gage, clamps down tightly upon the kettle

With heating elements placed in the seams between the bark and logs, this electric heater looks like the real thing

A cone-shaped colander with a hollow central column, which serves for a sleeve in the upper end of the handle. An inner grating shreds the food, forcing it through the perforations in the outer cone

A combined rack and serving tray for the sick-room. The rack is for cloths, towels, and the like, while the tray extends over the patient's bed and may be used for serving meals or for holding books or other articles

A canvas-covered substitute for the old wash-board

The latest thing in clothes-closet decoration is to line the walls with cretonne to match the hangings in a room

With this device the gas-jet heats as well as lights the room

This metal-polisher looks like a pencil-eraser; it removes rust by merely rubbing the metal surface

This little closet is built at a convenient height in the wall to house the telephone

A grate stove that is made to fit in any fireplace opening: it uses kerosene

The candle in this style holder is held flush with the tube by spring pressure, preventing drip

Several Useful New Ideas for Housekeepers

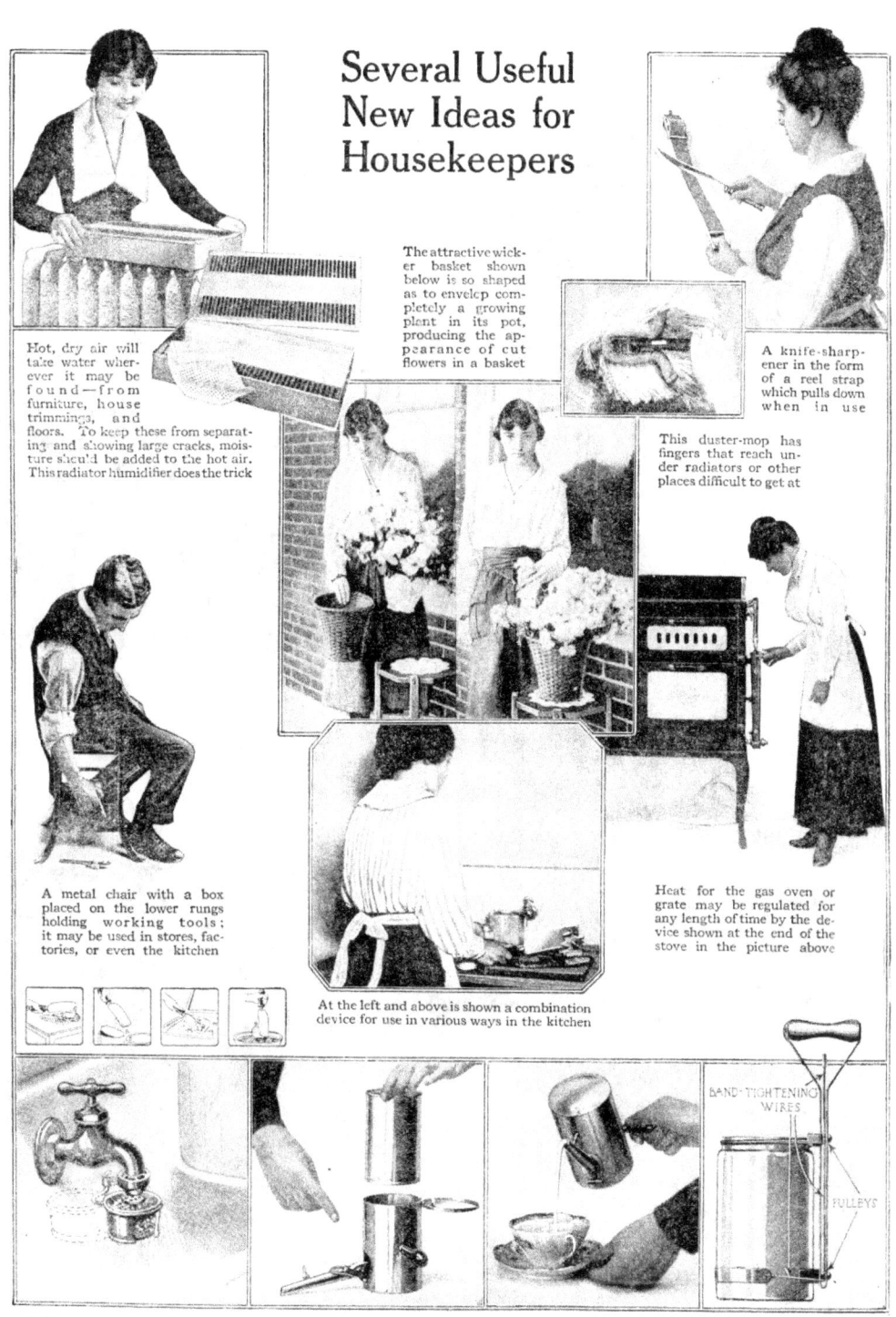

Hot, dry air will take water wherever it may be found—from furniture, house trimmings, and floors. To keep these from separating and showing large cracks, moisture should be added to the hot air. This radiator humidifier does the trick

The attractive wicker basket shown below is so shaped as to envelop completely a growing plant in its pot, producing the appearance of cut flowers in a basket

A knife-sharpener in the form of a reel strap which pulls down when in use

This duster-mop has fingers that reach under radiators or other places difficult to get at

A metal chair with a box placed on the lower rungs holding working tools: it may be used in stores, factories, or even the kitchen

Heat for the gas oven or grate may be regulated for any length of time by the device shown at the end of the stove in the picture above

At the left and above is shown a combination device for use in various ways in the kitchen

A faucet receptacle for soap. It swings aside when not in use

When the condensed milk can is inserted in this retainer, the spout pulled up punctures the can, making it ready for instant use

A handle for lifting preserve jars out of boiling water

If your hot-air register refuses to work, it is probably clogged with dirt. Let your vacuum cleaner suck the dirt away. Hot air will then shoot forth

No wonder she "cleaned up." She drew their Red Cross money from them with a vacuum cleaner

© Underwood & Underwood

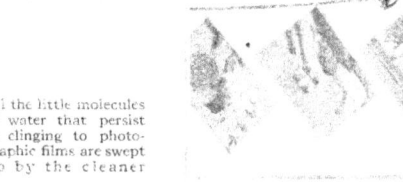

It is much easier to clean the inside of a box by inserting a tube than one's head, shoulders, arms, and a dust-rag

An ever-present cleaner hanging on your belt will lift the road dust from your automobile —if you own one

What chance do frost and ice stand of clinging to a window-pane when a vacuum cleaner gets after them?

All the little molecules of water that persist in clinging to photographic films are swept up by the cleaner

Vacuum Cleaner Will Do

Armatures, like pies, require an even oven. This oven for armatures was made even by a vacuum cleaner

Vacuum-clean the dust off your old boxes and sell your stock as new

Cellars are damp and cellar floors usually wet. This one was, and she went after it with a vacuum cleaner

Screen doors gather huge quantities of dust from the street, and pass it along indoors every time the door bangs. If you pick the dust off the screen with a vacuum cleaner, you prevent dust and bacteria from spreading

As a reward for his faithful service Dobbin received the V. C.

It's much better than hanging your head out of the window or airing it on the roof

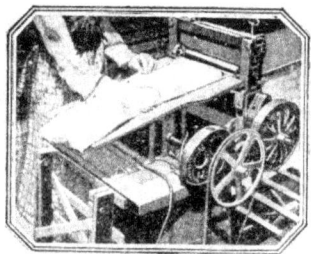

Inventions that Will Help Make the Housewife's Job Less Hard

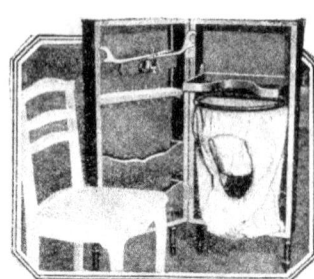

Here is an electrically heated clothes-mangle: attached to the washing-machine wringer; it is said to do the family ironing in one third the time formerly taken

This little sewing-stand, with its hanger places for spools, scissors, and sewing material, folds up exactly as would a book, and hides behind the door when not needed

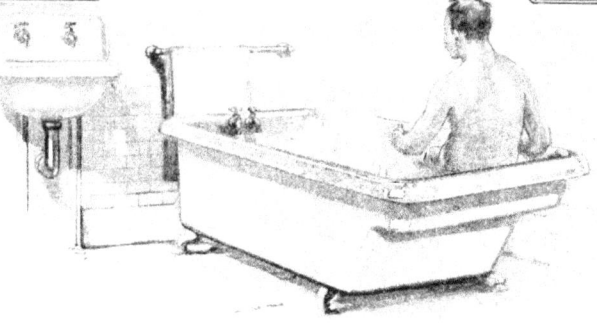

A new bath-tub is so designed that it offers the occupant either a head-rest or a seat at one end of the tub

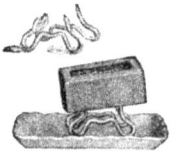

An unusual but very practical soap-holder is shown here. The cake of soap is simply set upon the prongs, which prevent it from slipping

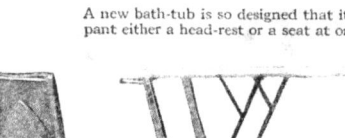

For the theater and the home is this automatically closing bag, hat, or program holder. When not in use the spring lug snaps back into the pear-shaped holder

The combination chair, step-ladder, and ironing-board is for the two-by-four apartment. It is entirely practical and saves buying three separate kitchen articles

The fumes of burning benzine cause the metal in the perforated box to glow for eight hours at a time, making this little hand-warmer a useful thing in the cold weather

The ashes from the kitchen range are placed in a sifter built almost like an ice-cream freezer: The sifter is made air-tight

This folding stool for the bath-room is set compactly against the wall, out of the way, when not in use. Very useful in small bath-rooms

Push down the pin, insert the rubber cork into the neck of the bottle, release the pin, and the cork will expand sufficiently to cork it: It will fit any bottle of reasonable size

The Daily Grind for Our Daily Bread

You can buy your own grains and be your own miller

"Time to get up," you think, when you hear the grinding of a coffee-mill. But the particular coffee-mill shown here is not grinding the morning coffee. In fact, it isn't grinding coffee at all: but wheat for bread. The thrifty housewife thereby saves the price of grinding and gets double service out of her coffee-mill

A few years ago this mortar and pestle pounded out things in the back room of a drug store; now the mortar and pestle have been transferred to the kitchen, where they are pounding out grain instead. Sufficient patience and power in using them will result in reducing grain to a powder fine enough for bread-making

By means of a meat-chopper the grains of wheat in the top picture were turned into whole-wheat flour for the food in the lower picture

The principle of the tub and poles for pounding grain is the same as that of the mortar and pestle shown above, but the proportions are different. There, a large woman pounded with a small pole; here, small children pound with large poles. The children are Filipinos and they are pounding rice

Boys who long for a country where they won't have to go to school are recommended to parts of China

Surely the rolling stone in the picture above gathers no moss, though it may be gathering corn flour. Two firm brown hands are rolling it on another stone, and between the stones grains of corn are being crushed. The hands belong to a young Indian girl of New Mexico.

This Chinese boy does not go to school, but he does have to spend several hours a day grinding corn

Things that Add to Household and Office Comfort

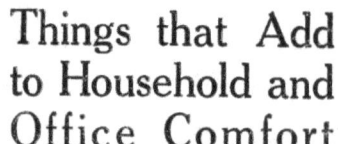

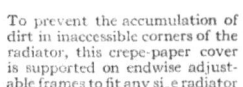

And now, as you finish washing your hands, your foot touches a lever and warm air quickly dries them, doing away with the old unsanitary towel

Having but one leg, constructed in such a manner as to produce absolute stability, this ironing-board may be attached to any closet or to a window-sill

The last word in writing-desks is the built-in type. It comes in a variety of styles. If desired, it can be so disguised as to appear like a part of the wall panel

Unlike the usual moth-bag, this portable folding wardrobe is odorless. It is closed down one side by a patent interlocking device. It may be folded up small and packed in a suitcase

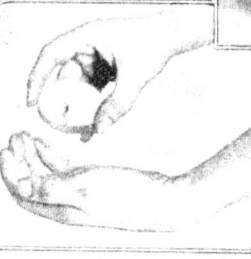

Something new is a soap tablet that upon being punctured permits liquid soap to ooze out, thus adding greater cleansing qualities than those of ordinary soap

Plants will not require constant attention if this device is attached to the flower-pot. It drops water directly at the roots of the plant, insuring an even distribution without overflowing

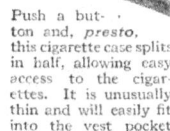

Push a button and, *presto*, this cigarette case splits in half, allowing easy access to the cigarettes. It is unusually thin and will easily fit into the vest pocket

To prevent the accumulation of dirt in inaccessible corners of the radiator, this crepe-paper cover is supported on endwise adjustable frames to fit any size radiator

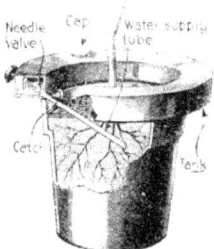

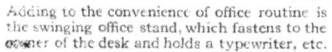

Adding to the convenience of office routine is the swinging office stand, which fastens to the corner of the desk and holds a typewriter, etc.

Things New for All the Members of the Household

Why not introduce some of these labor-savers?

An armful of small bundles that are always slipping is no longer necessary, now that many grocers supply their customers with heavy wrapping-paper shopping-bags that have strong cord handles

When cleaning house, put birdie in the yard

Attach this sponge-dryer beneath your soap-dish

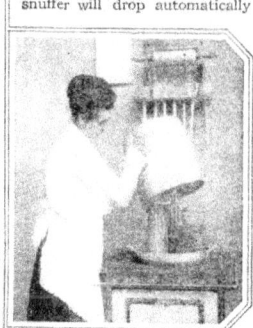

You are tempted to buy more when you can carry the things so easily

© Press Illustrating Service

Replace the bristles in an old shaving-brush with a small sponge: it takes readily to lather

Pour some paraffin in a glass, sink into it a wire cage containing absorbent cotton and a wick, and use for a night-light

You can leave this candle lighted. When it has burned low, the snuffer will drop automatically

This broom's Japanese inventor claims that the joint enables the user to dig in corners conveniently

"Bake in a hot oven," says your recipe, and you turn up the gas. But how can you be sure that your oven won't get too hot? Use this automatic temperature regulator

A German preserving outfit provides a water-jacket with a steam cover. The jars are easily handled

Making Things Easier for the Busy Housekeeper

The key-fastener is secured to the door-knob. It is so constructed that it snaps to the key in the lock and prevents the door from being opened on the outside

"When shall I give the next dose?" doesn't bother the nurse who has this medicine bottle: it has a dial and pointer which designates the time for the medicine to be given again

Made from a steering column and wheel of an automobile, this lamp is a distinct novelty in that it is the only one of its kind

After the hair is washed, this frame is placed under it, allowing whatever air is moving to circulate through the hair and dry it in half the usual time

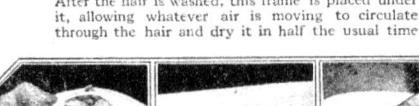

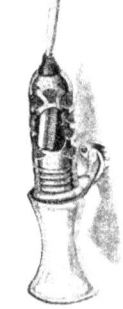

When the owner of this tooth-brush finishes cleaning his teeth, he dips it into the sterilizing liquid and sets it into its holder

A decided novelty in collar buttons is this new one, which, when in position, folds neatly against the collar-band, doing away with that unsightly bunch at the back of the collar

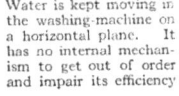

Just big enough for two is this picnic basket, which accommodates a small piece of ice for keeping the food and liquids cold

Water is kept moving in the washing-machine on a horizontal plane. It has no internal mechanism to get out of order and impair its efficiency

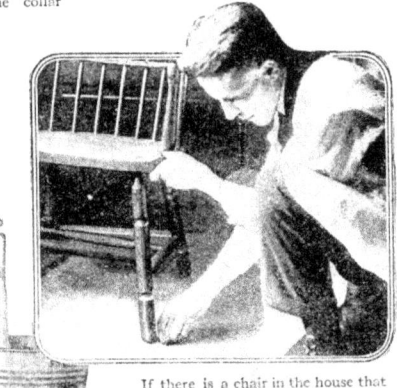

If there is a chair in the house that is wabbly, then why not attach a tin bottle cap underneath the bottom of the short leg to steady it?

The water is forced through the fabric of the clothing and cleans without injuring it. The motor and driving mechanism are entirely enclosed

SPORTS

Racing for the "America's" Cup

When sport becomes a science

By Joseph Brinker

SIR THOMAS LIPTON has come to this country with one of the most remarkable and sensationally radical challengers that ever crossed the Atlantic. Yachting experts who have seen the hull of the *Shamrock IV* agree that she is perhaps the lightest and yet the most powerful British racing creation that has ever visited our shores.

The features of design that stand out conspicuously are the extremely long keel; the full form of the hull in the bow and stern sections; the bulge of the sides of the hull inward at the top or deck; and the unusually lofty sail rig. The hull is of the lightest composite construction, the planking being laid in three plies with the two inner layers placed diagonally. The outer layer of the planking runs longitudinally.

Because of the long keel, with the lead placed low, yachting experts predict that the best chance of success of the *Shamrock IV* lies in a strong offshore breeze. Then her great sail spread will drive her through the water at a high speed. In light breezes the chances of winning are not considered so good, because her large keel will present a big area of wetted surface, and the "skin" friction between it and the water will tend to decrease her speed.

Against the *Shamrock IV* either the *Resolute* or the *Vanitie* will race. These yachts resemble each other to a far greater extent than either one resembles the *Shamrock IV*. Both of the American boats are approximately 75 feet on the water-line, the *Vanitie* being the larger when measured by the extent of the part that overhangs the water-line dimensions. The *Vanitie* has the larger sail area, carrying 9,465 square feet of canvas, compared with 8,188 square feet of sail carried by the *Resolute*, which is slightly narrower, but has a body that is more full beneath the water-line where the hull joins the vertical sides of the keel. All three of the contestants are provided with centerboards set in the bottom of the keels.

No matter which boat is selected to defend the Cup, she will compare in general design more closely to her foreign rival than did America's earliest cup contestant, the *America*. In the span of seventy-nine years from the *America* to the *Resolute* and the *Shamrock IV*, there has been a remarkable evolution in the design of the racing yacht.

The *America* was built to beat the sloop *Maria*, then the fastest pilot-boat in New York harbor. While she did not beat the *Maria*, she proved to be so fast for a schooner that the yachtsmen who had ordered the boat accepted her and made ready for the trip across the Atlantic. The route was to Havre. It was made in seventeen and one half days.

After watching the *America*, with her widest beam amidships and fine lines fore and aft, the British accepted the American type of clipper

© Edwin Levick

The racers coming head-on. From left to right are the *Resolute*, the *Vanitie*, and the *Defiance*. The *Vanitie* has a greater sail area than the *Resolute*

bow. But, because of the deep British waters, the hulls were deep and narrow. In America the conditions favored boats of greater beam and less draft because of the shallow waters. The British type is exemplified in the *Genesta* of 1885, as shown in one of the accompanying illustrations, and the American type in the *Mischief*, which was built in 1881.

Coming development was seen in the *Thistle*, the challenger for the Cup in 1887. She was wider in proportion to her depth than any previous contestant. Then came the *Valkyrie I* in 1893, with a fixed, finlike keel instead of a centerboard, because a fixed keel can carry from sixty to ninety tons of lead. Finally the British type of narrow, deep hull gave way to the wide but shoal hull with a deep fin keel to carry lead ballast. The *Reliance*, an example of this type, with a water-line length of about 90 feet and an over-all length of 140 feet, carrying 16,000 square feet of sail—the largest ever carried by a Cup yacht—was perhaps the acme of the highly developed racing machine.

But the *Reliance* and the *Shamrock III* were rule-beating freaks, which, as soon as the Cup races were over, were broken up on the junk-pile because of their extreme design, their unseaworthiness, and the inability to get other boats to race against them in club regattas. Because these boats had developed into freaks, the racing rules were changed. The *Resolute*, the *Vanitie*, and the *Shamrock IV*, all with a water-line length of 75 feet instead of the 90 feet of the *Reliance* are far more wholesome boats, and will probably see many years of regatta racing before they are discarded.

The old racing rule put a tax on water-line length and sail area, but on nothing else. Provided the yacht did not exceed 90 feet in length on the load water-line, she could be as broad and deep, and as long over all, as desired. The boats built under the old rule drew too much water for cruisers sailing in the shallow American harbors. To save weight the hulls themselves were made so shallow that there was insufficient headroom below the deck for comfortable accommodations. These reasons lie behind the fact that all of the racing machines of recent years were broken up for junk after the completion of the races.

The new rule and the formula by which the rating of the yacht is determined includes the factors of sail area, length, and displacement. The rating is determined by the formula:

$$\text{Rating} = 0.18\ \frac{L\ \sqrt{\text{Sail Area}}}{\sqrt[3]{\text{Displacement}}}$$

Because the displacement factor is the denominator of the fraction, and because the larger the denominator the smaller becomes the final fraction or rating, it is seen that the new rule favors boats of larger displacement. Other things being equal, the boat with the larger displacement will have the smaller rating and will receive a larger time allowance.

The *Vanitie* departs less from the old rule than the *Resolute*. The *Resolute* has a fuller and deeper under-water body and is built more closely to the rule, as shown by her sharp ends, deep V sections, and large displacement. The *Shamrock IV* is a compromise between the extremes of the *Shamrock III* and the *Reliance*. The *Shamrock IV* is full ended, with a large sail area and a deep keel having a large surface in contact with the water.

While the elements of yacht design are not expressed in so many words in the formula by which the rating of the Cup yachts is determined, they are present just the same. The resistance that a vessel encounters in passing through water is made up of three kinds—frictional resistance, wave-making resistance, and eddy-making resistance. Frictional resistance is that due to the friction of the water on the under-water surface of the vessel. It depends upon the area of the surface and the nature and shape of the surface. This resistance is known as skin friction, and forms a large part of the total resistance at low speeds. It is, of course, decreased by cutting down the area of the hull in contact with the water. This area is commonly called the "wetted surface." The speed of the boat depends upon the ratio of the sail area to the wetted surface. Without unduly reducing the area of wetted surface, it is the task of the yacht designer properly to proportion the ratio of the sail area to the wetted surface. Because the

The first winner of the cup, the *America*. Though a swift schooner in her day, the boat does not compare in speed with the modern racing yacht built with a full hull and extremely long keel. The *Resolute* and the *Vanitie* could sail around her in circles, because of their modern construction

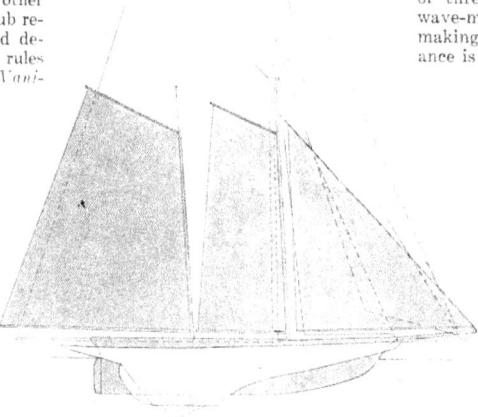

Compare the *America*, which won the race in 1851, with the *Resolute*. While the *America* was about 13 feet longer than the *Resolute* on the water-line, she was less in length over all, on account of the extent of that part of her construction that "overhangs" the water-line measurement. The *America* is indicated by the darkened portion, to show the difference in the shape of the hull and the relative sail area

skin friction is the most important resistance to the progress of the yacht through the water, that vessel with the larger ratio of sail area to wetted surface will be faster, other things being equal.

In designing the *Shamrock IV*, Nicholson gave her an extremely long keel. This greatly increased the area of the wetted surface, but also enabled him to spread the lead in the keel out longitudinally instead of building it up vertically. Hence the center of gravity of the lead lies lower, and this in turn means greater sail-carrying capacity for the same weight. Because of this low-placed lead and great sail area, which increases her speed in strong breezes, yachtsmen contend that the *Shamrock IV* will have her best chances of winning in strong winds.

At lower speeds in light airs, where the wetted surface and its skin friction are the most important considerations, the Lipton yacht will be at a great disadvantage compared with either the *Resolute* or the *Vanitie*, which have a much smaller area of wetted surface. The bottom length of the keel of the *Resolute* is only about half

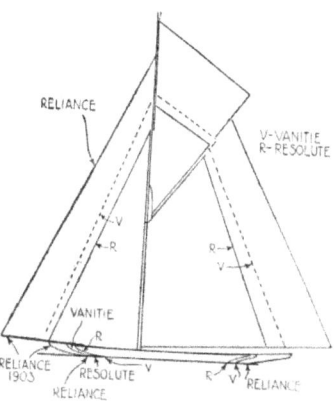

A comparison of the rig and sail plan of the *Resolute* and the *Vanitie*, with that of the *Reliance*, which defeated the *Shamrock III* in 1903. The *Reliance* was 90 feet long on the water-line and 140 feet overall. The water-line length of both the *Resolute* and the *Vanitie* is 75 feet. The *Reliance* carried 16,000 square feet of canvas. The *Resolute* carries 8,188 square feet of sail, the *Vanitie*, 9,465 square feet

that of the *Shamrock IV*. Although changes may be made in the rig of the *Shamrock IV* before she crosses the starting line on July 15, she may carry approximately 10,000 square feet of sail area as compared with the 8,188 square feet of the *Resolute*. If these ratios hold true, the *Shamrock IV* will probably have to give the *Resolute* three or four minutes of time allowance. This is an important factor, for on at least one occasion the American defender of the Cup beat the British challenger on time allowance. That was the race between *Columbia* and *Shamrock II* on October 4, 1901, when the *Shamrock II* actually beat the *Columbia* by two seconds on elapsed time, but lost the race by forty-one seconds because she had to give the *Columbia* a time allowance of forty-three seconds.

Some idea of the tremendous sail spread of the *Resolute* may be gained from the fact that if her sails were made from ordinary bed-sheets, fifty-four inches wide by eighty-one inches long, it would require 270 of these sheets, sewed end on end, to give the equivalent sail area.

The *Resolute* at the left and the *Vanitie*, at the right, cutting gracefully through the water. The "lines" of the sails are calculated to lie perfectly straight in the wind in the speed of the race

Famous Contestants for the "America's" Cup

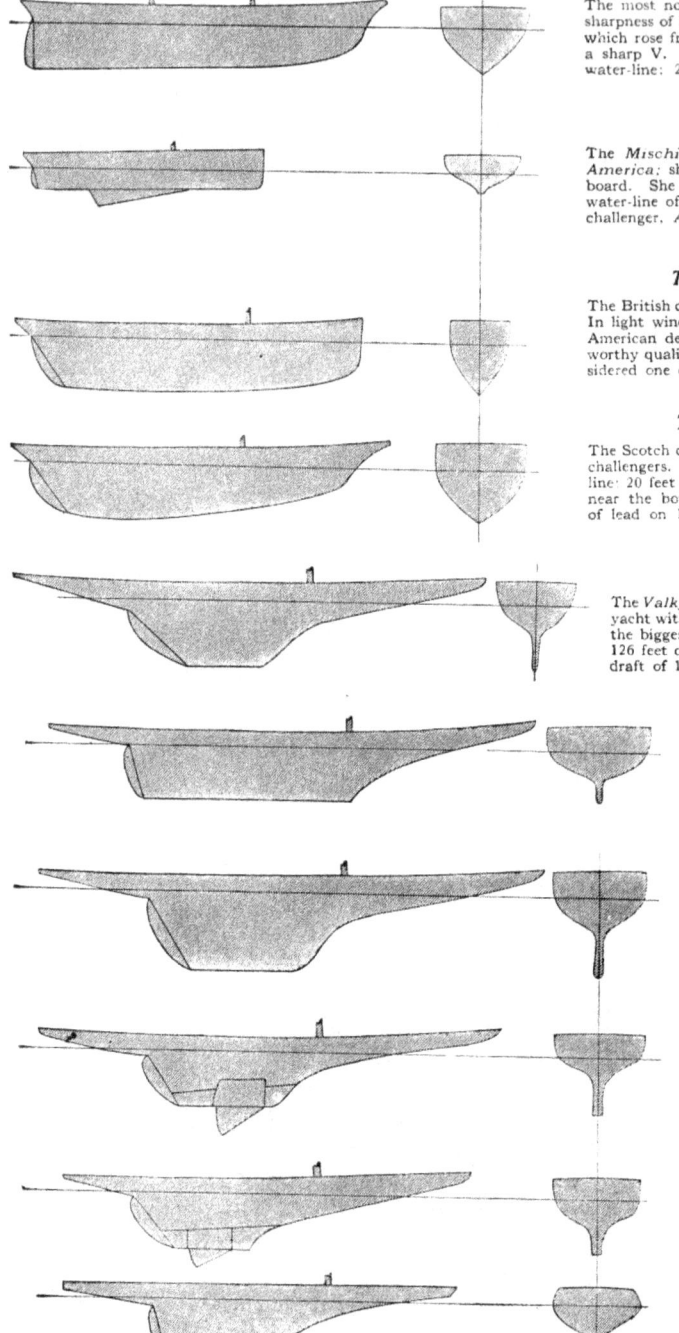

The America

The most notable peculiarities of the *America* were in the sharpness of her bow and in the shape of her hull on the sides which rose from the keel in straight lines, each pair forming a sharp V. She was only 94 feet over all; 88 feet on the water-line; 22 feet beam, and had 11¼ feet maximum draft

The Mischief

The *Mischief* was an iron sloop, much smaller than the *America*; she had a draft of only 5 feet, but a deep center-board. She had a beam of 20 feet, with a length on the water-line of 61 feet. The *Mischief* defeated the Canadian challenger, *Atalanta*, in two out of the three races sailed

The British Cutter Genesta

The British cutter *Genesta* was built along entirely new lines. In light winds she was an easy mark for the *Puritan*, the American defender, but in a heavy blow showed such sea-worthy qualities that her last race with the *Puritan* was considered one of the best Cup races of any sailed up to 1885

The Scotch Cutter Thistle

The Scotch cutter *Thistle* marked a new departure in British challengers. She was 108 feet over all; 86 feet on the water-line; 20 feet beam and 13¾ feet draft. Her underwater hull near the bow was well cut away and she carried 55 tons of lead on her keel. Inside she carried ten tons of lead

The Valkyrie

The *Valkyrie I* was the prototype of the present day racing yacht with long overhangs at the bow and stern. She was the biggest of the challenging sloops up to 1893 and was 126 feet over all, 85 feet on the water-line, and she had a draft of 16½ feet. She carried 10,042 square feet of sail

The Vigilant

The *Vigilant* which defeated the *Valkyrie I* in two races in 1893, was deeper and wider than any cup defender built up to that time. She was designed and built by Nat Herreshoff. Her sail area was 11,272 square feet

The Columbia

The *Columbia*, which defeated Sir Thomas Lipton's *Shamrock I* in 1899, was the most pronounced skimming-dish type of hull up to that year. The hull proper with a beam of just 24 feet had a depth of only 7 or 8 feet

The Vanitie

One of the contestants for the honor of defending the *America*'s cup this year, the *Vanitie*, has 65 tons of lead in her keel and draws 13¾ feet of water exclusive of a small centerboard. She is 119 feet in length over all, but only 75 feet on the load water-line

The Resolute

The *Resolute*, the second candidate for the defense of the Cup this year, is similar to the *Vanitie* in design except that she has fuller underwater body lines and shorter overhangs

The Shamrock IV

In his fourth attempt to win the Cup, Sir Thomas Lipton has brought over in *Shamrock IV* one of the most remarkable challengers built. Her keel is extraordinarily long, measuring about 35 feet along the bottom

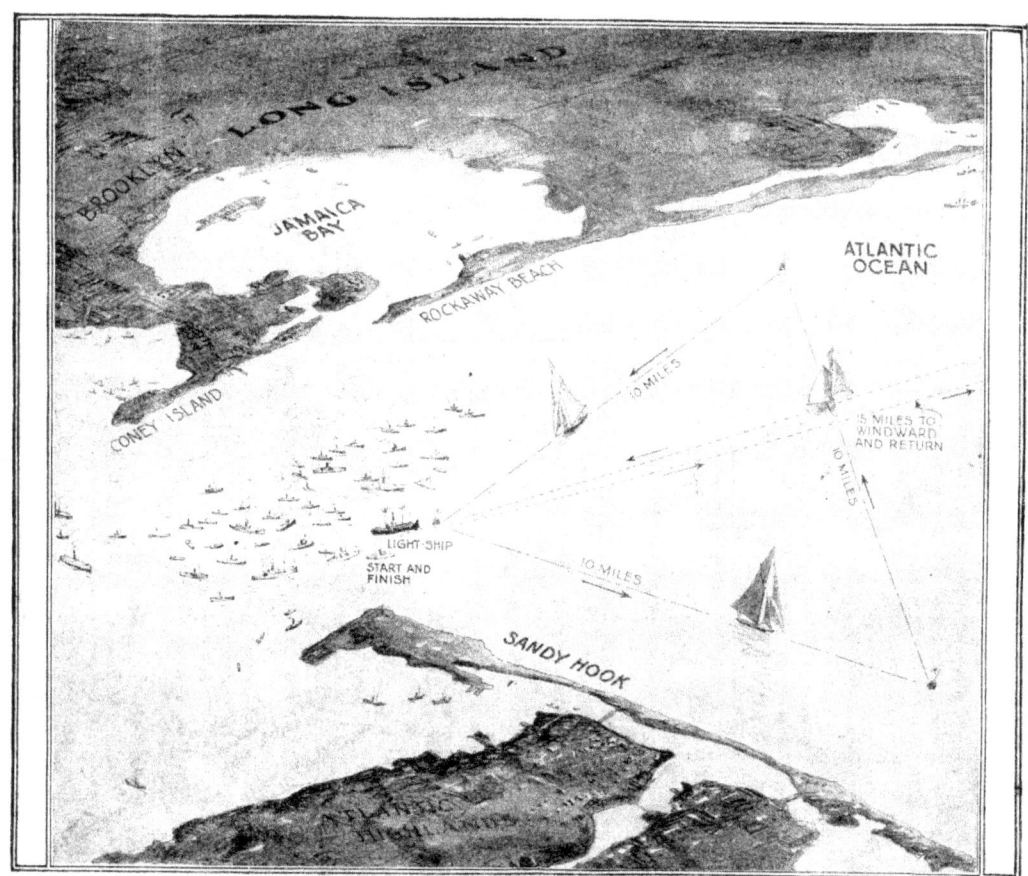

The race-course for the yachts off Sandy Hook. The first race for the *America's* Cup will be held on July 15. It will be a race fifteen miles to windward and return, as indicated by the dotted lines in the center of the triangle. Three out of five races must be won. The second and alternate races will follow the triangle, each side of which is ten miles long, making the total length thirty miles. These races will be sailed on Tuesdays, Thursdays, and Saturdays until the winner is decided

While the present racing rule gives a bigger rating the longer the water-line length and the larger the sail area, the length L in the formula is not the water-line length, nor does the rating increase in the direct ratio of the sail area. The sail area rating increases as the square root of the area, and it follows as a matter of course that four times the sail area would double the rating. The cube root of the displacement, being the denominator of the rating fraction, helps to give a smaller rating as the displacement increases, and this means greater seaworthiness of the boat. The length L in the formula is not the water-line length, but a corrected length

Sail area, displacement, and water-line length are considered in determining the rating of a racing yacht, with the time allowance that must be given a vessel of smaller rating. The water-line length is not measured on the load water-line, but is a corrected length

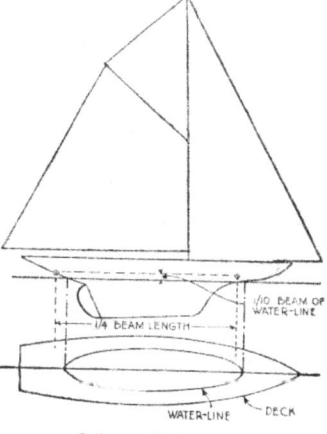

Sail area, displacement, and water-line length

which includes the load water-line length plus one-half the excess of the quarterbeam length over the percentage of the load water-line length given by the formula:

$$P = 100 - \sqrt{\text{load water-line length}}$$

The formula quarter-beam length is determined as shown in the accompanying sketch, and is employed to prevent beating the rule by freak designs in the bow and stern overhangs.

The rating, as determined by the above formula, which includes the load water-line length plus one-half of the excess of the quarter-beam length, divided by a certain percentage of the water-line length determined by another formula. The quarter-beam length is measured as indicated in the sketch at the left gives the time allowance, which depends upon the assumption that a yacht of racing measurement R will sail a nautical mile in the number of seconds shown by the formula:

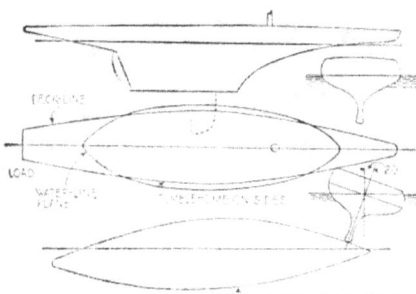

The drawings of the *Shamrock IV* show the tumble-home of sides of the hull in the center. The form of the hull at bow and stern is shown by the increase in area of the water-line plane when the *Shamrock IV* is heeled over to an angle of twenty degrees in the water

$$\frac{2160}{R} = 188.92$$

For a yacht of a different rating r, the allowance per mile between the two yachts will therefore be expressed by the formula:

$$\frac{2160}{r} - \frac{2160}{R}$$

in which R is the rating measurement of the larger yacht and r that of the smaller one. Tables have been worked out to give the time allowance in seconds for any given difference in rating.

In these days of airplanes, motor-boats, and swift automobiles, the racing of yachts seems a tame sort of sport. But the modern yacht race is indeed the keenest of sports. To those who like the touch of the salt air, and who delight in the sight of a trim sailing-craft leaning in the wind, the race for the Cup has lost none of its former charm.

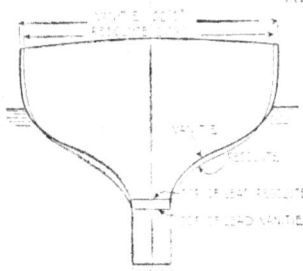

The chart shows the narrower width of the *Resolute* compared with the *Vanitie*, and the fullness of lines of the *Resolute* under water. The lead keel is higher than that of the *Vanitie*. The keel of each boat is three feet wide at the base

What more could be desired, on the day when the race begins, than a "spanking" breeze, a sun-glistening sea with waves of just the right size to add spirit to the scene? If the present race is like its predecessors, a procession of tugs, launches, motor-boats, excursion steamers, and all manner of craft will pass through the Narrows to take the throng of yachting enthusiasts

Experts agree that the *Shamrock IV* is the lightest and most powerful British racing boat ever built. The chances of winning the race will depend upon a strong offshore wind

and "good sports" out to a point of vantage. It will be a gala occasion, a scene of color and life.

In the staked-off course the beautiful yachts, their sails presenting lines that are straight and beautiful in the sunlight, will be seen. Tilted in the wind, cutting through the green water, glistening where the waves have split around their bows, the yachts will speed from buoy to buoy. Thousands of eyes will be focused upon them through glasses from deck and shore, wherever a glimpse of the course can be obtained.

Whether the *Resolute*, the *Vanitie*, or the *Shamrock IV* will be first to reach the final stake, this yacht race perhaps more than others will arouse international interest in yachting. But the rules of the race will make the crowd hold its decision in

reserve. It may be that no one will know, for a while, who the winner may be. If one boat skims the line a fraction of a minute before the other, this does not mean that it has actually won the race. When the rules of the "game" are applied and the proper "weights" allowed, the judges will announce the winner. Not until then will the curiosity of the crowd be satisfied.

But cheering will not be hindered on account of that. It is often well for those on each side to enjoy the pleasures of success, and then to suffer the pangs of disappointment; for then everyone is given a chance to prove himself a good sport. But the man who laughs last is the one who wins the bet, and he will have to await the decision of the judges who have considered every factor and applied the mathematics of the rules.

Further zest is added to this year's races because they mark the thirteenth attempt to lift the cup, twelve made by British challengers and one by the Canadians. If Sir Thomas believes 13 is his lucky number, who can say but that when the mist lifts off the Sandy Hook course some fine morning in July, the cup will be on its way back to England? It has remained continuously in this country for 69 years or since that memorable day in 1851 when the fleet *America* outsailed the fastest that Britain could produce.

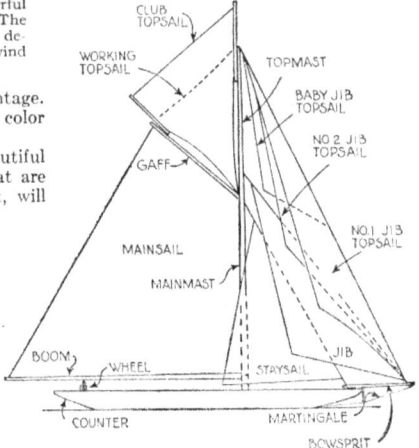

A diagram of the sails of a racing yacht. The shape of these sheets of canvas are cleverly designed to respond to the wind and air resistance. It is the manipulation of the sails combined, with their area, that speeds the craft through the water. The mainsail of the *Vanitie* weighs about one ton and the top of the mainmast towers more than one hundred feet in the air. The amount of canvas carried is almost twice that carried by the *America*

Amazons in the Making

It appears to be a happy as well as healthful operation

Ready to dance? Not in those shoes; they are heavily weighted in the toe and are used for exercising the legs. Such shoes are officially known as "pedal dumb-bells"

In her hands she holds a sand-bag, thirty inches long and five pounds heavy. When her arms grow tired she will slip her foot into the strap on top of the bag and raise it up and down

Now she treats the sand-bag as if it were a Roman candle—whirling it skyward through the air. The pull she must exert in order to keep it from flying off at a tangent is a good muscle - developer

Though it looks very much like an iron handle, it is really a new kind of dumb-bell. It must be held tightly while in action or it will topple over; for the weight is concentrated in the upper half

While her right hand holds the bag on high, She balances by resting her left hand on her hip. In this position she manipulates the bag easily

Five pounds of sand hanging from her foot disturbs this young woman not at all She takes to it as easily as if she were going upstairs

The Home Gymnast
Entertains His Friends

By R. P. Staats, 2d

BEFORE trying any of these tricks, *make sure of your chair*. It must be a heavy, stout one that can't wabble. The kind often described as a kitchen chair will generally fill the bill; and a straight-backed, heavy "Mission" chair is ideal. The rear legs should form the sides of the chair's back—that is, they should extend up through the chair seat. The cross rail at the top of the chair's back should be straight and horizontal. Personally, I prefer a solid wooden seat.

Next, be sure that your chair is standing on a rug or carpet, which will keep it from slipping and at the same time serve as a mat. While learning, it is well to wear light shoes or "sneakers" and as little clothing as time and place permit. Make sure, too, that you have a clear space of four or five feet around your chair; furniture close to you is a mental as well as physical handicap.

Short-Arm Stand. Take your position facing a side of the chair, and from twelve to eighteen inches away from it. Now lean forward so that the "heels" of your hands rest upon the near edge of the chair; to do this you must hold your elbows close together and turn back the hands so that your finger-tips point toward your body. Your hands should rest on the chair about three or four inches apart, and your thumbs rest on the chair bottom, pointing slightly in from its edge, while your fingers grip the edge and under side of the seat. Now bend the arms at the elbows so that the pit of your stomach rests upon your elbows.

By stretching out your neck and expanding your chest, you will find the body is in sufficient balance to permit the legs being raised to the horizontal position shown in the figure. Equilibrium is readily maintained by regulating the position of the head, and by down pressure of the thumbs. To hold the body in this rigid pose, parallel to the plane of the floor and at right angles to the forearms, will be a tax on the muscles. Don't hold it too long.

The one disadvantage of the Short-Arm Stand is that it is an awkward trick to get in and out of. The performer is too apt to fall forward on his hands clumsily, and make the raising of his body into line seem simply a "muscle-grind" and not a graceful acrobatic trick.

Head Stand on Chair. This looks dangerous, but isn't. Until you have mastered it, however, it is wise to have a friend steady the chair to guard against the possibility of an upset. Directions for this trick are given with its two pictures on page 57.

Chair-Roll. Sit a-stride the chair, facing front, toes resting on floor close beside the chair's back legs. Now bend forward and grab the front legs at the very end, so that the forefinger and thumb of each hand will be *next to the floor*. This hand hold is important, **for** you must be able to press the sides of your clenched

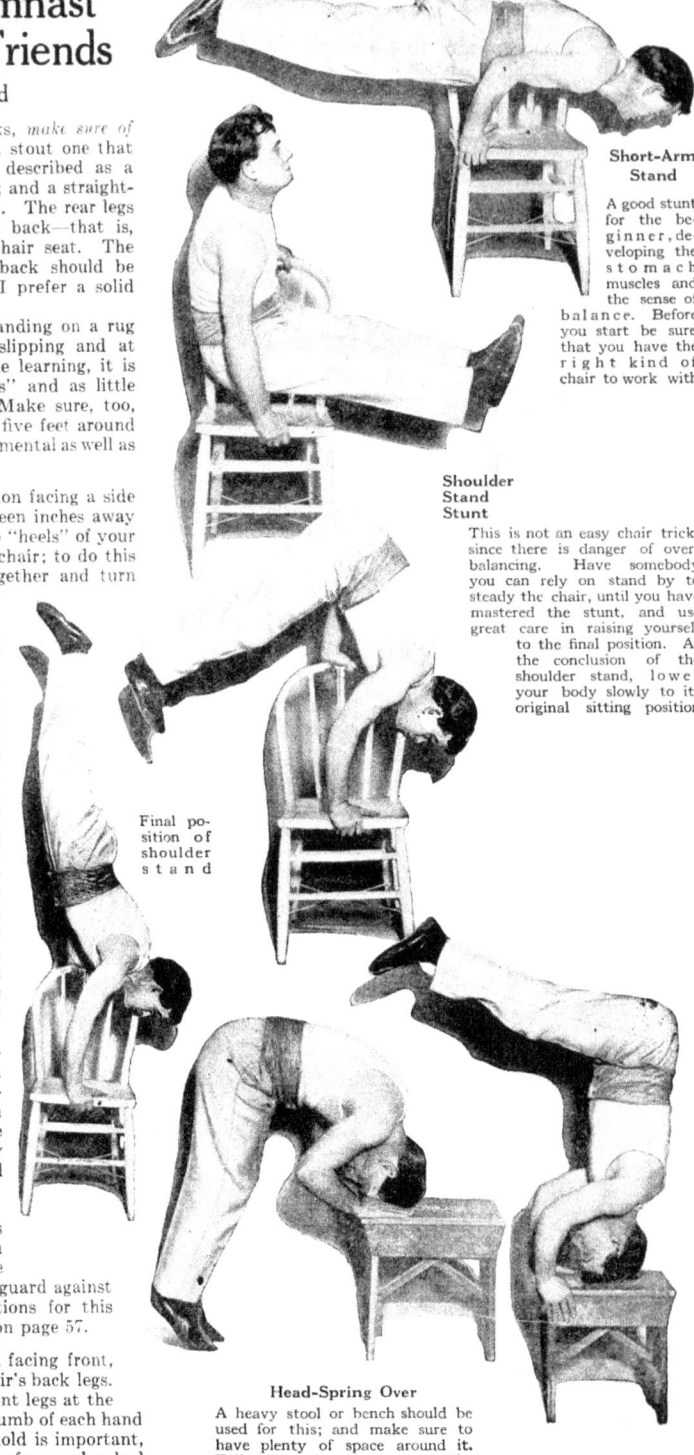

Short-Arm Stand

A good stunt for the beginner, developing the stomach muscles and the sense of balance. Before you start be sure that you have the right kind of chair to work with

Shoulder Stand Stunt

This is not an easy chair trick, since there is danger of over-balancing. Have somebody you can rely on stand by to steady the chair, until you have mastered the stunt, and use great care in raising yourself to the final position. At the conclusion of the shoulder stand, lower your body slowly to its original sitting position

Final position of shoulder stand

Head-Spring Over

A heavy stool or bench should be used for this; and make sure to have plenty of space around it. Follow the directions carefully

A Kitchen Chair Is All the Gymnasium He Needs

Illustrated with photographs of Bluch
Landolf of the New York Hippodrome

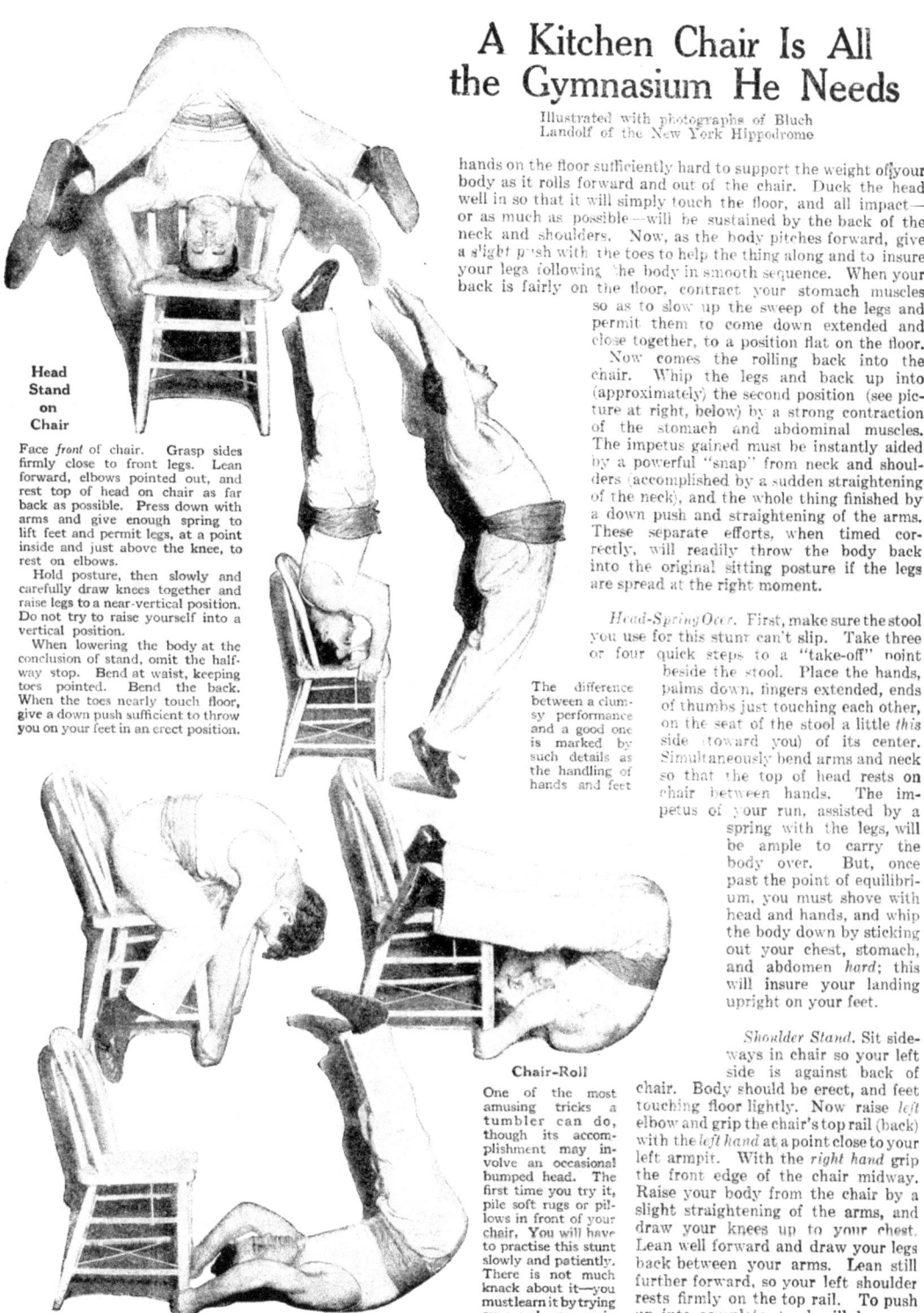

Head Stand on Chair

Face *front* of chair. Grasp sides firmly close to front legs. Lean forward, elbows pointed out, and rest top of head on chair as far back as possible. Press down with arms and give enough spring to lift feet and permit legs, at a point inside and just above the knee, to rest on elbows.

Hold posture, then slowly and carefully draw knees together and raise legs to a near-vertical position. Do not try to raise yourself into a vertical position.

When lowering the body at the conclusion of stand, omit the half-way stop. Bend at waist, keeping toes pointed. Bend the back. When the toes nearly touch floor, give a down push sufficient to throw you on your feet in an erect position.

The difference between a clumsy performance and a good one is marked by such details as the handling of hands and feet

Chair-Roll

One of the most amusing tricks a tumbler can do, though its accomplishment may involve an occasional bumped head. The first time you try it, pile soft rugs or pillows in front of your chair. You will have to practise this stunt slowly and patiently. There is not much knack about it—you must learn it by trying over and over again

hands on the floor sufficiently hard to support the weight of your body as it rolls forward and out of the chair. Duck the head well in so that it will simply touch the floor, and all impact—or as much as possible—will be sustained by the back of the neck and shoulders. Now, as the body pitches forward, give a slight push with the toes to help the thing along and to insure your legs following the body in smooth sequence. When your back is fairly on the floor, contract your stomach muscles so as to slow up the sweep of the legs and permit them to come down extended and close together, to a position flat on the floor.

Now comes the rolling back into the chair. Whip the legs and back up into (approximately) the second position (see picture at right, below) by a strong contraction of the stomach and abdominal muscles. The impetus gained must be instantly aided by a powerful "snap" from neck and shoulders (accomplished by a sudden straightening of the neck), and the whole thing finished by a down push and straightening of the arms. These separate efforts, when timed correctly, will readily throw the body back into the original sitting posture if the legs are spread at the right moment.

Head-Spring Over. First, make sure the stool you use for this stunt can't slip. Take three or four quick steps to a "take-off" point beside the stool. Place the hands, palms down, fingers extended, ends of thumbs just touching each other, on the seat of the stool a little *this* side (toward you) of its center. Simultaneously bend arms and neck so that the top of head rests on chair between hands. The impetus of your run, assisted by a spring with the legs, will be ample to carry the body over. But, once past the point of equilibrium, you must shove with head and hands, and whip the body down by sticking out your chest, stomach, and abdomen *hard*; this will insure your landing upright on your feet.

Shoulder Stand. Sit sideways in chair so your left side is against back of chair. Body should be erect, and feet touching floor lightly. Now raise *left* elbow and grip the chair's top rail (back) with the *left hand* at a point close to your left armpit. With the *right hand* grip the front edge of the chair midway. Raise your body from the chair by a slight straightening of the arms, and draw your knees up to your chest. Lean well forward and draw your legs back between your arms. Lean still further forward, so your left shoulder rests firmly on the top rail. To push up into complete stand will be easy.

Hurdling with a Canoe

Some clever tricks an expert has taught his birch-bark steed

Photographs by
Universal Film
Company

Canoe-hurdling is the original sport of Mr. Charles H. Clark. When he has mounted the hurdle, he balances a moment across the other canoe before taking the downward plunge

When he's ready for the plunge, he tips his body forward and the canoe slides off into the water. He can complete the entire hurdling operation in less than three minutes

Taking a canoe hurdle seems an easy feat for the rider, and one that is pretty to watch. The hurdling canoe rises gracefully at the bow, and slides over the one beneath

A canoe is like a balky horse: its behavior depends on the skill of the rider. Mr. Clark can ride dangerous swells with his paddling partner perched on the bow

Mr. Clark finds that in the wake of a steamer he absorbs some of its speed. He runs down a hill of water at a speed of twelve miles an hour

"They're sinking!" you exclaim. Not at all. These two people are merely showing you that a canoe will stay up and even move toward shore when filled with water

If the Bell Rings You're a Bad Shot

Instead of announcing a bull's-eye, the *ting-a-ling* says you're a trigger-jerker

By Captain Edward C. Crossman, U. S. A.

ABOUT ninety-five per cent of successful rifle or pistol shooting lies in the smooth release of the striking mechanism of the arm that fires the cartridge. Trigger "squeeze" instead of jerk is the hardest thing to teach a beginner, and it is the first thing he forgets when he shoots hurriedly or under excitement. He is not alone in this; for the veteran shot gets spells of fudging or flinching, and they usually go hand in hand with convulsive pulls of the trigger.

The Novice's Dread of Noise

The novice jerks the trigger, displacing his aim and sending his shot wild, usually because he's afraid of the noise and the kick of the gun, and also because he tries to make the shot go when the sights are just right, and, in the parlance of the rifle shot, he "quits holding and pulls the trigger." The experienced shot can make his shot go when the sights touch the spot, but the recruit has to arrive at this stage by way of the gradual trigger squeeze.

Trigger-jerking is one of the most difficult tricks the instructor has to overcome. At the Small Arms Firing School of the Army, at Camp Perry, Ohio, the pistol instructors feel that, when they've got the student officer past this jerking period, he lacks only practice to make a good pistol shot.

Except by the wild shooting of the learner, the trick is hard to detect. If the gun is empty, he doesn't, as a rule, jerk the trigger. If it is loaded, the recoil usually covers up the movement produced by his yanking habit. Veteran instructors "get the goods on him" by the old trick of making him think he's pulling the trigger of a loaded gun when it is really empty, and noting whether the arm moved as the striker fell. This exposes without curing the trouble.

To Cure the Trigger-Jerker

The best way to cure it is to "catch 'em young," and to train them so carefully in the preliminary practice with empty rifle and with small-caliber gallery rifle that the careful trigger squeeze is accomplished without thought.

The most valuable accessory yet invented for detecting and demon-strating to the shooter himself bad trigger squeeze, is the Coleman aiming device, which infallibly and violently rings an electric bell when the shooter jerks the trigger instead of squeezing it. It is in use by the United States Army, having been adopted by the General Staff.

It consists, primarily, of a slidable frame carrying a steel peg a trifle smaller than the interior of the bore of the rifle. The steel peg is connected with one pole of a dry battery; the

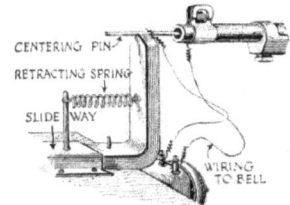

If he jerks the trigger or moves the rifle the merest trifle, a bell rings. The device that accomplishes this, the invention of an army colonel, is very easily made

other pole is connected by a temporary wire with the barrel of the rifle.

The shooter takes aim at a miniature target mounted on the apparatus —a target reduced to the proportion of the outdoor target at some given and common range; then, when he has steadied his aim, the instructor moves the frame carrying the steel pin or peg until it slips into the muzzle of the rifle. It is a trifle smaller all round, and as long as the rifle is held steadily it does not touch the bore. With the steel pin protruding into the bore, and the rifle held steadily as the rifleman aims, the instructor closes the electric circuit, save for the gap between the steel pin—one side of the circuit—and the barrel, the other side. The least jerk of the rifle muzzle will move the bore against the steel pin, close

the circuit, and ring an electric bell that is part of the apparatus. Then, with all this cooked up for the benefit of the learner, he is told to aim carefully and squeeze the trigger as if firing an actual shot at the target.

Ringing the Bell with New Meaning

If he has mastered the trigger squeeze, and lets off as he must to do accurate shooting on a real range, the muzzle does not move, the circuit is not closed, and the bell does not ring. If, however, he jerks the trigger or moves the rifle the least bit in letting off, the bell buzzes merrily, and bystanders hand him the merry ha-ha. There is no alibi —no bad cartridge to blame, no puff of wind, no cant, no mirage. He didn't get off with a smooth trigger squeeze, and the bell told the tale—told the same tale the target-marker would have told, without allowing the various excuses usually dug up for bad shots when in actual range practice.

After a séance of preliminary training with this device, with its "bawling-out" bell to tell to all men the tale of the trigger-jerker, the recruit either learns proper trigger squeeze beyond all chance of forgetting it, or else he's without the pale of possibilities as a rifle shot.

The relation between the steel pin and the bore depends on the skill of the firer. The inexperienced marksman couldn't even hold steadily enough to avoid closing the contact with a pin as near the size of the bore as that which the crack shot could use without trouble; but, once he's learned how to hold steadily, the pin can be so large as nearly to touch the bore, and still never close the circuit when the trigger is let off.

The device, the invention of a colonel in the army who understands the difficulties of rifle training, is cheap to make up and simple to operate, the total accessories necessary for the production of rough but serviceable copies being merely a dry battery, a bell, a little wire, a nail of the right size, a piece of 1x12-inch pine board. Add a simple switch for closing the circuit when the aim has steadied down, and you are all ready to begin the test.

Contests—But Not for Money

Speed and endurance often mean bread and butter as well as sport

A bonus waits for the men who can put together a first-class ship in record time; these men are trying to win it. As soon as the plate is in place the riveters begin

The man who can scramble to the top of this great ball and stay there wins. The winner is here perched on top of the ball, while the men who are holding it keep turning it and throwing it up and down, the man on top never losing his place

Here are three one-armed men in a race to see which can dress first. At the Walter Reed Hospital, in Washington, all kinds of queer contests are held by convalescent soldiers

On the last Fourth of July, before a large audience, two gangs of husky convalescents, each carrying a spouting fire-hose, approached each other. The first to let go would lose the battle for his team, so they all held on as long as they could

Three patient victims, ready to take part in the barbers' contest. They are about to be given the quickest shave and hair-cut they ever had. The winning barber did his bit in six minutes and fifteen seconds

This miner won a contest by drilling a hole fifty-six inches deep in seven minutes. At the word "Go" he made his pipe connections, turned on the air and water, jumped back to his machine, and began to drill

MAKING OVER CRIPPLED MEN

Crippled But Undaunted

Which shall it be for the crippled veteran of thirty — playing cards in a home or earning his living?

By Waldemar Kaempffert and A. M. Jungmann

If you were to read his letter, you would never know that it was written with a pencil held in a special chuck

IN one of the battles of the Boer War, Piper Findlater, a Scotch Highlander, had his legs shot off. He lay on the ground, a bleeding wreck of a man, clutching his bagpipes and playing "The Cock o' the North." A whole regiment was fired by the man's grit. The next day Findlater's name was on every lip; newspapers and magazines made much of him; music-hall ditties typified him as the stuff of which Britain's army is made. He was the hero of the hour, and deservedly so.

After that hour had passed, and the decorations that had been pinned on Findlater's breast had begun to tarnish, the legless hero who cheered on his regiment dropped completely out of sight.

But England was destined to hear of him again. His name reappeared—this time on the posters of music-halls; he was to play his bagpipes, sandwiched in between the trained seals and the Japanese acrobats.

England was horrified—not at itself, but at Findlater. No one stopped to investigate the social and economic conditions that compelled poor Findlater to make a living in a field which he had probably never dreamed of

Yes, they even box

entering. Instead of holding out a tin cup on a street corner, he was doing his best to earn an honorable living, and to relieve the State of caring for him.

Forty Thousand Cripples Out of an Army of a Million

Peace is not yet in sight, but already every government has seen to it that there shall be none of the thoughtlessness that marked the return of Findlater to private life. The privilege of selling shoestrings on the streets of New York or Chicago after having been blinded by German gas or having an arm torn off by a German shell—is that to be the reward of fighting for democracy? Or to watch the flies buzzing against the panes of some whitewashed home at the age of twenty-five or thirty? Or to gaze on the world as if one were no longer a part of it?

Until the war is ended the total number of men to be provided for is a matter of mere estimate. If we use the word "disabled" in its broadest possible sense, we may expect the return of 120,000 for each million men sent overseas.

It is unthinkable that these men shall be permitted to become objects of a demoralizing charity. And so, following the example set by France and England, this country has laid its plans to preserve the self-respect of men who will return to us blind, legless, armless, and otherwise handicapped. The Red Cross Institute, the Federal Board for Vocational Education, the Surgeon-General's office, and various societies and industrial firms will see to it that there shall be no Piper Findlaters.

Federal Board for Vocational Education

Since fully 80 per cent of the disabled will be capable of earning a livelihood, according to Major J. L. Todd, a medical officer connected with the Board of Pension Commissioners of Canada, it will be no light task to find places for

Here is an example of what the chuck will enable a man to do

Ping-pong for the armless! Who could have foreseen that at the outbreak of the war?

An artificial arm should be made, not for looks, but for service

No doorkeeper's or watchman's job for him. He wears the chuck type of arm, made to receive different tools

It is easier to design a leg than an arm; after all, it is used simply for the purpose of support

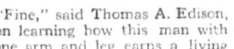

It is not necessary for a man, because he is a cripple, to give up tennis and other sports

"Fine," said Thomas A. Edison, on learning how this man with one arm and leg earns a living

men who are still young and who have been taught to make the most of the physical equipment that still remains. In addition to the thousands of amputation cases with which the United States must deal, it must also make some provision for tens of thousands of men afflicted with severe rheumatism, paralysis, shattered muscles, stiff joints.

On the Surgeon-General of the United States Army will devolve the duty of healing the disabled man and of giving him some preliminary training. Then he passes to the jurisdiction of the Federal Board for Vocational Education. He will earn something while he is learning a trade anew, and a family allowance will be paid. When he has completed the prescribed course, he will be turned over to some accredited organization, which will see to it that he finds employment at the trade that he has learned. No man is compelled to attend a re-education school, but his pride will be deliberately touched, so that he will have no desire to become an object of sympathy or charity.

Some Men Who Have Been Re-educated

How does this work out in practice? Lieutenant Frederick Holmes, Ontario officer in charge of Industrial Surveys, recently wrote of a Canadian who had lost one eye and both legs, and who was turned over to a jewelry house for a course in silver-polishing. In six months the man was not only an expert silver-polisher, but had also learned gilding and soldering.

Then there is the case of Captain Edward Baker, who went with the 4th Canadian Field Company to France, won the Military Cross and the Croix de Guerre, and lost both his eyes. He had been an electrical engineer, but was so disabled that he could not practise his profession.

His old firm—a hydro-electric company—took him back. He now operates a typewriter, takes trouble reports along the lines between Niagara and Ontario, makes out reports on a sheet of paper with twelve rulings, and, by an ingenious spacing arrangement invented by himself, typewrites in the proper space the date, time of day, the trouble, the

France re-educated him and showed him how he could work his farm

peak load, the low load, the number of minutes, and his own name—all as quickly and as accurately as if he had his sight.

Teaching a Man to Use Untrained Muscles

The first task is to determine what a disabled man can do with the natural tools still left to him. A tailor or a novelist who has lost both his legs is a subject for sympathy, but not for re-education. If a man is blind, he does not have to content himself with becoming either a piano-tuner or a weaver of baskets and chair bottoms —the traditional occupations of the sightless. A hundred callings are open to him — winding the wire on induction coils and armatures, assembling airplanes and parts of automobiles, feeding material to automatic machines.

If he has lost his right arm, he is taught how to use his left. Even if both arms are gone his case is not industrially hopeless: with the aid of appliances, he may still hold his own in competition with an able-bodied man. A hand is a tool—one of the most wonderful, versatile, complicated tools that

There are dozens of ways of moving the spacing-bar of a typewriter. This man uses a string attached to a cam mechanism

The machine must sometimes be adapted to the man. Here is one cripple's way of working the space-bar of a typewriter

nature has devised. Study it. Bend and stiffen its five fingers, and you have a powerful claw. Clench the fist and you have a hammer. With the aid of the thumb you can form a pair of pincers with any one of four fingers. The arm is a lever by means of which this marvelous tool is moved to a place indicated by the brain. The invention of a serviceable substitute arm and hand is a far more difficult problem than the invention of a leg.

Remarkable results have been obtained by fitting to an arm-stump special appliances to be used for special purposes. Thus a chuck may be attached to the stump in which a knife, fork, spoon, or tool can be interchangeably inserted. For some special cases, devices must be invented which the cripple may attach to parts of his body, not to take the place of missing limbs, but to serve as new ones. A ring or loop is attached to the suspenders or belt to assist a one-armed man to handle a shovel; or yokes, special belts, and grasping devices are operated by pressure of the body against the work-bench; or a "third thumb" holds a magnifying-glass.

How the Disabled Will Be Re-educated

What is the process of restoring a crippled soldier to a living? First of all, the hospital surgeon must heal the man, so far as that is possible—loosen his stiff joints, soften his scars. Occupation is one of the best agents that the surgeon has at his disposal.

When the surgeon has done all that can be asked of his skill, the patient is ready for vocational guidance. He receives free a leg or an

How it looks from the front

arm—sometimes two legs or two arms. The arms are implements and not deceptions. They are intended to serve a man in earning his living, not to trick the community into believing that he is whole. Just as the natural arm is a tool manipulated by a wonderful system of levers, joints, and tendons, so the artificial arm is an artificial set of muscles—steel wires, rawhide cords: the whole linked up to form a new muscular combination with which the man must familiarize himself. It is amazing what can

Legless but undaunted

be accomplished with such substitutes.

Since the hook and the clamp are the most useful of all devices, it is no difficult matter to provide a useful artificial hand which consists of nothing more. Only when the necessity of providing a tool is considered, regardless of its appearance, is real mechanical success attained. The simplest form of artificial hand is one so constructed that the fingers are bent in a fixed position to form a hook, and the thumb is made to move against them to form a clamp. A spring gives the thumb grasping power enough to clutch small objects. If large objects are to be seized, some form of locking device must be provided, since grasping power is limited by the spring.

For Use, Not for Show

Indeed, the modification and combinations of the hook and clamp are almost numberless. Perhaps the most useful of all is a combination of two fingers with a thumb—the fingers fixed and the thumb moving between them. Open the hand and you have a hook; close it and you have a means of holding nearly all ordinary tools.

If drawing-room considerations must be abandoned to help the one-armed, they are even less significant in restoring the armless to a place in the industrial world. Here the best plan seems to be that of adapting the mechanism provided to the particular kind of work to be performed. With the aid of a strap or two, for example, a man can dress and feed himself. The simplest devices, aided by perseverance and intelligence, accomplish wonders. The old-fashioned "peg-leg" and the simple hook and claw are by far the most popular contrivances still, despite all the thought that engineers have given to the problem of the crippled soldier.

Long before the United States entered the war, Frank B. Gilbreth, the American efficiency engineer and motion-study expert, gave much thought to the problem of the crippled soldier, partly through humanitarian motives, partly at the request of European governments. Mr. Gilbreth demonstrated that it is possible for a cripple to adapt himself to specified work, if the motions required to perform that work are analyzed and tabulated in proper sequence. Slight changes in machines enable many a cripple to practise his old calling.

Adapting Tools to Cripples

Gilbreth has accomplished marvels in adapting the typewriter, for example, to the needs of the armless. He has shown how a legless man can travel from machine to machine in a shop on a little seat running on a track; he has demonstrated that a one-eyed, one-armed man can practise dentistry, if his patients will but assist him by holding down gums. He has pointed out the necessity of training cripples for the work in which they themselves are most interested, and of teaching the public what attitude should be assumed toward the crippled. He would show the cripple how to adapt himself to existing vocations, so that he will actually compete with an able-bodied worker; would create entirely new occupations where that may be necessary; and, lastly, would reserve certain places for cripples. Gilbreth maintains:

This hammer has been invented for one-armed cripples. On pressing the trigger with the forefinger, a nail drops from the magazine, to be carried to the proper place until it is driven in place by a blow. When the finger is removed from the trigger, the holder drops back, and leaves the hammer free to finish driving in the nail

Every one of us is in some degree a cripple, either through being actually maimed or through having some power or faculty which has not been developed or used to its fullest extent. . . . From an efficiency standpoint, a policeman with corns on his feet, or a golfer with gout in his toe, is more of a cripple during his working hours than a legless man while operating a typewriter. We can then think of every member of the community as having been a cripple, as being a cripple, or as a potential cripple.

Nearly all of us are one-armed in the sense that we use the right arm almost exclusively; the left is merely an auxiliary used as a weight to hold a sheet of paper or a piece of work. That simplifies the re-education of those who have lost a left hand or arm enormously. And if the right is lost, the left is trained to acquire its functions. Indeed, it is by no means unusual for a one-armed man to discard all artificial aid. The remaining stump serves to hold down the piece of work.

The Future in Industry

No past war has ever made such demands on populations as this. While, therefore, we may be prosecuting this war with more foresight than our fathers showed, we must not forget that the very character of the conflict compels the exercise of generalship in industry as well as on the battlefield. Past wars did not cripple men by tens of thousands.

The men who will return from France sightless and maimed will constitute a veritable army. For that reason, the warring countries have made it their business, even while cannon are still belching death, to give back to the crippled veteran his place in the world.

Frank Gilbreth, the motion-study expert, sent us this picture to show what persistence can accomplish

At twenty-three this man became a cripple from rheumatism. He learned the machinist's trade, and is now employed by a big typewriting concern

After all, anybody can brush one hand at a time. We must use the tools that nature has given us

Healing Burns without Pain or Disfiguring Scars

REMEMBER the excruciating pain which even a slight burn causes you, and then try to imagine the suffering resulting from a severe burn by which a considerable area of skin is scorched and destroyed. Such severe burns occur sometimes in factories, foundries, and mines, and are classed among the most serious injuries to which workers are exposed.

When the Germans introduced "liquid fire" as one of their weapons of attack during the great war, the Allied surgeons in the field hospitals found it difficult to cope with the task of adequately treating burns. At that critical moment Dr. Barthe de Sandfort, a retired French Navy surgeon, who had invented a method of treating burns by coating the scorched surface with an air-tight covering of medicated paraffin wax. presented his invention to the French military authorities and offered his services in administering the treatment and teaching others the proper method of applying it. He was given charge of a small hospital at Issy-les-Moulineaux, in Paris, to which all soldiers suffering from burns were sent for treatment. The results obtained from this treatment were so remarkable that it was adopted by the hospital staffs of all the other Allied armies.

The compound that Dr. Sandfort

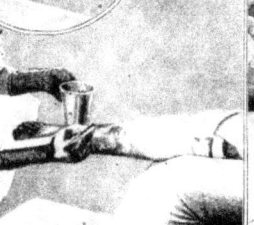

(Left) Before the protective wax coating is sprayed on the burned surface the scorched tissue is dried by an electric dryer

(Right) The melted paraffin compound, kept liquid by the hot water in the jacket of the atomizer, is sprayed on the burned surface and forms a protecting cover

A thin layer of cotton is placed over the paraffin coating of the burned surface, and by painting it with melted wax it is made to adhere to the coating

When the dressing is to be changed it is ripped open with the scissors and it can then be peeled off without difficulty and without giving the patient any pain

employs for covering burned surfaces consists of neutral paraffin, resins, and certain disinfectants. It has a low melting-point, and, when heated to 140° or 150° F., forms an amber-colored liquid which is either brushed over the surface of the burn or sprayed on. The hot liquid hardens quickly, and forms a pliable yet air-tight covering of the burned area. A thin layer of cotton is placed on the first coating of wax, and is made to adhere to it by painting it liberally with the melted wax. For further protection the coated part is swathed in cotton. After the first twenty-four hours the patient, in most cases, is free from pain.

During the first stages of the healing process the injured tissues secrete a large quantity of lymph, which collects under the shell of the dressing and prevents it from adhering to the tissues. The wax dressing may easily be lifted or peeled off without causing the slightest pain to the patient. The burns should be dressed every twenty-four hours for the first week or so.

Workshops that Travel on Wheels

Workshop trains like this, compactly arranged and equipped with electric power and machinery, were used by the Allies behind the Flanders front

DURING the intensive military operations along the Flanders front it was a matter of the greatest importance to keep the enormous technical fighting apparatus, the wonderfully efficient war machinery, in good condition. Repairs were constantly needed, and time was too precious to permit the delay that would have been caused by long transportation over difficult roads. The British War Office supplied six completely equipped military workshop trains which were operated on the narrow-gage military railways behind the front.

Each of these six trains consisted of six cars, including a generating car, two machine-shop cars, a tool van, a stores van, and an officers' car. The cars were all of the same dimensions, 17 feet 8 inches long, 5 feet 5 inches wide, and a trifle more than 9 feet high. The generating car was equipped with two 15 to 20 horsepower gasoline engines directly coupled to 10-kilowatt generators. The machine-shop cars carried drills, lathes, shaping machines, etc.

Making the Dumb to Speak

The latest methods of specialists in treating victims of shell shock

By Lucien Fournier

COMPLETE loss of speech was one of the serious results that developed in many soldiers suffering from shell shock. It afflicted most frequently men whom the horrors of the Great War had brought to a state of nervous tension which predisposed them to this particular pathological condition. Drs. G. Liebaut and E. Coissard, two eminent French nerve specialists, studied the disease, and organized a clinic for restoring to these soldiers the power of speech in accordance with the methods employed in schools for the treatment of deaf-mutes.

Among the patients in that clinic there are none who have lost their ability to speak through any physical injury of the organs of speech. Nearly all of those who undergo treatment are eventually cured or at least greatly improved by a systematic training of the organs of speech, which include the lungs, the larynx, the pharynx, the mouth, nose, and lips.

Careful Examination of Each Patient

Before beginning treatment the doctors carefully examine each patient to ascertain his lung capacity, method of breathing, and any abnormal peculiarities in the mechanism of drawing in and expelling air. The lung is measured by a simple yet accurate apparatus devised by Dr. Coissard. The apparatus is constructed like a gasometer. A cylindrical tank is filled with water. Within that tank, but closed at the top, is a smaller cylindrical container, held in position by counter-weights connected with it by cords running over pulleys.

The air exhaled by the patient is conducted by a tube into the floating cylinder, and, by displacing the water contained in it, causes the inverted container to rise. A scale on the side of the tank indicates the volume of air exhaled. The chest expansion is measured with a tape-measure, and the movements of the diaphragm are recorded by two radiographs taken of the extreme positions of elevation and depression during the process of breathing. The force of exhalation is measured by a pulsometer, and the time of

inhalation and exhalation by a stop-watch. Then the functioning of the larynx, lips, and other speech organs is carefully observed and noted.

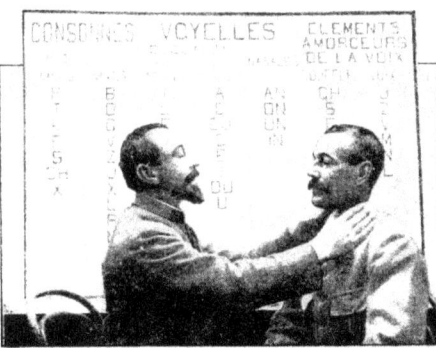

The doctors give individual attention to each case, a detail of the treatment being massage of the throat by hand or by an electric vibrator

Each case is studied individually by the doctors. After a thorough consideration of the facts in each case, the method of treatment best suited to the condition of the patient is determined.

Breathing exercises are given first, to correct any defects developed by the nervous affection of the patients in their respiration. These exercises are supported by gymnastics of the arms and body.

To restore the normal functions of the vocal cords is far more difficult. After the patient has been taught to breathe properly, his throat is massaged by hand or by an electrical vibrator. By slow degrees the shell-shock victim is taught to emit sound at the time of exhaling.

First the Vowels

The patient begins with the vowel *o*. If he does not succeed in emitting the simple sound, his throat is gently manipulated by the doctor, who depresses the larynx of the patient with his fingers to stimulate and assist that organ in the effort of resuming its normal functions. These exercises can be taken for only a brief time each day.

After the patient has learned to pronounce the vowel *o* without assistance, he begins to learn how to pronounce the vowel *e*, which in French is pronounced like "ay" in the English word "bay." Then follow in the same order the vowel sounds *eu, ou, u, a,* and *i* as pronounced in French.

The last phase of the treatment includes the careful and systematic training of the muscles of the cheeks, tongue, lips, pharynx, and nose which present symptoms of paralysis. These exercises are performed by the patients in front of a mirror.

After they have learned to respond in an almost normal manner, the patients essay the pronunciation of simple combinations of vowels and consonants.

In most of these cases speech is restored in a few months.

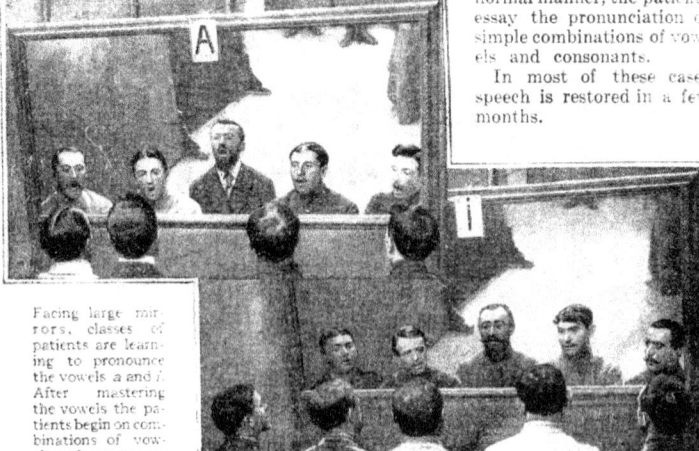

Facing large mirrors, classes of patients are learning to pronounce the vowels *a* and *i*. After mastering the vowels the patients begin on combinations of vowels and consonants

GENERAL SCIENCE

Ripples of Life and Waves of Death

By Alfred J. Lotka

YOU may whisper in her ear or you may wireless to her in Honolulu; it makes no difference—it is only a ripple, anyway.

I am not speaking of the *substance* of your message, but of the *means* by which it is conveyed. In either case, it is carried to its destination by a wave. Of course, there are waves and waves. A whisper is a sound wave measuring only a few inches from crest to crest, while the electro-magnetic waves used in wireless telegraphy may be several thousand feet long.

The world is full of waves—waves visible and waves invisible; waves inconceivably small; waves of ordinary dimensions; waves so long that the imagination stands aghast in the effort to picture them; waves in the water, waves in the air; waves in the solid earth, and waves in the empty space between the stars.

The most familiar of these are the waves on water. Look at water waves and you can form a mental picture of some others that occur in nature—most of them either quite invisible, or else visible only under very special circumstances.

If you throw a stone into a pond, an ever-widening train of circular waves

If you throw a stone in a pond, an ever-widening train of waves spreads in all directions. The waves have two dimensions only. A somewhat similar wave action takes place when you strike a bell

spreads in all directions. The waves, then, in this case are circular and are propagated in two dimensions only, for they cannot leave the surface of the water and travel out into space.

Making Your Eyes See Sounds

Something similar happens if you strike a bell, for example. The bell is thrown into vibration by the blow; you discover that if you touch the bell with your finger immediately after striking it. The vibrations of the bell are communicated to the air around it, and a wave ripples out in all directions. If it reaches your ear you say: "I heard the bell strike."

In this case, if there are no obstacles, the sound spreads approximately evenly in all directions; in other words, the waves are not circles, but spheres. Though your ear detects these waves, they are ordinarily quite invisible. Yet you can be made to see them by special devices. Professor Foley, of Indiana University, has obtained some exceptionally fine photographs of sound waves. As a sound wave in air moves at a speed of about 1,100 feet a second, or 750 miles an hour, and as it is quite impossible to point

Probably not a day passes without an earth wave. Fortunately, most of these are very slight. This picture shows car-tracks and cobblestones torn up by a recent earthquake

your camera and work a shutter by following the wave with the eye, it requires special ingenuity and skill to secure a photograph. The feat is accomplished by pointing the camera at the source from which the sound proceeds, and making a flash of intense light immediately after the sound is produced. The particular sound used in the experiment is the loud crack of a powerful electric spark, and the flash of light is given a very small fraction of a second later by another electric spark sprung between magnesium terminals to render it intensely luminous.

Try It on Your Piano

Just as the waves on water vary in height or amplitude, and also in length as measured from crest to crest, so sound

The earth writes its autograph with this instrument —a seismograph. Daily waves in the earth are thus written down for the scientist to study

waves differ in amplitude or loudness and also in wave-length or musical pitch. An ordinary piano keyboard extends from the note A three octaves and a half above the so-called "middle C," the note on the first ledger line below the staff in the violin clef and four octaves above this to the top note of the piano, again a C. The string that sounds the middle C vibrates 273 times a second at concert pitch and sends out waves measuring $\frac{1100}{273} =$ 4 feet in length. The wave length is halved every time we go up one octave. Hence the wave length of the topmost note on the piano, four octaves above the middle C, is $\frac{4}{2^4} = \frac{4}{16} = \frac{1}{4}$ ft. = 3 inches. Similarly, the wave length of the lowest note (A) on an ordinary upright piano is nearly 39 feet. The lowest note on some of the largest organs is the C an octave and a half below this, sounded by a 64-foot open pipe. Its wave length is 128 feet, and it gives only 8½ vibrations a second.

It is doubtful whether the ear truly recognizes the pitch of the notes at the extreme ends, both above and below, of the full range of a great organ. Even on the piano it is difficult to recognize a melody played, say, on the three lowermost notes. And if we go a little further down the scale of vibrations, we cease to recognize them as sounds at all at most, we may become conscious of a kind of rumble. At the upper end of the scale, too, a point is first reached where we are unable to distinguish musical pitch, where we can still hear a note, but are unable to place it in the musical scale. Then beyond this come vibrations that the human ear does not perceive at all. This limit of audibility varies considerably in different persons, but corresponds to a frequency of about 24,000 vibrations a second or a wave length of one half inch, a note two or three octaves above the highest note on the piano. Cats can hear sounds beyond this limit, and a whistle can be constructed by which you can call a cat, although its note is quite inaudible to a human being.

Sounds that You Can't Hear

Audible sounds, then, are comprised within some ten or eleven octaves of the musical scale. Both above and below this range we have a region of what may be termed "inaudible sounds." That is to say, there are waves that, just like the audible sound wave, consist of alternating layers of compressed and rarefied air traveling out from a vibrating body, but to whose wave length our ears are not attuned.

Is there any limit also to these inaudible sound waves? The answer is that there is, at any rate, an upper limit, for a wave only one tenth of an inch long can be started, but by the time it has traveled about an inch its intensity ("loudness") has fallen off to only one hundredth of its initial value. Such exceedingly short waves, then, if they occur at all, can never travel more than a very short distance. In contrast with this, sounds of ordinary wave length have been heard over very long distances. Thunder may sometimes carry as far as twenty miles. In the World War the boom of big guns was heard

at a distance of more than three hundred miles. In the year 1883 the top of the Krakatoa mountain in the East Indies was blown off by a terrific volcanic explosion which was heard for fifteen hundred miles.

The water upon the earth and the air of the atmosphere are very mobile bodies, and perhaps we would natur-

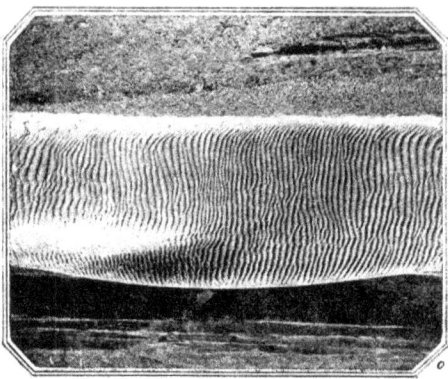

This looks like the back of some great wriggling reptile. In reality it is nothing more than a harmless stretch of sand ruffled up by the wind. The waves occur with almost unbelievable regularity

ally expect that they should be constantly thrown into vibration or wave motion.

Even the solid earth itself is not exempt, although powerful earth waves are fortunately of rare occurrence in most localities; I say fortunately, because when they do occur they are absolutely the most deadly of all catastrophes in nature. It is estimated that in the earthquake of Messina in 1908 about 100,000 people were killed, while

300,000 are said to have perished in the Indian earthquake of 1737.

While such catastrophic convulsions as these are fortunately rare, minor earth tremors are very common, especially in certain localities. They range from slight but easily perceptible shocks to fine vibrations detected only by delicate instruments, so-called seismographs, constructed especially for this purpose. Even in New York, whose immunity from serious earthquakes is attested with picturesque eloquence by its sky-line of tall buildings, the seismograph of the Museum of Natural History is seldom idle. Probably not a day passes without an earth tremor, such as a delicate seismograph would record, taking place somewhere on our globe. It is hoped that in the course of time the observations gathered with seismographs and by other methods will enable us to foretell the advent of severe earth-shocks, and to warn the occupants of the threatened territory.

A Bridge that Was Destroyed by Walking Horses

The subject of vibrations in buildings and other engineering structures is of interest also quite apart from the often only remote possibility of earthquake disturbances. The sway of a tall structure, such as a factory chimney, under the wind may attain an amplitude of several inches. Nor does it necessarily require a large force, such as high wind pressure, to produce very serious oscillations. You know that by properly timed pushes a small child can gradually set into

Even fog comes in waves. You do not see the waves when you are in the midst of it—but if you stand above it on a mountain-peak, the top of the fog will look like a rippling ocean. There are waves in water, in air, in the solid earth, and in the ether space between the stars

Water waves vary from the fleeting ripple to the mighty ocean billow, mounting thirty or forty feet from trough to crest, and rolling on for hours after the gale that gave it birth has entirely subsided

motion a swing with an adult sitting in it. This is a typical example of what is called resonance. The swing has a certain natural period of vibration, just like the pendulum of a clock, which makes one, two, or more beats every second, according to its length and the construction of the clock.

Now, this is a perfectly general property of bodies or systems capable of vibration—that they have a definite period of oscillation, depending on their dimensions, and that, if the proper kind of force is applied at regular intervals corresponding with their natural period of vibration, even a small force is capable of producing very powerful effects. This is the reason why troops marching over a bridge are always ordered to break step. Were they to march in unison it might happen that their footsteps kept time with one of the natural periods of vibration of the bridge or some of its members. Serious results might then follow. A suspension bridge at Manchester, England, many years ago actually collapsed under the hoofs of a troop of cavalry.

Danger of Oscillation

Similarly, a building may be perfectly safe for ordinary use, and its floors able to support the weight of heavy machinery; yet this same building might be thrown into violent vibrations if its natural period of oscillation should happen to coincide with the rate of revolution of a motor installed within it. The same kind of effect is liable to occur and has to be guarded against in the design of electric installations. If the dimensions of the electric circuit happen to stand in a certain numerical relation, powerful surges are liable to arise, which would either burn out the conducting lines or else break through the insulation; the period of revolution of the dynamo, and hence the fre-

quency of the alternating current, coincides with the natural period of oscillation of the circuit.

Perhaps the most familiar example of resonance, the one after which the phenomenon has been named, is that of a body set into vibration by sound waves striking upon it. If you lift all the dampers off the strings of a piano by pressing down the "loud pedal," and sing any note of the musical scale into the body of the instrument, after the sound of your voice has died out you will hear certain of the strings vibrating and giving out their note, namely those strings one of whose natural periods of vibrations coincides with the note you sang into the piano.

Waves in the water, waves in the air, waves in the earth—there we have at least something tangible. But waves in the empty space between the stars—how can that be? The question is hard to answer. Modern developments seem to indicate that our diffi-

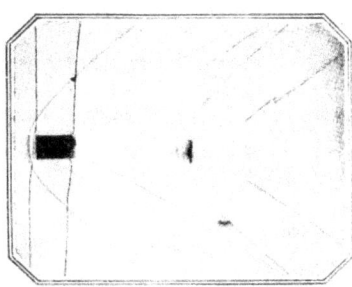

A brass bullet with a hole through the center was shot from a rifle and photographed in the act. The high velocity of the bullet did not permit much air to pass through the hole, and thus the front air wave does not differ from that of a flat-headed bullet. But the air that did pass through hit the vacuum at the base of the bullet and started a sound wave. The friction of the air along the sides of the bullet is seen

culties in answering it are psychological at least in part. Just as the ear is deaf to sounds outside a certain compass of frequency, so the human mind may be incompetent fully to grasp certain facts and relations of nature. Perhaps the human race will have to recast its conceptions of time and space before it is able to gather into the realm of understanding certain aspects of such phenomena as the journey of light from sun to earth. But this we know: that light does travel through what is for all ordinary purposes empty (airless) space, and, in fact, travels through such empty space with greater ease and at a higher speed than even through the most transparent glass. And we know, further, that in its travel light displays all the characteristics of a wave disturbance; that it travels through space with the perfectly definite velocity of 186,000 miles a second, identically the same as that of the electric waves employed in wireless telegraphy. It is one of the feats of the genius of Clerk Maxwell, that he predicted the existence of these waves on theoretical grounds alone long before they had ever been observed, and that he identified light with them.

The wave lengths of visible radiation, the "light" we see, range from about sixteen one millionths of an inch for violet rays to about thirty one millionths of an inch for red rays.

Where the Eye Fails

Just as there are sounds too "high" and too "low" for the human ear to perceive, so there is light "too violet" and "too red" for the eye to see. But the photographic plate still detects ultra-violet rays with a wave length of about four one millionths of an inch. On the other hand, though a body may not be hot enough to give out visible radiation (as at red or white heat), it may still be emitting "heat rays" perceptible to the hand or observable by means of special instruments. The wave length of these may be taken to extend to about twenty-four ten thousandths of an inch.

Above the ultra-violet waves the scale is continued in X-rays, with a wave length of about two one billionths of an inch. Below the infra-red waves comes a long series of electromagnetic waves produced experimentally and ranging from about one tenth of an inch to many thousands of feet in length.

Lastly, the earth being negatively charged, and revolving around the sun, which probably carries a positive electric charge, must send out an annual wave having the stupendous length of six million million miles!

Once They Would Have Burned Him at the Stake

MEN have been tortured and burned at the stake for upsetting ideas. Galileo and Copernicus had a hard time of it in their day. Now comes a revolutionist, an idea-upsetter as great as any. He is Professor Albert Einstein, and his idea-upsetting "theory of relativity" will make it necessary to rewrite Newton's laws of gravitation and every text-book on physics.

Einstein starts with two suppositions. One is that all motion is relative. In the cabin of a ship you cannot tell whether you are moving or not without looking out of the window. Is a man who is walking from stem to stern of a ship at the same rate that the ship is moving in the opposite direction standing still or not? If astronomers could not see the stars, they would not know that the earth is moving.

Einstein's second supposition is this: The speed of light (186,000 miles a second) cannot be increased or diminished. This is like saying that the speed of a rifle bullet is always the same when it is fired forward or backward from an airplane going at one hundred and twenty miles an hour. Nothing can be faster than light. Newton says that gravitation acts instantaneously throughout space. "No," says Einstein, "the action of gravitation is not instantaneous; it cannot exceed the velocity of light."

What is your size? What is your shape? "Tell me how fast you are moving and in what direction," says Einstein, "and I will answer. If you are traveling vertically upward at the rate of 136,000 miles a second, you are not six feet tall, as you supposed, but just three. But horizontally you will measure six feet. It is no use to bring in a standard yard-stick and start to measure, for that too contracts and measures only a half a yard vertically held."

Light is a form of energy. Therefore, says Einstein, it must have mass and must be affected by gravitation. In other words, you ought to be able to weigh light. But how? It moves so fast that it cannot be weighed on the earth. If we could see a star close up to the edge of the sun, a ray of light coming from the star would bend under its own weight, and the star would be seen, not where it actually is, but a little bit to one side. During the last solar eclipse exactly what Einstein predicted happened. A light from a star was deflected, just as a bullet fired from a gun gradually curves toward the earth. We can safely speak of a pound of light now. As a matter of fact, the sun showers on the earth 160 tons of light daily.

Other astonishing consequences follow as soon as Einstein's theory of relativity is accepted. Straight lines do not exist. They are parts of gigantic curves. Travel fast enough on a straight line, and you will come back to your starting-point. It would take a beam of light 30,000,000 years to describe a complete circle.

All this seems like sheer nonsense. *And yet, Einstein's statements have been proved to be true by experiments!* You have been living in a dream world. Your conception of time and space are true only within limits. "Wake up," says Einstein, "and acquaint yourself with the real world."

The Diamond and Its Bloody Story

All the revel, riot, recklessness, quick dramas, and dazzling riches of all the world's mining rushes and gold stampedes are crowded into the mad romance of the diamond

By Walter Noble Burns

ALL the diamonds in the world could be packed in your wife's clothes closet. They could be stored in a kitchen pantry, where, in the dim light, the cook might mistake them for navy beans and attempt a puree. They would form a pile about as big as the pile of coal the truckman dumps on the sidewalk at the basement entrance to your apartment building. If the pile had a base diameter of eight feet and were rounded into a cone, it would be five feet high. A pile of coal of equal size—and coal by every tie of chemical relationship is the diamond's first cousin—would cost $28. The pile of diamonds, reckoned at $100 a carat, would have a value of $4,635,547,480. If figured at current diamond prices, it would be worth from three to five times that much.

There are, it is estimated, 46,355,474 carats of cut and polished diamonds in existence. In terms of avoirdupois they would weigh $10\frac{1}{2}$ tons. The total includes possibly the first diamond ever found on earth—who knows?—and the last gem picked from the chimneys of South Africa; the little twinkler that the shop girl wears on her finger and the Kuh-i-Nur that blazes in Great Britain's crown.

The War Advanced Diamond Prices

One hundred dollars a carat, used as a basis in the estimate, is perhaps below the average cost of diamonds throughout history. Diamond prices have been subject to wide variations. The war advanced the price about one third. Present prices are about one hundred per cent. higher than those of fifty years ago, and they undoubtedly will go higher in the next few years. But every diamond is an individual problem as far as price is concerned. The price always depends on the stone's color, comparative flawlessness, inherent brilliancy, and cutting.

A one-eighth carat diamond sells at present for from $12.50 to $20; one-fourth carat from $37.50 to $62.50; one-half carat from $100 to $200; three-fourths carat from $187.50 to $337.50; one carat from $300 to $500. Importers buy rough diamonds in foreign markets for about $90 a carat. A rough crystal of $2\frac{1}{2}$ carats, which will cut to a gem of one carat, costs $225. Import duty is 10 per cent.; 1 per cent. is to be added for insurance and brokerage charges; the labor of cutting may be figured at $15. The polished one-carat gem thus represents an outlay of about $250. If this diamond turns out to be a gem of first quality, it will retail at from $500 to $550.

But such quotations are not wholly dependable. Some blue-white one-carat stones sell for $2,500, while you can buy a one-carat yellow diamond for $150. Blue-white diamonds bring the highest price in the market. But many connoisseurs prefer as more beautiful the snow-white gem often found among river diamonds, whose sharp, cold brilliancy is like that of clear ice gleaming in winter sunshine.

Democratic Uncle Sam and His Diamonds

The United States in recent years has become the greatest diamond-buying nation on the globe. For years it absorbed from fifty to sixty per cent., and during the war 85 per cent. of the output of the South African mines, which supply 98 per cent. of all the diamonds in the world's markets. A recent estimate placed the value of the diamonds in this country today at $1,350,000,000. Of this $500,000,000 was set as the value of the stones in the country in 1900. Importations since 1900 have amounted to $506,000,000, this including

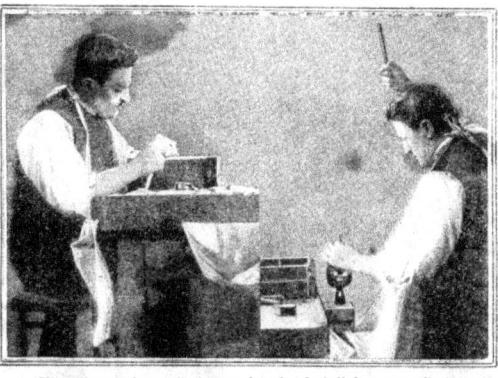

The diamond was a clumsy jewel of dull luster until the art of cutting and polishing it was discovered in the fifteenth century. If the stone is large the first step is to make an incision and then to cleave it with a mallet, which two steps are illustrated by this picture

Some of the Diamonds that Have Had a Thrilling Past

A   B   C   D   E

The Regent, or Pitt (A), weighed 410 carats, and was bought for about $120,000 by Pitt, Governor of Madras. The Duke of Orleans, Regent of France, paid $400,000 for it. It was cut to nearly 137 carats, and was stolen during the Revolution, but was recovered and is still in France. The Kuh-i-Nur (B) led one Indian potentate to kill his three brothers and imprison his father. It has been cut and recut. It weighs now 125 carats and has been valued rather fancifully at $1,000,000. The

Pigott (C) weighed 82 carats. It was last heard of in Egypt. It is valued at $150,000, rather little for a stone of such size. The Empress Eugénie (D) weighs 51 carats, and is the property of the famous Gaikwar of Baroda. Two centuries before it shone on the bosom of the proud Eugénie it was given by a peasant to a blacksmith for mending a plow. The Duke of Westminster owns the Nassak (E), weighing something under 79 carats. Little is known about it

$175,000,000 worth of rough stones which were doubled in value by cutting.

Prosperity has no better barometer than the diamond trade, and the increase in diamond buying year by year has reflected the nation's growing wealth. Yearly importations of cut diamonds increased from $1,317,420 in 1867 to $27,000,000 in 1913. They fell to $18,000,000 the first year of the war, and to $9,000,000 in 1915. They showed a reaction from war conditions in 1916, when they reached $20,567,222, and were $21,855,735 in 1917 and $13,925,772 in the first eight months of 1918. The diamond-cutting industry in the United States began in 1873, when $176,426 worth of rough stones were imported. Importations amounted to more than half the importations of cut stones in 1916 and 1917, and had become almost equal in 1918.

Time was when they adorned only the princes of the earth and sparkled only in palaces. But they have become a democratic gem in the great democracy of the West. The stenographer emits Kimberley sparkles. Faint Dutoitspan gleams show in the ears of the pretty waitress. No cook or housemaid can hold up her head without a diamond among her jewels.

Diamond Cutting—a Distinctly Modern Art

The perfectly cut and brilliant diamond the world knows today is not more than fifty years old. The ancient world knew little of diamonds. From the first pharaoh to the last, through all the pageantry of thirty-one dynasties, diamonds were unknown in Egypt. From the dawn of history, Babylon remained unfamiliar with them for forty centuries. The pioneering conquest of Alexander across the Indus in

327 B.C. acquainted Greece vaguely with their existence. The patricians of Rome in the days of the early empire rarely owned them. Byzantine supremacy, the rise of Venice to maritime power, the Moorish conquest of Spain, brought only a trickle of diamonds into western Europe. A fashionable jewelry store in America today carries more diamonds in stock than were in all Europe when Columbus sailed from Palos.

© Topical

The Cullinan diamond was divided into nine large stones and a number of small brilliants. Here is the biggest of the stones. It weighs 516½ carats and is the largest brilliant in the world

To the right is a brilliant weighing more than 309 carats. It is the second largest stone cut from the famous Cullinan

The earliest cutters used their wheels deftly enough, but they neglected their mathematics. Bringing out a diamond's full brilliancy is a mathematical problem. Increase of facets adds to surface area and surface glitter. But the angle of total reflection must be considered in relation to the angle of incidence, and the facets so arranged that a ray of entering light will be reflected from the inner facet surfaces and returned in refracted rainbow sparkles through the top of the stone.

Making the Diamond Sparkle—a Mathematical Problem

Henry D. Morse, of Boston, in the last century was the first to discover the balanced proportions that developed a diamond's highest reflective and refractive possibilities. Since brilliancy is the crowning glory of a diamond, he did not hesitate to sacrifice whatever weight was necessary to achieve it. Retaining the fifty-eight facets of the earlier cutters, he found that a diamond is at its sharpest climax of brilliancy when its depth from table to culet is six tenths of its diameter, and a little more than two thirds below. Cut in this style, a diamond not only flashes light from every polished facet surface, but seems alive with coruscating inner fires.

Morse's proportions are the rule of the world today, and they mark the final

To the left is the Cullinan diamond as it appeared in the rough. It weighed in this state 3,025¾ carats (1½ pounds) and was as white as water. The stone was purchased from the Transvaal Government in 1907 and presented to King Edward VII

To the right is the Excelsior diamond, found in 1893 at the Jagersfontein by a native while loading a truck. It weighed 971 carats in this rough state, and was ultimately cut into ten stones weighing from 68 to 13 carats

A   B   C   D   E

Swallowed by a faithful serving-man to save it from the robber who slew him, the Sancy (*A*) was sliced from his stomach to adorn the royal person of Henry of France and Navarre. The Orloff (*B*) was stolen by a French soldier from the eye of an idol in a Brahmin temple, stolen again from him by a ship's captain, bought by Prince Orloff for $450,000, and given to the Empress Catherine II. It weighs nearly 105 carats, and was one of the Russian crown jewels. The great Mogul (*C*), most magnificent of Indian gems, disappeared from history, never definitely to reappear. It has a bloody history going back to the year 1665. Its fame lured Nadir Shah to the sack of Delhi. This is a glass reproduction made from extant descriptions. It probably weighed after cutting, 280 carats. The Akbar Shah (*D*) was originally a stone of 116 carats with Arabic inscriptions upon it. After being cut down to 71 carats it was bought by the Gaikwar of Baroda for $150,000. The Polar Star (*E*), a magnificent stone weighing 40 carats, belongs to the Princess Youssoupoff

triumph of art in the achievement of the perfect modern jewel.

Fiction in its maddest moods never invented romance more bewildering than the stories of the great diamonds of India. For these baubles wars have been waged, nations devastated, thrones and dynasties overturned, men slaughtered by tens of thousands. For gems men have plotted, intrigued, robbed, murdered, committed every cruelty and treachery, stained their souls with every crime.

The fame of the Great Mogul lured Nadir Shah to the sack of Delhi. Desire to possess the Kuh-i-Nur was woven into the complex motives that led Aurung-zeb to deluge India with blood, slay his three brothers, and dethrone and imprison Shah Jehan, his father.

The Orloff, stolen from the eye of a temple idol and sold overseas, was presented to Catherine of Russia by her princely paramour to patch a lovers' quarrel. Swallowed by a faithful serving man to save it from robbers who slew him, the Sancy was sliced from his stomach to adorn the royal person of Henry of France and Navarre.

The Great Mogul, the most magnificent gem of the Indian mines, disappeared from history, never definitely to reappear, its fate a riddle of the centuries.

The Baleful Gleam of the Hope Diamond

The Hope blue diamond—stone of tragic fame—is the only one of the great historic diamonds to come to the United States. When Louis XIV bought it in the seventeenth century, it was a gem of 67½ carats. It disappeared during the French Revolution, and remained lost until 1830, when it reappeared as a jewel of 44¼ carats.

© Harris & Ewing

The ill-starred Hope diamond eventually passed into the possession of Mrs. Edward Beale McLean, whose husband bought it for $300,000. She wore it on one occasion together with the Star of Este, the two stones together being worth $500,000. The occasion was a dinner which, a curious statistician figured, cost about $166 a minute

From its first appearance in Europe, a superstition has clung to it that it brought disaster to all whoever owned or wore it. Certainly it has been associated with a long list of tragedies. Tavernier, who brought it from India, failed in business, and died on his voyage back to the Orient to recoup his fortune. Madame de Monte-span, upon whom the Grand Monarque bestowed it, was supplanted in the king's affections by her rival, Madame de Maintenon. Nicholas Foquet, a courtier who borrowed it, was executed. Louis XVI and Marie Antoinette, who inherited it, lost their heads on the guillotine. Princess de Lamballe, of Marie Antoinette's entourage, was killed by a revolutionary mob.

The thieves who stole it were executed or deported to penal colonies. Wilhelm Fals, the gemsmith who cut it down for them, ended his life in poverty. Hendrik Fals, his son, who stole it from the thieves, committed suicide. Francis Beaulieu, last of its underworld owners, who sold it to Daniel Eliason, a London jeweler, died of starvation in a garret in Soho. Lord Francis Hope became a bankrupt, and was scandalized by the elopement of May Yohe, his American actress wife. At last accounts May Yohe was a scrubwoman in Tacoma.

Lorens Ladue was shot and killed by her infatuated admirer as she danced in the glare of the footlights with the diamond on her bosom. Her Russian cavalier, who had hung the jewel about her neck, was assassinated. Simon Montharides, who sold it to Sultan Abdul Hamid, was killed in an accident. Two of its Turkish custodians were murdered. Salma Subaya, the sultan's favorite, was shot while in the Yildiz Kiosk; and Abdul Hamid finally lost his throne.

Imported into the United States, it was bought by Edward B. McLean for $300,000. As beautiful as when, fresh from the mystic East, it dazzled the court of France, the diamond for years brought only happiness to its new owners. Then one day the little son of the McLeans, first-born of a happy marriage and heir to vast riches, was killed at play by an automobile. Instantly the tragic tradition recurred to the public.

A   B   C   D   E

The Florentine diamond (*A*), among the crown jewels of Austria, weighs 139½ carats and is valued at $525,000. It is a very pale yellow. It was picked up on a medieval battlefield and sold for two francs. The Hope (*B*), 44¼ carats, is believed to be a portion of a beautiful blue stone of 67 carats cut from a stone weighing over 112 carats, which was discovered in India, brought to Europe by Tavernier, and which was stolen from the French crown jewels. The Hope has the same color as the missing gem. The Kuh-i-Nur (*C*) eventually passed into the hands of the East India Company, and was presented by it to Queen Victoria in 1850. This is a picture of it recut to 106 carats. The Star of the South (*D*), perhaps the most famous of Brazilian stones, was found in 1853. It was cut from 254½ carats to 125 carats, and was bought by the Gaikwar of Baroda for $400,000. The Pasha of Egypt (*E*) weighs forty carats and is valued at $140,000

313

What Makes a Baseball Curve?

It is simply a matter of atmospheric pressure

By Lindley Pyle

HOLD up a sheet of paper and blow vigorously between the two separated ends. The leaves of paper move toward each other and cling together. Obviously, the moving stream of air exerts a suction effect upon adjacent bodies. In more precise language, the atmospheric pressure between the paper leaves is lowered by the air velocity action, and the unaltered pressure of the air on the outside crushes the paper leaves together.

Blow through a distended envelope that has been sealed and that has had its ends slit open. The side walls collapse and the envelope flattens, crushed by the greater pressure of the still air outside the envelope.

Suspend a cork from a silk thread, and with a "straw," such as those used at soda-fountains, blow a stream of air past the cork at a distance therefrom of at least three fourths of an inch. The air pressure on the breeze side of the cork is reduced, and the normal atmospheric pressure on the quiet side of the cork pushes the cork over and into the stream of air.

Not a New Discovery

Similar experiments may be performed with water as the fluid medium instead of air, using a submerged hose to furnish a high-velocity water current. Let the nozzle of a hose delivering a quarter-inch stream of water at high pressure be thrust under the surface of a body of water, and let the stream pass between the submerged hands, held with the palms an inch apart. Marked suction will be felt and sensible effort will be required to keep the palms apart.

Examples might be multiplied demonstrating that, where a portion of a fluid is moving faster than an adjacent part, the pressure in the higher-velocity region is less than in the lower-velocity region. This is not a new discovery in science: it has been known for nearly two hundred years. Nevertheless, it is of interest to all of us, since by this principle we may explain the curve of a baseball.

Tie one end of a six-foot length of string to a tennis-ball, and fasten the other end to the ceiling of a room free from air currents. Spin the ball for several minutes in one direction so as to twist the string, then

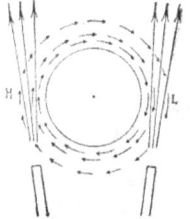

Fig. 1. Short arrows indicate the air rotating with the ball; long arrows, the air from two straws

allow the string to untwist, thus putting the ball into a high rate of rotation. The air close to the ball is

As soon as he stops blowing the sheets of paper will come together and cling. Why? Because the atmospheric pressure between them has been lowered by the blowing and it is overcome by the steady pressure from without

If you slit the ends of a closed envelope and blow through it the result will be the same as in the case of the paper

Blow through a straw at a suspended cork; the pressure of outside air forces the cork into the stream of air

dragged into motion along with the surface of the ball, as may be seen by bringing the smoke of a cigarette into close proximity. Now, while the ball is spinning (not swinging), place two

Fig. 2. The curve of a baseball, following the principle of air currents indicated in Fig. 1 above

soda "straws" in your mouth and blow a stream of air past each side of the ball.

In the diagram Fig. 1 the short arrows indicate the swirl of air rotating with the ball, while the long arrows represent the air that streams past the ball from the straws. It is plain that the net effect of the mingling of these air currents is to cut down the velocity effect at L (where the two currents are in opposite directions) and to maintain it at H (where the two currents are in the same direction). The preceding experiments have demonstrated that the air pressure is consequently greater at L than at H, and one is not surprised to see the ball *pushed over toward H.*

Discard the straws, and produce a breeze past the spinning ball by swinging it to and fro like a pendulum. The effect upon the spinning ball as it moves bodily through the air is the same as when the breeze past it is produced by blowing. In the diagram Fig. 2, when the spinning ball is pulled aside (by its string) to *a* and then swung toward *b*, it corresponds precisely to the case just described, except that the breeze past the ball is due to the bodily motion of the ball through the air. Instead of reaching *b* the side thrust on the ball carries it to *c*. On the return trip it goes to *d*, and on the next swing to *e*. If the direction of spin be reversed the plane of the pendular vibration rotates in the opposite direction.

The Ball "Follows Its Nose"

The front of the moving ball is often termed its "nose." It will be noted that the path of the moving ball deviates in such a direction as to lead one to say that the ball "follows its nose."

The experiments have led directly to the case of the curving baseball. The delivery of the pitcher is such that the ball leaves his hand with a rapid spin, and the ball follows its nose, whether it be in-shoot, drop, or what not—for a reason amply illustrated by the foregoing simple experiments.

Why Laundering Kills Clothes

A bit of science applied each Monday
will add weeks to the life of your linen

By I. Newton Kugeimass, Professor of Chemistry, Howard College

DIRT is matter in the wrong place. The business of the laundress is to remove it. The business of the chemist is to tell her how to do it. The life of clothes may be prolonged twenty-five per cent by scientific laundering.

The laundering process is started with soaking to loosen the dirt and save rubbing and thereby the goods, time, and energy. The great mistake made is to begin soaking with hot water. This coagulates the albuminous matter and starch, making them stick on the clothing with resultant blotches. Start with a cold-water bath, for cold water dissolves the starch and albuminous matter and gets rid of them for good.

The kind of water used should not be a matter of indifference. Woolens galore have been ruined by washing them in naturally hard water. The sticky soap settles in the pores of the wool fiber and materially reduces its wearing qualities. For safety and efficiency prepare the water before using it for washing. Add a minimum of ammonia, borax, soda-ash, or washing soda, enough to precipitate the objectionable minerals. Stir, let the water settle, and then allow the clear water to flow into the washing-tub.

With the water prepared, the next step is the actual washing operation,

which involves combined mechanical agitation and cleansing action of soap. To get maximum service from soap we must know how it works. Soap first dissolved in the water reacts chemically, giving a mild alkaline medium. This medium prepares the way. The rest of the soap is very finely divided into microscopic particles, all evenly distributed throughout the whole solution—all the water is soapy. Each soap particle is a worker—a dirt capturer! The more finely divided the particles and the greater the number,

Her soapsuds are cloth eaters, her bluing overused, the starch is full of germs; and your clothes are piled together with many other dirty clothes. Do you wonder that they don't last long?

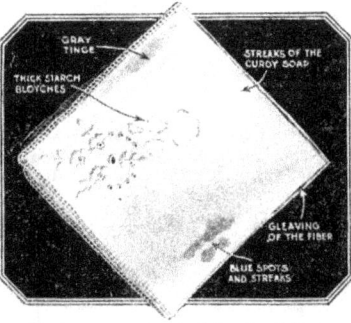

Hard water is very bad for the clothes; ammonia or washing soda should always be added; your handkerchief is likely to suffer from all the ailments shown above if you don't soften hard water

the more efficient the cleaning. The best condition is attained by slightly increasing the alkalinity with a mild alkali—soda. These dirt-fighters work best in a moderately alkaline field. Since dirt is held in soiled goods by grease, soap cleans in two operations. It first removes the grease from the materials by forming an *emulsion* with it. The dirt without any grease support on the clothes is now pulled in (absorbed) by the fighting soap particles. Every soap particle carries a dirt-load on its back and keeps the grease in emulsion form.

Many a laundress adds caustic soda to the soap solution. This gives an excessive alkalinity and ruins the strength, color, appearance, and wear of the clothing. Then, too, she does not invariably choose the best cleaning soap. It is "neutral soap," without free caustic, without fillers of water-glass, rosin, or peroxides, adulterations that loosen, weaken, and color the texture of the fiber. Neutral soap and a mild alkali together give the most efficient washing medium. The deadly policy of leaving the clothes overnight in the dirty soap bath "rots" them.

Using raw bleaching powder means more harmful effects on the clothes than hard water and caustic soda combined. Treat the bleaching powder with soda in a separate vessel. The sodium replaces the calcium, giving sodium hypochlorite, the bleach liquor, and precipitated chalk settles to the bottom and is rejected. The sodium hypochlorite is acted upon by the water, giving oxygen, caustic soda, and energy.

To bleach with little injury, use the least soda in making up the bleach liquor, so as not to have large alkalinity, keep the materials in the bleach a minimum length of time, heat the bleach bath gradually to prevent too rapid giving off of the oxygen, and rinse thoroughly, else the bleach liquor will "rot" the fabrics.

Rinsing should be thorough after each operation. Insufficient rinsing after the first suds decreases the soap efficiency in the second suds, after the bleach, ruins the clothes; before bluing, leaves the alkalinity to cause uneven setting of the blue; and also leaves the alkalinity to convert the starch into yellow decomposition products during ironing.

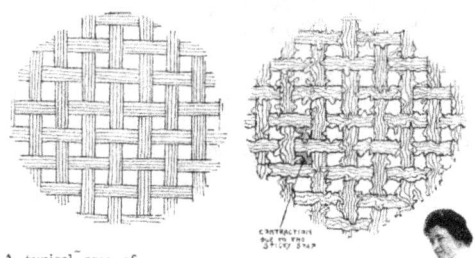

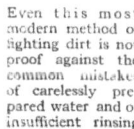

A typical case of before and after washing in hard water; the fibers are clogged with decomposed soap, which is insoluble

Even this most modern method of fighting dirt is not proof against the common mistakes of carelessly prepared water and of insufficient rinsing

They Monkeyed with the Buzz-Saw

And now it turns rough logs into finished lumber at the rate of one million feet a day

IT is probable that the first sawmill in the United States was erected at Jamestown in 1607. It was crude, and an improvement mechanically over the then common method, pit-sawing, only in that the work was done by simple machinery instead of, as formerly, by hand. In those first mills, wasteful and slow in operation and of light producing power, there was apparent but little progress toward better methods for very nearly two centuries, when a small circular saw supplanted the old "up-and-down" contrivance that had been in use. This, in turn, retained its place for many years, finally to be supplanted in the manufacture of lumber by the very efficient band-mill of today.

In this field the American was the pioneer, and it is said that his mills, driven by windmill or by water-power, later by the tides on the coast of New England, were cutting the virgin timber of America two hundred years before the running of a mill in England, where the first of them were broken up by mobs of men, loath to see the appearance of any new labor-saving contrivance.

The American sawmill was not always devoted exclusively to sawing timber; for wood was plentiful and ready at hand to those who needed it, and the exigencies of the times very frequently demanded food far more. Many of the earlier milling plants were for that reason a working combination of the saw- and grist-mill; and the miller who received the settler's corn and ground it into meal was also, in many settlements, the sawyer who converted logs into boards.

For many years after the advent of the first circular saws the mills showed few essential changes. Their development, if not entirely arrested, was very slow; old ways continued, as they still continue in certain sections of the country. Then (not more than a generation since) came the band-saw, and with the band-saw greater speed and skill, and a daily cut greater by far than that of which the old sawyers had ever dreamed.

Steam supplanted wind and water-power, to be later succeeded in some large-sized establishments by the electric motor. Crews of two men, the sawyer and his helper, disappeared with their circular saws, their overshot water-wheels and primitive log carriages. The larger timbers of the newer West and South made new demands, and native ingenuity proved equal to modern needs. Small logs and short carriages gave way to huge lengths of timber and carriages designed to travel at locomotive speed.

Though some work is still done by hand, and two or three men "ride the carriage" to fasten the logs in place after high-power machinery has put them there, the first small crews of two men, and sometimes even one, would today hardly suffice for a cut-off saw in the modern mill, with its thirty or forty men assigned to special tasks. The boards are not touched by hand when

The old-time sawyer found many stiff problems to be solved in the heavy timber of the Southern Appalachians

they leave the log. Endless chains carry them to the trimmers, graders, and loaders, while others carry the scraps and edgings to smaller machines, where they are turned to lath or other by-products.

The head sawyer is a high-salaried employee who could throw away his wages in half an hour of careless cutting; for thorough experience, sound judgment, and a quick, trained eye are necessary adjuncts to improved machinery and the increased production it makes possible.

In many of the mills a double-cut saw passes through the log as it goes out and back: the carriage runs forward, a board is cut, the log advanced an inch or two, and another cutting made on its return. Great logs are handled, and disappear, in the twinkling of an eye. The first-class band-mill is a model of efficiency.

A few hundred feet of rough, uneven lumber in a day was the output of the early mill; there are plants in operation in the state of Louisiana that have in excess of a million feet in twenty-four hours for their unit of production.

Methods like these in the hills of Kentucky remind one of days when pit- or whip-sawing was the only *modus operandi*

With skilled operators and fast, efficient band-saws, from log to finished boards takes but an instant, and the waste is at a minimum

Spy-Glasses on Guns

Bringing the game nearer with grandfather's old telescope

By Captain Edward C. Crossman, U. S. A.

DARKNESS was creeping into the grove. Forty feet overhead, in a crotch of a huge limb, something seemed to be snuggled that did not fit in with the usual outline of a black-walnut limb. Vainly the hunter tried to see it. Hastily he drew from a leather case by his side a thin steel tube, possibly a foot long and not much thicker than a man's thumb. He slipped it on the rifle and tightened a thumb-screw. Again he pitched the rifle to his shoulder.

Through the tube the great limb ceased to be merely a blur. In relief against the dark bark were two ghostly crossed hairs, their intersection at the exact center of the circle of light that came from the lenses. The tube crept along the limb to the crotch.

The something that the hunter suspected did not belong to the tree was suddenly revealed—a sharp ear, a bright eye, an impudent nose with the sensitive nostrils above, set in a background of light gray fur. The intersection of the hairs rested for an instant on the bright eye. Bright flame spurted from the black muzzle of the rifle. Something fell into the leaves at the foot of the tree.

If you took grandfather's old ship's telescope, focused it, and fastened it on a rifle, you would have the primary principle of the rifle telescopic sight. You could not hit anything with the rifle so fitted, because there would be nothing in the telescope to direct the aim.

If you carefully fitted a pair of crossed fine silk hairs or the finest possible steel wires in the focus of the lens at the eye end of the glass, so that the magnifying eye-lens enlarged the hairs as well as the image formed by the object lens at the front end of the telescope, you would have the second step of the telescopic rifle sight.

If you altered the fastenings of the

When the point where the hairs cross bears directly on the woodchuck's head, it is the moment to pull the trigger

The upper half of this diagram shows the old type of telescope sight; the lower half, a later type with achromatic lens. "Stop" means diaphragm. Like ordinary telescopes, the telescope rifle sight has a lens (the object glass) at the forward end which forms the image. At the eye end is another lens (the eyeglass) which is a simple form of microscope, and which takes the image from the object glass and magnifies it

instrument so the junction of the fine hairs and the point of the strike of the bullet agreed at some one desired distance away from the rifle, you would have the third step.

The power of the telescope rifle sight is low. This at first seems strange, because the more we enlarge an image the closer it seems and the plainer we can see the object. Nine times out of ten the ordinary American telescope sight is less than five power. Here's the reason. The rifle, in human hands, moves constantly. The viselike grip died with the heroes of Fenimore Cooper.

Point the telescope sight at a distant brick chimney. If without the telescope, we find that our front metal sight moves just the width of a brick in spite of all our efforts to hold steadily. With the telescope it still continues to move the width of the brick. As the image is enlarged five times, the rifle apparently has to move five times as fast and five times as much to continue to swing from side to side of the brick. The result is that the shooter beholds the cross-hairs galloping back and forth across a greatly magnified picture of a brick. With the ten-power glass the effect is just that much worse.

Two other reasons, minor but valid, are that the greater the magnification the less the light passed by the lenses—and we must have plenty of light.

The common form of American telescope sight is a steel tube from three fourths to a full inch in diameter and from ten to fifteen inches long. In the best make the lenses are known as "achromatic"—which means corrected for color dispersion. The old form of cross-hair is giving way to other and more useful forms. Probably the best form for hunting is a little slip of steel running to a sharp point just at the center of the tube. Sometimes, to strengthen this, another wire is run across the tube from side to side, intersecting the picket just below the point. Should this be our form, then whenever we touch our mark with the sharp point, our rifle is sighted for it, and the bullet will strike there.

The telescope sight as mounted on a sporting rifle. Prismatic glasses will add half an hour to the evening shooting

The American type. It is good for target work, but not for war. Its low power makes it excellent for twilight shooting. The power must be low because, in human hands, the rifle is constantly on the move

German telescope sight, showing how to detach from rifle. War also has proved the value of the telescope sight

The Story of a Splash

What happens when the drop falls in the bucket

By H. E. Howe

"JUST a drop in the bucket" is a familiar and useful simile, but it remained for an English schoolboy to wonder what happens when the drop strikes in the bucket. He tried to find out by causing drops to fall upon carefully smoked glass, after which he could study the prints made by the drop upon the carbon. Some twenty years later Professor A. M. Worthington became interested in drops, and in 1894, under the title "The Splash of a Drop," delivered an address setting forth his observations. This was afterward printed by the Royal Institution.

It was certain that any changes that a drop undergoes must be of exceedingly short duration. Hence the first problem, as in most investigations, was that involved in providing suitable apparatus. A drop of definite size was to be allowed to fall from a definite height in darkness, and any desired stage in the phenomenon illuminated by a spark of such exceedingly short duration that all other stages would remain unobserved and not blur the picture.

Many Observations Necessary

These conditions were eventually satisfied by providing two similar levers released simultaneously by an electromagnet. From one lever the drop was allowed to fall from a small watch glass smoked to prevent wetting, and the other let fall a marble that struck an adjustable releasing mechanism, which in turn broke an electrical connection, causing a short spark near the spot where the drop struck. By regulating the release mechanism, and therefore the distance through which the marble fell, intervals separated by very small fractions of a second could be secured.

At the time the first experiments were made the photographic plate was not sufficiently rapid to record the image, so that Professor Worthington found it necessary to make a very large number of observations of each stage until he could accurately sketch what he had seen. Later on he had the satisfaction of being able to confirm his early work, using more modern photographic apparatus and fast emulsions.

The illustrations reproduced here are from a series of drawings, and illustrate the exercise a drop of

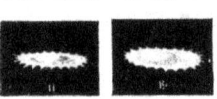

As the drop spreads it forms distinct rays—sometimes as many as twenty-four (Fig. 9)

Surface tension causes the gymnastic exercises that a drop goes through when it falls

Forming pairs of lobes. Fig. 17, with twelve pairs, is the climax of this stage

mercury takes when one .15 inch in diameter falls three inches upon a smooth glass plate.

Time, One Twentieth of a Second

Surface tension is responsible for the movements and patterns observed. The distinct rays shown in number nine are twenty-four in number, as determined by counting the droplets that squeeze off the end of these rays. Next, an annulus with a very thin area between center and rim is formed, followed by the rim forming pairs of lobes. These are drawn together until

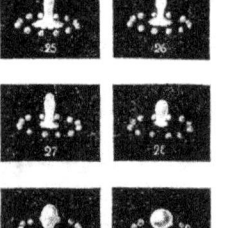

The surface tension checks the spread of the drop and causes the center to rise

The rebound. Only one twentieth of a second intervened between Figs. 1 and 30

in stage seventeen we find twelve lobes.

Surface tension checks the spread, and then a rapid and powerful contraction takes place, causing the mercury to rise in the center almost to the breaking-point, as in twenty-six. This sudden drawing in of the lobes causes droplets to become detached and marks the beginning of the rebound. At the end we may have two circles of different-sized drops surrounding the remainder of the original drop, and the entire series of thirty stages takes place in a total time of about one twentieth of a second—an interval of one six hundredth of a second for each stage.

At certain stages, such as numbers fourteen, fifteen, and sixteen, the mercury is sometimes covered with beautiful concentric rings.

The phenomenon depends upon four factors:

1. The size of the drop.
2. The height of the fall.
3. The surface tension of the liquid in question.
4. The viscosity of the liquid.

The possibilities of this interesting study have by no means become exhausted. Study drops of ink or of milk falling upon water, or indeed the phenomenon of large drops of water falling upon water, giving rise to the beautiful bubbles so often seen in a hard rain. Solids falling into liquids are another possibility.

Value of Modern Photography

Modern photography — remember that Professor Worthington did his work before 1894—ought to record unusual features. He used a Leyden jar spark for his photography, the duration of such a spark being probably less than ten millionths of a second.

It would be interesting to determine whether our more recent methods are any more satisfactory, so far as results are concerned, than the careful work of a patient trained observer.

Butter from the Coal-Oil Cow

You've never seen a coal-oil cow, but you may hope to see one

By John Walker Harrington

THE village pump has long competed with Bossy. Now comes the derrick to substitute for the churn. For butter can be made from petroleum.

As yet, this artificial petroleum butter does not possess the desirable new grass taste; it savors more of the flavor of axle-grease. Dr. Gustave Egloff, a well known chemist who has been experimenting with it, does not recommend it for the table, or even for automobile luncheons by the wayside. But the day will dawn when the oil refinery will compete with the creamery.

A Problem for the Chemist

Petroleum is a highly complex liquid composed chiefly of hydrogen and carbon in chemical combination. Hence chemists call petroleum a "hydrocarbon." Many of our foods, including butter, are also combinations of hydrogen and carbon—but different. You can build with brick hundreds of houses that bear no architectural resemblance to one another; you can build up from hydrogen and carbon atoms thousands of substances as different as coal-tar dyes and potatoes.

So this problem of making good butter out of a vile oil that oozes from the earth resolves itself into a rearrangement of its atoms. That is not an easy problem; for petroleum is composed of some thirty different chemical compounds classed as hydro-carbons, and is impregnated, besides, with soluble nitrogen and sulphur. We have many hydro-carbons that are good to eat, such as the starch which is an important constituent of wheat and corn and potatoes.

Butter is a solid fat consisting of a group of acids. Of these acids the principal one is called "butyric" acid. It is made by agitating or beating milk so as to break up the globules of fat and to bring them into a solid mass. The problem of the chemist is to change the hydro-carbon of petroleum into the pleasing acid of good creamery butter.

Hydro-carbons consist of hydrogen and carbon. The first step in the transformation is to "chlorinate" the petroleum. This is done by forcing chlorine into it by an electric current which is turned on while it is confined in a closed mixing vessel. The chlorine combines with atoms of hydrogen and produces hydrochloric acid and chlorides of the hydro-carbons.

It Looks All Right

Next this mixture is boiled with caustic soda, a chemical resulting from the union of hydrogen and soda, and technically known as a hydrate of that element. Chlorine reacts with soda and forms a chloride of soda which is common salt. Thus we have our derrick brand of butter literally salted in the making!

There is present in the mixture, also, a combination of the carbon, the hydrogen, and the oxygen, which have all been brought into new relations. They constitute a form of alcohol. By intricate chemical processes more oxygen is added, so that

Turning the oil from this "gusher" into butter for our bread is merely a matter of rearranging atoms

the compound is changed into a group of acids which may be assimilated by the human system exactly the same as are those fatty acids which we call butter.

Dr. Egloff's early experiments were made with a light colored fuel-oil of the kind used under the boilers of ocean liners. This experimental butter had the proper yellow tint, but, owing to the fact that many things had not been eliminated from it, it had a taste that was far from palatable. By bleaching and filtration it is possible to obtain a bland, colorless, and tasteless petroleum. Such a product is now sold by every drug store as an internal lubricant.

When the chemist starts with a bright, pure oil, he can undoubtedly produce a most edible substitute for butter.

Butter substitutes are legion. The best known of them is the oleomargarine, which is a chemically pure mixture of animal oils and stearin. Butterine is oleomargarine flavored with real butter. Both of these products are made under government supervision, and are accepted as valuable foods.

Theoretically, the way is open for the production of a food adjunct of great economic value.

Chemistry in a Transition State

Industrial chemistry is in a state of transition. The impossible of the present becomes the inevitable of the future. It was only a short while ago that the molecule was regarded as indivisible; yet, in the chemistry of petroleum alone, this belief has been repeatedly shattered. By the "cracking" of molecules of kerosene, the supply of gasoline, in this country, has been appreciably increased.

By treating oils obtained from cotton-seed, cocoanuts, and peanuts, with hydrogen gas in the presence of nickel or iron, hard fats result which are acceptable substitutes for lard. Petroleum butter may take its place with these lard substitutes, and come to be regarded as a household necessity, while the bovine variety will pass into the list of luxuries.

There are very good reasons, as Dr. Egloff has pointed out, why the use of fats from petroleum will someday be considered a matter of prime importance.

Wonderful Sea Creatures Spun from Glass

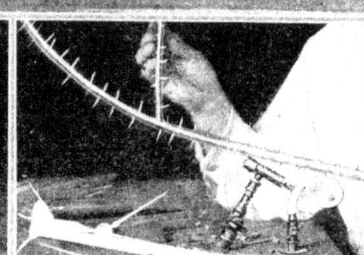

Herman Miller, expert glass-blower at the American Museum of Natural History, in New York, at work on a glass model of a sea-spider enlarged to twenty-five times its natural size

His tools are a pair of forceps, a carbon pencil, and a flame; with these and glass tubes he copies nature's work exactly

Unusual specimens of the glass-blower's art, which form a part of the wonderful collection at the Museum. This group represents six months' work

Assembling the parts of a very complicated model of sea life made up of thousands of delicate parts. This piece of work represents the result of six weeks of labor

Glass model of a branch of a marine plant (magnified many times) being held in the gas-flame while the small spines are carefully attached to their places

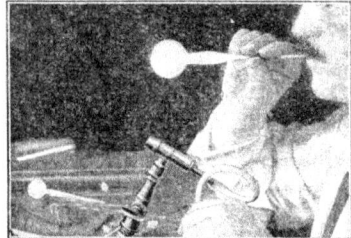

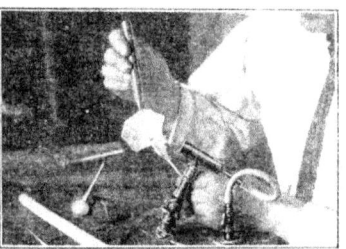

Making a glass flower. First the glass is heated in the center and drawn out to form a bulb at one end

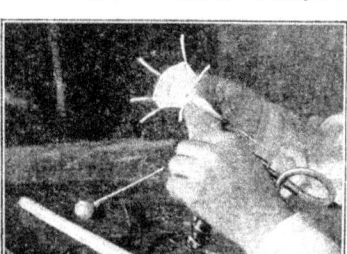

Then the glass bulb is blown out to the desired size, broken open, and shaped by a carbon pencil

The sides of the flower are formed by means of a carbon pencil as the glass is held in the flame

These bits are attached by touching them to the edges of the petals at points heated red-hot

www.ingramcontent.com/pod-product-compliance
Lightning Source LLC
Chambersburg PA
CBHW081106170526
45165CB00008B/2348